Neuropsychopharmacology of the Trace Amines

Experimental and Clinical Neuroscience

Neuropsycopharmacology of the Trace Amines, edited by ***A. A. Boulton, P. R. Bieck, L. Maitre,*** and ***P. Riederer,*** *1985*

Neurobiology of the Trace Amines, edited by ***A. A. Boulton, G.B. Baker, W. G. Dewhurst,*** and ***M. Sandler,*** *1984*

Neural Membranes, edited by ***Grace Y. Sun, Nicolas Bazan, Jang-Yen Wu, Giuseppe Porcellati,*** and ***Albert Y. Sun,*** *1983*

NEUROPSYCHOPHARMACOLOGY OF THE TRACE AMINES

Experimental and Clinical Aspects

Edited by

A. A. BOULTON, L. MAITRE, P. R. BIECK, and P. RIEDERER

Humana Press · Clifton, New Jersey

Library of Congress Cataloging in Publication Data

Main entry under title:

Neuropsychopharmacology of the trace amines.

(Experimental and clinical neuroscience)
Based on the 2nd Trace Amines Symposium held at Weitenburg Castle May 15-19, 1985.
Includes index.
1. Biogenic amines—Physiological effect—Congresses.
2. Biogenic amines—Therapeutic use—Congresses.
3. Neuropsychopharmacology—Congresses. I. Boulton, A. A. (Alan A.) II. Trace Amines Symposium (2nd : 1985 : Weitenburg Castle) III. Series. [DNLM: 1. Amines—congresses. 2. Behavior—drug effects—congresses.
3. Psychopharmacology—congresses. QU 60 N4946 1985]
QP801.B66N486 1985 615'.78 85-24904
ISBN 0-89603-099-7

Crescent Manor
PO Box 2148
Clifton, NJ 07015

Printed in the United States of America

PREFACE

A probable neurobiological role for the trace amines began to emerge as soon as techniques sufficiently sophisticated to detect them were developed. Techniques for quantitative analyses included radioenzymatic procedures and, more recently, HPLC–EC and automated GC–MS. The methods are applied after separation of the substances to be analyzed and after their purification by physicochemical procedures that are becoming more and more efficient.

The identification and quantitative analysis of minute amounts of trace amines in biological fluids or tissues is in itself a remarkable technological achievement. The enormous task that several teams of research workers have set themselves is to investigate the origin of the trace amines, to understand their metabolism, and to discover whether or not they possess an important biological role. The 2nd Trace Amines Symposium at Weitenburg Castle near Tübingen from the 15th to the 19th May 1985 was held to gain a somewhat clearer insight into the present state of knowledge. Since the first meeting in 1983 our knowledge has increased and interest is deepening. Hence the idea proposed at that first meeting, of organizing similar gatherings at regular intervals, has come to fruition.

Covered in the Neuropsychopharmacology section of this book are studiesof the effects of certain trace amines on different forms of behavior, on neurotransmission mediated by the classical neurotransmitters, on their biosynthesis or on catabolism, on their possible occurrence as a result of alternative metabolic pathways for the amino acids, and in some instances their neuroanatomical distributions.

In addition, certain psychotropic and psychotomimetic agents have been shown to influence neurobiochemical and/or behavioral parameters associated with the trace amines. By contrast, the association of certain of the trace amines with specific receptor sites, compared with the affinity sites for the classical neurotransmitters, are questions that have been touched upon only in a few instances to date.

Presented in the Clinical Pharmacology section are analyses of some of the trace amines and their acidic metabolites in certain pathophysiological states and also in healthy volunteers. Also described are the effects of MAO inhibitors—either of the "classical" or "new" types—on the concentration, elimination, or physiological effects of the trace amines in humans.

It appears to us, therefore, that the combined efforts presently being made to determine the importance of the trace amines both in experimental biology and in human clinical pharmacology promise to open up quite a large field of potential application. In addition, new approaches might well be offered by which to study the pathophysiological bases of certain of the neurological and psychiatric disorders.

A. A. Boulton, *Saskatoon*
P. R. Bieck, *Tübingen*
L. Maitre, *Basel*
P. Riederer, *Vienna*

ACKNOWLEDGMENTS

In 1983 at the conclusion of the successful First Trace Amines Meeting in Edmonton it was suggested by our Canadian hosts that a second meeting should be held, preferably in Europe. The Heads of research of the Pharma Division of CIBA-GEIGY in Basle and the director of the Medical Department of CIBA-GEIGY in Germany decided to provide some financial help for such a Second Trace Amines Meeting. We wish, therefore, to express our thanks to CIBA-GEIGY for their generous and philanthropic support. Financial help from the International Society of Neurochemistry is also greatly appreciated and was most helpful in covering part of the costs of these proceedings. Finally we thank all the members of the Human Pharmacology Institute, Ciba-Geigy Tübingen, who helped to make this meeting the scientific and social success that it was.

CONTRIBUTORS

ANTONIN, K. H. • Human Pharmacology Institute, CIBA-GEIGY GmbH, Tübingen, Federal Republic of Germany

ARBILLA, S. • Department of Biology, Laboratoires d'Etudes et de Recherches Synthelabo, Paris, France

BAILEY, B. A. • Psychiatric Research Division, University of Saskatchewan, Saskatoon, Canada

BAKER, G. B. • Neurochemical Research Unit, University of Alberta, Edmonton, Canada

BIECK, P. R. • Human Pharmacology Institute, CIBA-GEIGY GmbH, Tübingen, Federal Republic of Germany

BOULTON, A. A. • Psychiatric Research Division, Saskatchewan Health, University of Saskatchewan, Saskatoon, Canada

BROWN, M. J. • Department of Clinical Pharmacology, Hammersmith Hospital, London, UK

COOPER, S. J. • Department of Psychology, University of Birmingham, Birmingham, UK

COUTTS, R. T. • Neurochemical Research Unit, University of Alberta, Edmonton, Canada

DAVID, J. C. • Laboratoire de Biochimie du Development LA CNRS 256, Universite de Rennes, Paris, France

DELISI, L. E. • National Institute of Mental Health, Bethesda, Maryland, USA

DOWNER, R. G. H. • University of Waterloo, Department of Biology, Waterloo, Ontario, Canada

DURDEN, D. A. • Psychiatric Research Division, Saskatchewan Health, University of Saskatchewan, Saskatoon, Canada

DYCK, L. E. • Psychiatric Research Division, University of Saskatchewan, Saskatoon, Canada

EDWARDS, D. J. • Department of Pharmacology-Physiology, University of Pittsburgh, Pittsburgh, Pennsylvania, USA

GILBERT, D. B. • Department of Psychology, University of Birmingham, Birmingham, UK

GOUDIE, A. J. • Psychology Department, University of Liverpool, Liverpool, UK

GREENSHAW, A. J. • Psychiatric Research Division, University of Saskatchewan, Saskatoon, Canada

GUFFROY, C. • Laboratoires Fournier, Centre de Recherches, Fontaine les Dijon, France

HARRIS, J. • Chemistry Department, Arizona State University, Tempe, Arizona, USA

HETEY, L. • Institute of Pharmacology and Toxicology, Humboldt University, Berlin, Federal Republic of Germany

JUORIO, A. V. • Psychiatric Research Division, Saskatchewan Health, University of Saskatchewan, Saskatoon, Canada

KAROUM, F. • Neuropsychiatry Branch, NIMH Saint Elizabeths Hospital, Washington, DC, USA

KIENZL, E. • Ludwig-Boltzmann Institut für Klinische Neurobiologie, Krankenhaus Wien-Lainz, Wien, Austria

KINEMUCHI, H. • Department of Pharmacology, Showa University, Tokyo, Japan

KISARA, K. • Department of Pharmacology, Tohuko College of Pharmacy, Japan

LANGER, S. Z. • Department of Biology, Laboratoires d'Etudes et de Recherches Synthelabo, Paris, France

LARSON, A. A. • Department of Veterinary Biology, College of Veterinary Medicine, Saint-Paul, Minnesota, USA

LOCOCK, R. A. • Neurochemical Research Unit, University of Alberta, Edmonton, Canada

MAITRE, L. • Research Department, CIBA-GEIGY AG, Basel, Switzerland

MCQUADE, P. • Douglas Hospital Research Centre, Verdun, Quebec, Canada

MOISES, H. W. • Central Institute of Mental Health, Mannheim, Federal Republic of Germany

MOSNAIM, A. D. • Chicago Medical School, North Chicago Veteran's Administration Medical Center and Loyola University, North Chicago, Illinois, USA

MURPHY, D. L. • Clinical Neuropharmacology Branch, NIMH, NIH Clinical Center, Bethesda, Maryland, USA

NAZARALI, A. J. • Neurochemical Research Unit, University of Alberta, Edmonton, Canada

ORELAND, L. • Department of Pharmacology, University of Gothenburg, Gothenburg, Sweden

ORTMANN, R. • Research Department, CIBA-GEIGY AG, Basel, Switzerland

OSBORNE, N. N. • Nuffield Laboratory of Ophthalmology, University of Oxford, Oxford, UK

PHILIPS, S. R. • Psychiatric Research Division, Saskatchewan Health, University of Saskatchewan, Saskatoon, Canada

REINHARD, J. F. • The Wellcome Research Laboratories, North California, California, USA

REYNOLDS, G. P. • Department of Pathology, University of Nottingham, Nottingham, UK

RIEDERER, P. • Ludwig-Boltzmann Institut für klinische Neurobiologie, Krankenhaus Wien-Lainz, Wien, Austria

ROMMELSPACHER, H. • Department of Neuropsychopharmacology, Freie Universität Berlin, Berlin, Federal Republic of Germany

SANDLER, M. • Bernhard Baron Memorial Research Laboratoires, Queen Charlotte's Maternity Hospital, London, UK

STULA-DELINI, A. • Research Department, CIBA-GEIGY AG, Basel, Switzerland

TADANO, T. • Department of Pharmacology, Tohuko College of Pharmacy, Japan

WALDMEIER, P. C. • Research Department, CIBA-GEIGY AG, Basel, Switzerland

WEIL-FUGAZZA, J. • Unite de Neurophysiologie Pharmacologique, Paris, France

WEINER, N. • Physiologisch-Chemisches Institut, Philipps Universität, Marburg/Lahn, Federal Republic of Germany

WOLLMANN, H. • Human Pharmacology Institute, CIBA-GEIGY GmbH, Tübingen, Federal Republic of Germany

YEUNG, J. M. • Neurochemical Research Unit, University of Alberta, Edmonton, Canada

YU, P. H. • Psychiatric Research Division, Saskatchewan Health, University of Saskatchewan, Saskatoon, Canada

CONTENTS

Free Communications

CLINICAL PHARMACOLOGY

Invited Communications

Free Communications

INTRODUCTION

THE TRACE AMINES: RECENT OVERVIEW AND FUTURE POINTERS

Alan A. Boulton

Psychiatric Research Division
Cancer and Medical Research Building
University of Saskatchewan
Saskatoon, Saskatchewan
Canada S7N OWO

At the 1st International Conference on the Trace Amines held in Edmonton, Canada in 1983, [Proceedings of which have been published Boulton et al. (1984)], not only was the new work up to the end of 1982 described but in addition, the general properties of the trace amines and their identification and quantitative analysis was reviewed. This 2nd meeting held in Tübingen, West Germany covers work completed since then. Clearly everything could not be included and so the organizers have chosen to concentrate on the neuro- and psycho-pharmacology of the trace amines and some bioclinical findings. Not all the trace amines could be included, for example, histamine, the polyamines, aliphatic amines and the classical biogenic amines when present in 'trace' concentrations, were ommitted as was the burgeoning role of the trace amines in invertebrate systems; rather, as in 1983, it was the arylalkylamines on which we focussed.

The trace amines, m- and p-tyramine, phenylethylamine, tryptamine and their acidic metabolites and several other related substances (the octopamines, synephrines, phenylethanolamine, etc.) are ubiquitously distributed. They have been found in all tissues analysed to date and they are heterogenously distributed in the brain and across the neurone, although with respect to the brain, this heterogeneity is based on relatively crude large area dissections. The time has come to undertake more precise neuroanatomical studies, utilising finer dissections, the Palkovits 'punch' technique (Palkovits, 1985) and strategic lesions. Simultaneously more sensitive analytical procedures [different trace amine and acid derivatives, high pressure liquid chromatographic and mass spectrometric procedures including chemical ionisation (CI), negative CI, metastable techniques, etc.], (Durden and Boulton, 1982; Durden, 1985) and newly developed immunological procedures (see Osborne, this book) should be introduced. In fact the use of monoclonal antibodies coupled with autoradiography holds out the prospect for precise mapping of phenylethylaminergic, tryptaminergic, tyraminergic and octopaminergic fibre tracts now that indirect evidence for such pathways has been assembled (see Juorio *et al*, this book).

The enormously fast turnover rates for phenylethylamine and tryptamine (Durden and Philips, 1980) and probably the tyramines (Wu and Boulton, 1975), the reciprocal relationship that exists between the tyramines and dopamine (see Table 1) and the relatively high levels of some trace acid metabolites in urine (see Table 2) now indicate that a better understanding of the synthesis and catabolism of

Table 1. Reciprocal relationship between *p*-tyramine and *m*-tyramine concentrations and dopamine turnover in the caudate nucleus of the brain.

	p-TA concentration	*m*-TA concentration	DA turnover
DA antagonists	↓	⇡	↑
DA agonists	↑	↓	↓
DA stimulation of synthesis	↓	↑	↑
DA inhibition of synthesis	↑	↓	↓
DA releasors	↓	⇡	↑
Stress	↓	↑	↑

↑ increase; ↓ decrease; ⇡ tendency to increase.
Data from (Juorio 1979a, 1979b).

some of the trace amines is necessary. As Karoum (this book) argues, the trace acids probably do not arise simply by decarboxylation of their parent amino acids followed by oxidative deamination of the formed amines; the question of the role and magnitude of transamination must be sorted out. In addition, there are hints that there may be more than one decarboxylase (Boulton and Juorio, 1983; Juorio, 1983) and the role of conjugases and monoamine oxidases must be sorted out. Since meta-tyramine (Philips et al, 1975), meta-octopamine (Danielson et al, 1977; Ibrahim et al, 1985) and probably meta-synephrine along with their respective meta acidic metabolites exist in tissues and body fluids it follows that m-tyrosine must almost certainly exist as an endogenous tissue constituent since this is the most parsimonious explanation for the biosynthesis of these meta isomers, especially since the amine hydroxylases and dehydroxylases have now been shown to be of limited quantitative significance (Boulton and Dyck, 1974; Boulton et al, 1974; Juorio and Boulton, 1982; Dyck et al, 1983).

That the trace amines release the catecholamines and vastly potentiate the latter's properties when they are iontophoresed simultaneously is now established (Jones, 1984) as is the fact that some of the trace amines themselves are taken up and released (Dyck, 1984). In fact the release of the tyramines can be distinguished from that of dopamine using a variety of stimulants and aminotetralins (Dyck et al, 1980). That the trace amines, in addition to some of them acting as inhibitory neurotransmitters in their own right, possess a neuromodulatory role and interact with several of the classical neurotransmitters including some of the peptides is also now established (Jones, 1982, 1983, 1984; see also Langer et al, this book). The demonstration of their in vivo release awaits the development of more sensitive methods whilst a more detailed understanding of their actions at the level of the membrane requires the involvement of patch clamp and intracellular recording techniques. The existence of specific receptors for tryptamine has now been established by several groups (Kellar and Cascio, 1982; Cascio and Kellar, 1982, 1983; Bruning and Rommelspacher, 1984; Wood et al and Kienzl et al, this book).

As will be seen from these proceedings, characterisation of the behavioural pharmacology of some of the trace amines continues apace with studies on phenylethylamine and

Table 2. Urinary levels of some trace acids in the human.

	Unconjugated (mg/24h)	Conjugated (mg/24h)
Phenylacetic acid	4.10±0.81 (0.7-13.5)	96.4±5.6
p-Hydroxyphenylacetic acid	10.00±1.47 (5.6-54)	<1.2
m-Hydroxyphenylacetic acid	1.72±0.25 (0.3-15)	<0.1
Mandelic acid	0.09±0.009(0.04-0.19)	nm
p-Hydroxymandelic acid	0.76±0.09 (0.37-3.60)	nm
m-Hydroxymandelic acid[9]	0.060±0.008(0.015-0.127)	nm
Indoleacetic acid	4.05±0.66 (1.30-43)	nm

Actual values (mean ± SEM) in above table are from Davis, Durden, Yu, Dawson and Boulton (unpublished) except where otherwise indicated. The range of values (in parentheses) are from Davis and Boulton (1981a), Sandler *et al* (1979), Davis *et al* (1985), Domino *et al* (1979), Midgley *et al* (1979) and Davis and Boulton (1981b). nm = not measured.

tryptamine being in the forefront due to their ability to cross the blood brain barrier after peripheral administration. For the future, it is going to be necessary to investigate specific behavioural paradigms following intracerebral introduction of specific amines into targeted neuroanatomical locations. Coupled with such studies the quantitative assessment of regional brain levels of the biogenic amines in the presence and absence of specific (where possible) and catecholic and serotonergic agonists and antagonists will permit a clearer understanding of the roles, both direct and indirect, of the various trace amines. What is now urgently required from our pharmaceutical colleagues is a selection of drugs whose primary actions are directed at certain of the trace amines as opposed to dopamine, noradrenaline and 5-hydroxytryptamine. Some start on this has occurred with respect to some of the newer monoamine oxidase inhibitors.

Two years ago, I alluded to the effects of strategic deuterium substitution on the properties of the trace and other amines. By replacing the hydrogens on the side chain α-carbon atoms with deuterium it became possible to reduce considerably their oxidative deamination by monoamine oxidase (Yu et al, 1981, 1982). After injection of suitable deuterated amines, certain components of spontaneous locomotor activity and stereotypy were vastly potentiated (Dyck *et al*, 1983; Dourish *et al*, 1983a, 1983b). Since then we have investigated this phenomenon further and shown, using dopamine (Yu *et al*, 1985), that it is the R α-^{2}H enantiomers of dopamine that are critical since this is the bond that is cleaved by monoamine oxidase. Table 3 summarizes kinetic data for the various enantiomers of dopamine. It is expected that all monoamine oxidase catalysed amine substrates will exhibit this phenomenon which is in contradistinction to benzylamine and polyamines where the cleaved bond is S α-^{2}H via diamine oxidase (Richards and Spenser, 1983). As Dr. Dyck will describe later in this

Table 3. Oxidative deamination of several deuterated dopamine (DA) enantiomers.

DA	V_{max}	K_m	V_H/V_D	$(V/K)_H/(V/K)_D$
proteo	0.76	2.42	-	-
α,α-$^2H_2$-DA	0.30	2.59	2.53	2.71
R(α-$^2H_1$)DA	0.29	2.53	2.62	2.74
S(α-$^2H_1$)DA	0.77	3.37	0.99	1.37

MAO-B as present in human platelets. V_{max} (maximal velocities, nmoles/ng protein/min); K_m (Michaelis-Menten constant 1 + 10^{-4}M). Data taken from Yu *et al* (1985).

meeting, when dideutero-phenylethylamine or dideutero-tryptamine are injected intraperitoneally into rats, along with an equal amount of the proteoamine, considerably more of the deuterated forms of the amine persist in the brain as opposed to the plasma or liver for as long as two hours after the injection. This appears to be the result of the reduced ability of monoamine oxidase to deaminate oxidatively the amine coupled with a peripheral "first-pass" effect. That it is the consequence of the monoamine oxidase rather than an increased ability of the deuterated amine to penetrate into the brain is inferred from Dr. Durden's studies (this book) on ionization constants and partition coefficients. It could be argued that if the lipophilic amines phenylethylamine and tryptamine when deuterated penetrate into the brain more easily than their proteo analogues, then this ought to show up in their pK_a values and partition coefficients. In neither case was this so; in fact the pK_a values of the deuterated amine were increased by about 0.05 units and their partition coefficients were about 95% those of the non-deuterated amine. If anything this is likely to reduce the ability of these deuterated amines to penetrate into the brain.

As has been argued by Foster (1984) strategic deuteration of existing drugs and substances; as well, of course as the synthesis of new ones, could prove to be very beneficial to the drug industry. The data described above certainly support this suggestion, anything that can cause a 67-fold increase over conventional brain levels one hour after peripheral injection is worthy of further study.

In the clinical sphere, which Dr. Sandler will review, evidence in support of some sort of involvement of the trace amines in some psychiatric and neurological disorders continues to mount. Sabelli et al (1983) have claimed that a low total urinary phenylacetic acid may be a state marker for the diagnosis of some forms of unipolar major depressive disorders and Sandler et al (this book) claim a consistent and significant decrease in the urinary output of conjugated p-tyramine after an oral challenge; they suggest that this abnormal conjugation can be construed as a trait marker for depression.

These claims have been confirmed to some extent by Harrison et al (1983, 1984) and our own group (Yu et al, 1983). Others have implicated phenylethylamine, as well as some of the other trace amines and their acidic metabolites, in aggression, schizophrenia, epilepsy, migraine, Parkinson's disease, phenylketonuria, hypertensive crises and hepatic coma (see Davis et al, 1982, for summary). At this meeting, data implicating the trace amines and their metabolites with stress, personality, aggression, depression, attention deficit disorder, hypertension and schizophrenia is described as well as assessments of the excretion of the trace amines and their metabolites in healthy volunteers who have been treated with new experimental drugs, particularly monoamine oxidase inhibitors.

ACKNOWLEDGEMENT

The author thanks Saskatchewan Health and the Canadian MRC for continuing financial support.

REFERENCES

Boulton A.A., Baker G.B., Dewhurst W.G. and Sandler M. (eds.) (1984) Neurobiology of the Trace Amines. Humana Press, Clifton, New Jersey.

Boulton A.A. and Dyck L.E. (1974) Biosynthesis and excretion of meta- and para-tyramine in the rat. Life Sci. 14, 2497-2506.

Boulton A.A., Dyck L.E. and Durden D.A. (1974) Hydroxylation of β-phenylethylamine in the rat. Life Sci. 15, 1673-1683.

Boulton A.A. and Juorio A.V. (1983) Cerebral decarboxylation of meta- and para-tyrosine. Experientia 39, 130-134.

Brüning G. and Rommelspacher H. (1984) High affinity [^{3}H]tryptamine binding sites in various organs of the rat. Life Sci. 34, 1441-1446.

Cascio C.S. and Kellar K.J. (1982) Tetrahydro-β-carbolines: affinities for tryptamine and serotonergic binding sites. Neuropharmacology 21, 1219-1221.

Cascio C.S. and Kellar K.J. (1983) Characterisation of [^{3}H]Tryptamine binding sites in brain. Eur. J. Pharmacol. 95, 31-39.

Danielson T.J., Boulton A.A. and Robertson H.A. (1977) meta-Octopamine, para-octopamine and phenylethanolamine in rat brain: a sensitive specific assay and the effect of some drugs. J. Neurochem. 29, 1131-1135.

Davis B.A. and Boulton A.A. (1981a) Longitudinal urinary excretion of some 'trace' acids in a human male. J. Chromatog.: Biomed. Appln. 222, 161-169.

Davis B.A. and Boulton A.A. (1981b) Excretion of meta-hydroxymandelic acid in human urine. J. Chromatog.: Biomed Appln. 222, 271-275.

Davis B.A., Durden D.A. and Boulton A.A. (1985) Simultaneous analysis of 12 biogenic amine metabolites in plasma, cerebrospinal fluid and urine by capillary column gas chromatography-high resolution mass spectrometry with selected ion monitoring. J. Chromatog.: Biomed. Appln. (in press).

Davis B.A., Yu P.H., Carlson K., O'Sullivan K. and Boulton A.A. (1982) Plasma levels of phenylacetic acid, m- and p-hydroxyphenylacetic acid and platelet MAO activity in schizophrenic and other patients. Psychiat. Res. 6, 97-105.

Domino E.F., Mathews B.N. and Tait S.K. (1979) Urinary neurotransmitter metabolites in drug-free schizophrenic patients measured by gas chromatography selected position ion monitoring. Biomed. Mass Spec. 6, 331-334.

Dourish C.T., Dewar K.M., Dyck L.E. and Boulton A.A. (1983b) Potentiation of the behavioural effects of the antidepressant phenelzine by deuterium substitution. Psychopharmacol. 81, 122-125.

Dourish C.T., Greenshaw A.J. and Boulton A.A. (1983a) Deuterium substitution enhances the effects of phenylethylamine on spontaneous motor activity in the rat. Pharmacol. Biochem. Behav. 19, 471-475.

Durden D.A. (1985) High resolution and metastable mass spectrometry of trace amines and metabolites. In, NEUROMETHODS, Vol 2. Amines and Their Metabolites (A.A. Boulton, G.B. Baker and J.M. Baker, eds.), Humana Press, Clifton, New Jersey (in press).

Durden D.A. and Boulton A.A. (1982) Mass spectrometric analysis of some neurotransmitters and their precursors and metabolites. In, Handbook of Neurochemistry (A. Lathja, ed.), Plenum Press, New York, pp. 397-428.

Durden D.A. and Philips S.R. (1980) Kinetic measurements of the turnover rates of phenylethylamine and tryptamine in vivo in the rat brain. J. Neurochem. 34, 1725-1732.

Dyck L.E. (1984) Neuronal transport of trace amines: an overview. In, Neurobiology of the Trace Amines (A.A. Boulton, G.B. Baker, W.G. Dewhurst and M. Sandler, eds.), Humana Press, Clifton, New Jersey, pp. 185-204.

Dyck L.E., Boulton A.A. and Jones R.S.G. (1980) A comparison of the effects of methylphenidate and amphetamine on the simultaneous release of radiolabelled dopamine and m- and p-tyramine from rat striatal slices. Eur. J. Pharmacol. 68, 33-40.

Dyck L.E., Durden D.A., Yu P.H., Davis B.A. and Boulton A.A. (1983) Potentiation of the biochemical effects of β-phenylhydrazine by deuterium substitution. Biochem. Pharmacol. 32, 1519-1522.
Dyck L.E., Yang C.R. and Boulton A.A. (1983) The biosynthesis of p-tyramine, m-tyramine and β-phenylethylamine by rat striatal slices. J. Neurosci. Res. 10, 211-220.
Foster A.B. (1984) Deuterium isotope effects in studies of drug metabolism. TIPS 5, 524-527.
Harrison W.M., Cooper T.B., Quitkin F.M., Libowitz W.R., McGrath P.J., Stewart J.W. and Klein D.F. (1983) Tyramine sulfate excretion test in depressive illness. Psychopharm. Bull. 19, 503-504.
Harrison W.M., Cooper T.B., Stewart J.W., Quitkin F.M., McGrath P.J., Liebowitz M.R., Rabkin J.R., Mankowitz J.S. and Klein D.F. (1984) The tyramine challenge test as a marker for melancholia. Arch. Gen. Psychiat. 41, 681-685.
Ibrahim K.E., Couch M.W., Williams C.M., Fregly M.J. and Midgley J.M. (1985) m-Octopamine: normal occurrence with p-octopamine in mammalian sympathetic nerves. J. Neurochem. 44, 1862-1867.
Jones R.S.G. (1982) Tryptamine: a neuromodulator or neurotransmitter in mammalian brain? Progress in Neurobiology 19, 117-139.
Jones R.S.G. (1983) Trace biogenc amines: a possible functional role in the CNS. TIPS 4, 426-429.
Jones R.S.G. (1984) Electrophysiological studies of the possible role of trace amines in synaptic function. In, Neurobiology of the Trace Amines (A.A. Boulton, G.B. Baker, W.G. Dewhurst and M. Sandler, eds.), Humana Press, Clifton, New Jersey, pp. 205-223.
Juorio A.V. (1979a) Drug-induced changes in the formation, storage and metabolism of tyramine in the mouse. Brit. J. Pharmacol. 66, 377-384.
Juorio A.V. (1979b) Effect of stress and L-dopa administration on mouse striatal tyramine and homovanillic acid levels. Brain Res. 179, 186-189.
Juorio A.V. (1983) The effect of some decarboxylase inhibitors on mouse striatal tyramines. Neuropharmacology 22, 71-73.
Juorio A.V. and Boulton A.A. (1982) The effect of some precursor amino acids and enzyme inhibitors on the mouse striatal concentration of tyramines and homovanillic acid. J. Neurochem. 39, 859-863.

Kellar K.J. and Cascio C.S. (1982) [^{3}H]Tryptamine: high affinity binding sites in rat brain. Eur. J. Pharmacol. 78, 475-478.
Midgley J.M., Couch M.W., Crowley J.R. and Williams C.M. (1979) Identification and quantitative determination of o- and m-hydroxymandelic acid in human urine. Biomed. Mass Spec. 6, 485-490.
Palkovits M. (1985) Microdissection of individual brain nuclei and areas. In, NEUROMETHODS, Vol 1. General Techniques (in press).
Philips S.R., Davis B.A., Durden D.A. and Boulton A.A. (1975) Identification and distribution of meta-tyramine in the rat. Can. J. Biochem. 53, 65-69.
Richards J.C. and Spenser I.D. (1983) ^{2}H NMR Spectroscopy as a probe of the stereochemistry of enzymic reactions at prochiral centres. Tetrahedron 39, 3549-3568.
Sabelli H.C., Fawcett J., Gusovsky F., Javaid J., Edwards J. and Jeffriess (1983) Urinary phenylacetate: a diagnostic test for depression. Science 220, 1187-1188.
Sandler M., Ruthven C.R.J., Goodwin B.L., Reynolds G.P., Rao V.A.R. and Coppen A. (1979) Deficient production of tyramine and dopamine in cases of depression. Nature 278, 357-358.
Wu P.H. and Boulton A.A. (1975) Metabolism, distribution and disappearance of injected β-phenylethylamine in the rat. Can. J. Biochem. 52, 374-381.
Yu P.H., Bailey B.A., Durden D.A. and Boulton A.A. (1985) Stereospecific deuterium substitution at the α-carbon position of dopamine and its effects on oxidative deamination catalyzed by MAO-A and MAO-B from different tissues. Biochem. Pharmacol. (in press).
Yu P.H., Barclay S., Davis B.A. and Boulton A.A. (1981) Deuterium isotope effects on the enzymatic oxidative deamination of trace amines. Biochem. Pharmacol. 30, 3089-3094.
Yu P.H., Bowen R.C., Davis B.A. and Boulton A.A. (1983) Platelet MAO activity and trace acid levels in plasma of agoraphobic patients. Acta Psychiat. Scand. 67, 188-194.
Yu P.H., Kazakoff C.W., Davis B.A. and Boulton A.A. (1982) Deuterium iosotope effects on the enzymatic oxidation of dopamine and serotonin. Biochem. Pharmacol. 31, 3697-3698.

NEUROPSYCHOPHARMACOLOGY

Invited Communications

CEREBRAL 2-PHENYLETHYLAMINE CONCENTRATIONS AND TURNOVER NOT INFLUENCED BY VARIOUS PHARMACOLOGICAL TREATMENTS IN RATS

P.C. Waldmeier, A.-M. Buchle and J. Lauber

Research Department, Pharmaceuticals Division, CIBA-GEIGY Ltd., CH-4002 Basel, Switzerland.

INTRODUCTION

2-Phenylethylamine (PEA), a compound with stimulant properties when administered to animals, occurs in minute quantities in the normal mammalian brain (Durden et al. 1973). The turnover of this amine is high (Durden and Philips, 1980) and there is no evidence for storage, uptake or impulse-mediated release as yet. PEA, like other trace amines, has been regarded mainly as a metabolic accident in the past (see Boulton, 1984). However, a number of clinical studies with its precursor amino acid, phenylalanine, as well as investigations of its CSF concentrations or the urinary excretion of PEA or its metabolite phenylacetic acid in psychiatric patients seemed to indicate a possible role of the amine in depression, schizophrenia, and possibly also in extrapyramidal disorders and migraine (for reviews see e.g. Reynolds, 1979; Wolf and Mosnaim, 1983). Recently, it has been suggested that PEA at least partly mediates the antidepressant effect of maprotiline (Fawcett et al., 1983). We have consequently studied the effects of maprotiline and various other antidepressants on rat cerebral PEA concentrations and, as an index of PEA turnover, on their accumulation after inhibition of its degradation via B-type monoamine oxidase by pargyline. Since there is an obvious shortage of studies on possible effects of drugs affecting various cerebral transmitter systems which might interact with a hypothetical PEA system, we have also included compounds interacting with cholinergic, histaminic, serotonergic, dopaminergic and noradrenergic transmission in this study.

MATERIALS AND METHODS

Female Tif:RAIf(SPF) rats (Tierfarm Sisseln, Switzerland) weighing 160 - 200 g were used in these studies. (+)-Amphetamine sulfate, haloperidol and scopolamine HBr were purchased from Siegfried AG (Zofingen, Switzerland), Cilag AG (Schaffhausen, Switzerland), and Merck (Darmstadt, FRG), respectively. The following drugs were generously donated by their manufacturers: doxepine HCl (Pfizer & Co., Inc., New York NY, USA), iprindol HCl (Wyeth Labs., Philadelphia PA, USA), methysergide (Sandoz AG, Basle, Switzerland), nomifensine hydrogen maleinate (Farbwerke Hoechst, Frankfurt, FRG). Amitriptyline HCl, mianserin HCl, pargyline HCl, prazosin HCl x H_2O and trimipramine maleate were synthesized in our Chemical Department by Mr. H. Blattner, Dr. A. Storni, Dr. W. Bencze, Dr. Th. Leutert and Dr. H. Schroeter, resp. Imipramine HCl, methylphenidate HCl, maprotiline HCl and (+)- and (-)-oxaprotiline HCl are drugs of CIBA-GEIGY.

The drugs were given i.p. and the rats decapitated at the times indicated in text or legends. PEA was determined in the whole brains by GCMS as described previously (Lauber and Waldmeier, 1984). Accumulation rates were calculated according to equation (7) of Durden and Philips (1980).

RESULTS

Maprotiline at 50 mg/kg i.p. did not significantly ($p > 0.05$) influence the concentrations

Table 1 PEA Concentrations at different time intervals after administration of maprotiline

	time post adm.	PEA conc. (ng/g)	n
controls		2.2 ± 0.4	12
maprotiline	2 h	2.0 ± 0.2	5
	6 h	3.4 ± 1.0	4
	24 h	2.9 ± 0.6	5
	48 h	3.2 ± 0.4	5

Rats were treated with 50 mg/kg i.p. maprotiline and were decapitated at various times thereafter. PEA levels were measured in the whole brains. Data are means ± S.E.M. Statistical evaluation was done by means of Dunnett's t-test.

of PEA in the rat whole brain from 2 - 48 h after treatment (table 1). Moreover, none of the other antidepressants or other drugs interfering with various transmitter receptors or mechanisms showed a statistically significant effect on PEA concentrations 130 min after administration (tables 2 and 3); in particular, the absence of any effects of amphetamine and methylphenidate should be noted. In some cases there were considerable, though nonsignificant, changes; these drugs were reexamined in a further experiment, which invariably confirmed the absence of an effect.

The accumulation of PEA within 20 min after the administration of 100 mg/kg i.p. pargyline was linear (Fig. 1). Therefore, an interval of 10 min between pargyline treatment and death was chosen for the accumulation experiments in which the drugs were administered 2 h before pargyline. The results given in tables 2 and 3 demonstrate that none of the drugs significantly ($p > 0.05$) altered PEA accumulation induced by pargyline.

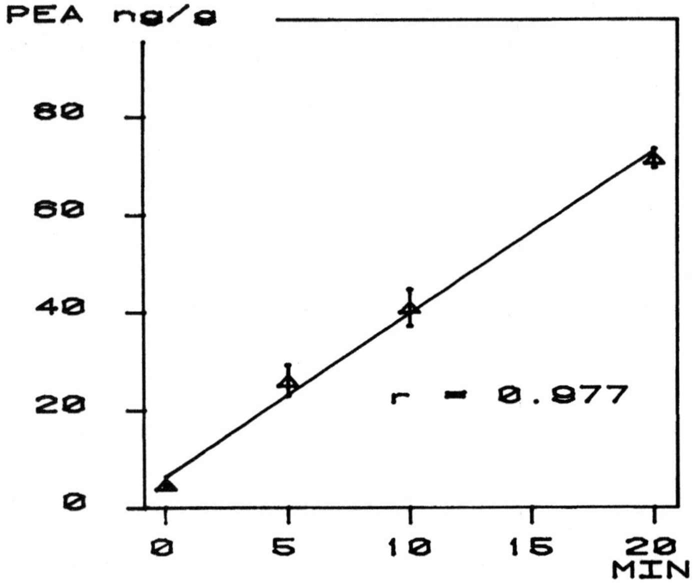

Fig. 1 Linearity of PEA accumulation in rat brain over 2Ø min after pargyline

Groups of 5 rats (controls = 9) received 1ØØ mg/kg i.p. pargyline and were decapitated 5, 1Ø or 2Ø min later. PEA levels were determined in the whole brains. Data are means ± S.E.M. The line was calculated by linear regression analysis.

Again, experiments were repeated in cases of doubt, and they invariably confirmed the absence of significant effects.

Table 4 compares the means of the PEA control levels and those obtained 1Ø min after pargyline, indicates the increase induced by the MAO inhibitor in percent of controls and the accumulation rate and half-life computed from each of the experiments referred to in tables 2 and 3 and of two additional experiments not shown in these tables. The considerable variation of the control levels from one experiment to the other (factor >

Table 2 PEA levels and their accumulation induced by pargyline after pretreatment with various antidepressants

treatment	dose mg/kg i.p.	controls ng/g	test drug ng/g	pargyline ng/g	test drug + pargyline ng/g
amitriptyline	50	4.4±0.7	4.9±0.4	36.5± 1.5	34.2±5.2
doxepine	50	2.6±0.4	1.7±0.3	66.7±12.2	48.9±9.9
imipramine	30	6.9±0.7	5.8±0.8	27.6± 3.0	27.7±1.5
iprindol	50	4.4±0.7	5.8±0.4	36.5± 1.5	25.9±7.3
	50	3.9±0.9	3.0±0.3	26.0± 3.3	21.6±1.5
maprotiline	20	3.9±0.9	4.1±0.4	26.0± 3.3	27.7±2.0
	50	3.9±0.9	3.5±0.3	26.0± 3.3	25.2±1.5
mianserin	30	6.9±0.7	4.6±0.4	27.6± 3.0	20.3±0.2
nomifensine	10	6.9±0.7	7.0±1.6	27.6± 3.0	28.2±7.0
(+)-oxaprotiline	10	4.4±0.7	5.7±0.3	36.5± 1.5	31.0±6.7
(-)-oxaprotiline	10	4.4±0.7	5.3±0.9	36.5± 1.5	29.2±5.6
trimipramine	50	4.4±0.2	11.0±2.0	36.5± 1.5	40.7±9.1
	50	5.2±0.4	7.6±0.8	38.9± 8.4	32.3±3.6

Groups of 5 rats (controls 9 - 10) received the test drugs 2 h before 100 mg/kg i.p. pargyline and were decapitated 10 min later. PEA levels were measured in the whole brains. Data are means ± S.E.M. Statistical comparisons between the groups receiving test drugs and controls or pargyline and pargyline plus test drugs, resp., were made using Dunnett's t-test.

Table 3 PEA levels and their accumulation induced by pargyline after pretreatment with substances interfering with various transmitter receptors and other mechanisms

treatment	dose mg/kg i.p.	controls ng/g	test drug ng/g	pargyline ng/g	test drug + pargyline ng/g
amphetamine	5	6.9±0.7	4.2±0.5	27.6± 3.0	26.6± 2.8
haloperidol	1	2.6±0.4	1.1±0.2	66.7±12.2	52.3±10.4
methylphenidate	10	6.9±0.7	5.2±0.6	27.6± 3.0	30.8± 3.1
methysergide	10	2.6±0.4	1.7±0.3	66.7±12.2	39.1± 6.4
	10	5.2±0.4	4.5±0.7	38.9± 8.4	32.9± 3.6
prazosin	10	2.6±0.4	1.3±0.4	66.7±12.2	39.3± 6.0
	10	5.2±0.4	5.2±0.6	38.9± 8.4	31.0± 3.1
scopolamine	10	2.6±0.4	1.6±0.2	66.7±12.2	35.6± 4.0
	10	5.2±0.4	11.4±1.2	38.9± 8.4	24.3± 1.6

Groups of 5 rats (controls 9 - 10) received the test drugs 2 h before 100 mg/kg i.p. pargyline and were decapitated 10 min later. PEA levels were measured in the whole brains. Data are means ± S.E.M. Statistical comparisons between the groups receiving test drugs and controls or pargyline and pargyline plus test drugs, resp., were made using Dunnett's t-test.

Table 4 Mean control PEA concentrations, their accumulation within 10 min after pargyline administration, and corresponding accumulation rates and half-lives from different experiments

controls ng/g	pargyl. ng/g	% incr.	accumul. rate nmol/g/h	half-life min
2.1±0.5	15.8± 1.5	752	0.68	1.53
4.8±0.5	41.0± 3.8	854	1.80	1.33
3.9±0.9	26.0± 3.3	667	1.10	1.76
2.6±0.4	66.7±12.2	2565	3.18	0.41
6.9±0.7	27.6± 3.0	400	1.03	3.33
5.2±0.4	38.9± 8.4	748	1.67	1.54
4.4±0.7	36.5± 1.5	830	1.59	1.37
4.2±0.3 (n = 64)	35.0± 3.6 (n = 34)		1.58±0.31 (n =7)	1.62±0.33 (n = 7)

Data are means ± S.E.M. from the experiments reported in tables 2 and 3 plus two additional ones not otherwise referred to in this study. The data in the last line are means of the values obtained in controls and pargyline treated rats; for the accumulation data, means were calculated from the data listed here.

3; see also Lauber and Waldmeier, 1984) should be noted. On the other hand, the accumulation after pargyline relative to the controls is remarkably constant, being 7 - 8 fold with 2 exceptions, one being higher and another lower. This is reflected in the accumulation rates and half-lives.

DISCUSSION

Earlier literature has suggested that imipramine and other antidepressants, amphetamine and methylphenidate increase brain PEA concentrations and synthesis (Fischer et al., 1972; Sabelli et al., 1973; Mosnaim et al., 1974; Borison et al. 1975; for review see Sabelli et al., 1978). However, it must be realized that the control concentrations found in these studies were orders of magnitude higher than those reported in later investigations (for a comparison see Lauber and Waldmeier, 1984), so that some doubts with respect to the specificity of these methods and consequently also to the above findings are certainly justified. In our experiments, using a method of which the specificity and accuracy has been clearly documented (Lauber and Waldmeier, 1984), we were unable to demonstrate any significant alterations of PEA concentrations in the rat brain by amphetamine, methylphenidate, antidepressants possessing noradrenaline and serotonin uptake inhibiting properties, anti-α-noradrenergic, antihistaminic and anticholinergic properties in varying proportions, and by antagonists of α_1-noradrenergic, 5-HT_2 serotonergic, muscarinic cholinergic and D_2 dopaminergic receptors.

Durden and Philips (1980) used 225 mg/kg i.p. pargyline, a rather toxic dose (LD50 = 175 mg/kg according to Usdin and Efron, 1972) to measure PEA synthesis by its accumulation after the MAO inhibitor, because they found a further increase with this dose over that observed with 100 mg/kg, 2 h after treatment. We wanted to avoid such a high dose and chose 100 mg/kg i.p. (a dose largely sufficient to totally block MAO in rat brain and liver for hours), which gave rise to a linear accumulation of PEA for 20 min. Accumulation rates calculated from the accumulation within 10 min were very similar to those reported by Durden and Philips (1980), suggesting that the degree of MAO B inhibition was sufficient to ensure complete blockade of PEA metabolism.

None of the antidepressants showed an effect on the accumulation of PEA after pargyline, suggesting that they did not alter PEA synthesis. This is at variance with the results of Sabelli et al. (1978), who reported imipramine and similar compounds to increase it. However, these data seem to suffer from the same methodological problems as those quoted above. Thus, our data do not provide a basis from animal experiments for the hypothesis of Fawcett et al. (1983), that PEA is involved in the antidepressant effect of maprotiline. Neither do they support the idea of a possible involvement of PEA in the antidepressant effect of tricyclic drugs in general.

If there existed a PEA neuromodulatory or even neurotransmitter system, as has been suggested by some authors (Sabelli et al., 1978; Boulton, 1979; Wolf and Mosnaim, 1983), one might expect it to interact with other transmitter systems in a way in which PEA synthesis and/or metabolism is affected by alterations of the activity of such other systems. In view of the low endogenous levels and the high turnover of this amine, even rather large changes might actually be anticipated. However, this study shows that pharmacological agents affecting a variety of transmitter receptors or mechanisms did not alter PEA concentrations nor its synthesis. The only treatments which seem to influence PEA levels are the increase of the availability of its precursor, phenylalanine, by its administration or by inhibition of its p-hydroxylation by p-chlorophenylalanine (Saavedra, 1974), the inhibition of its formation by blockade of aromatic amino acid decarboxylase (Silkaitis and Mosnaim, 1976) and inhibition of its degradation by MAO B inhibition (Philips and Boulton, 1979).

There is a long list of arguments against a neurotransmitter or neuromodulatory role of PEA, which includes

- -low and variable endogenous levels
- -absence of evidence for storage, uptake and release
- -the incongruity between the low endogenous le-

vels and the rather high doses of exogenously administered PEA necessary to elicit behavioural or motor effects (see e.g. Braestrup et al., 1975; Antelman et al., 1977; Borison et al., 1977; Ortmann et al., 1984)

-the fact that even high doses of MAO B inhibitors, which increase brain PEA severalfold, do not cause behavioural or motor changes comparable to those elicited by exogenously administered PEA (Ortmann, pers. commun.; see also Greenshaw, 1984).

This list is enlarged by the results of this study which suggest the absence of a regulation of PEA levels or synthesis by other transmitter systems.

REFERENCES

Antelman S.M., Edwards D.J. and Lin M. (1977) Phenylethylamine: evidence for a direct, postsynaptic dopamine-receptor stimulating action. Brain Res. 127, 317 - 322.

Borison R.L., Havdala H.S. and Diamond B.I. (1977) Chronic phenylethylamine stereotypy in rats: a new animal model for schizophrenia?. Life Sci. 21, 117 - 122.

Borison R.L., Mosnaim A.D. and Sabelli H.C. (1975) Brain 2-phenylethylamine as a major mediator for the central actions of amphetamine and methylphenidate. Life Sci. 17, 1331 - 1344.

Boulton A.A. (1984) Trace amines and the neurosciences: an overview, in Neurobiology of the trace amines, (Boulton A.A., Baker G.L., Dewhurst W.G. and Sandler M., eds), pp. 13 - 24, Humana Press, Clifton N.J..

Boulton A.A. (1979) Trace amines in the central nervous system, in International review of biochemistry, physiological and pharmacological biochemistry, 26, (Tipton K.F., eds), pp. 179 - 205, University Park Press, Baltimore.

Braestrup C., Andersen H. and Randrup A. (1975) The monoamine oxidase B inhibitor deprenyl potentiates phenylethylamine behaviour in rats without inhibition of catecholamine metabolite formation. Eur. J. Pharmacol. 34, 181 - 187.

Durden D.A. and Philips S.R. (1980) Kinetic measurements of the turnover rates of phenylethylamine and tryptamine in vivo in the rat brain. J. Neurochem. 34, 1725 - 1732.

Durden D.A., Philips S.R. and Boulton A.A. (1973) Identification and distribution of β-phenylethylamine in the rat. Can. J. Biochem. 51, 995 - 1002.

Fawcett J., Sabelli H., Gusovsky F., Epstein P., Javaid J. and Jeffriess H. (1983) Phenylethylaminic mechanisms in maprotiline antidepressant effect. Fed. Proc. 42, 1164 (abstr. 5110).

Fischer E., Spatz H., Heller B. and Reggiani H. (1972) Phenethylamine content of human urine and rat brain, its alterations in pathological conditions and after drug administration. Experientia 28, 307 - 308.

Greenshaw A.J. (1984) β-Phenylethylamine: a functional role at the behavioural level?, in Neurobiology of the trace amines, (Boulton A.A., Baker G.L., Dewhurst W.G. and Sandler M., eds), pp. 351 - 373, Humana Press, Clifton N.J.

Lauber J. and Waldmeier P.C. (1984) Determination of 2-phenylethylamine in rat brain after MAO inhibitors, and in human CSF and urine by capillary GC and chemical ionization MS. J. Neural Transm. 60, 247 - 264.

Mosnaim A.D., Inwang E.E. and Sabelli H.C. (1974) The influence of psychotropic drugs on the levels of endogenous 2-phenylethylamine in rabbit brain. Biol. Psychiatry 8, 227 - 234.

Ortmann R., Schaub M., Felner A., Lauber J., Christen P. and Waldmeier P.C. (1984) Phenylethylamine-induced stereotypies in the rat: a behavioral test system for assessment of MAO-B inhibitors. Psychopharmacology 84, 22 - 27.

Philips S.R. and Boulton A.A. (1979) The effects of monoamine oxidase inhibitors on some arylalkylamines in rat striatum. J. Neurochem. 33, 159 - 167.

Reynolds G.P. (1979) Phenylethylamine - a role in mental illness?. TINS 2, 265 - 268.

Saavedra J.M. (1974) Enzymatic icotopic assay for and presence of β-phenylethylamine in brain. J. Neurochem. 22, 211 - 216.

Sabelli H.C., Borison R.L., Diamond B.I., Havdala H.S. and Narasimhachari N. (1978) Phenylethylamine and brain function. Biochem. Pharmacol. 27, 1707 - 1711.

Sabelli H.C., Giardina W.J., Mosnaim A.D. and Sabelli N.H. (1973) A comparison of the functional roles of norepinephrine, dopamine and phenylethylamine in the central nervous system. Acta Psysiol. Pol. 24, 33 - 40.

Silkaitis R.P. and Mosnaim A.D. (1976) Pathways linking L-phenylalanine and 2-phenylethylamine with p-tyramine in rabbit brain. Brain Res. 114, 105 - 115.

Usdin E. and Efron D.H. (1972) Psychotropic drugs and related compounds. U.S. Dept. of Health, Education and Welfare, publication No. 72-9074, p. 401

Wolf M.E. and Mosnaim A.D. (1983) Phenylethylamine in neuropsychiatric disorders. Gen. Pharmac. 14, 385 - 390.

PHARMACOLOGICAL PROFILE OF β-PHENYLETHYLAMINE ON DOPAMINERGIC AND NORADRENERGIC NEUROTRANSMISSION IN RAT CEREBRAL SLICES : COMPARISON WITH AMPHETAMINE AND OTHER TRACE AMINES

Salomon Z. LANGER, Sonia ARBILLA, Ruth NIDDAM, Sanâa BENKIRANE and Patrice BAUD

Department of Biology,
Laboratoires d'Etudes et de Recherches Synthélabo (L.E.R.S.)
58, rue de la Glacière, 75013 Paris, France

INTRODUCTION

The locomotor stimulating properties of phenylethylamines have been attributed to their indirect dopamine and noradrenaline releasing action (Mogilnicka and Braestrup, 1976 ; Braestrup and Randrup, 1978), as well as to the direct stimulation of receptors for dopamine (Antelman et al., 1977) serotonin (Slotiver et al., 1980) or specific receptors for phenylethylamines (Hauger et al., 1982 ; Paul et al., 1982).

The receptor mediated modulation of the electrically-evoked release of ^{3}H-neurotransmitters from rat cerebral slices is a well-established model to study neurotransmitters interactions and the effects of drugs at the level of presynaptic receptors (Langer, 1981 ; Langer et al., 1980). In addition to the existence of presynaptic autoreceptors, modulatory heteroreceptors affecting the electrically-evoked release of several neurotransmitters have also been described (Langer and Arbilla, 1981). For example, the stimulation-evoked release of ^{3}H-acetylcholine (ACh) in the corpus striatum is modulated by an inhibitory dopamine receptor which is a target for dopamine receptor agonists (Lehmann et al., 1983a,b ; Arbilla et al., 1985), exogenous (Arbilla et al., 1985) or endogenously released dopamine (Baud et al., 1985).

Another example is the stimulation-evoked release of ^{3}H-5-hydroxytryptamine (5HT) which can be inhibited by pharmacological activation of α-adrenoceptors of the α_2-subtype (Frankhuyzen and Mulder, 1982).

At the level of the electrically-evoked release of ^{3}H-ACh from rat striatal slices, it has been demonstrated that amphetamine (AMPH) inhibits the release of ^{3}H-ACh by promoting the release of newly synthetised dopamine (Cantrill et al., 1983a,b ; Baud et al., 1985). Consequently, the modulation of the stimulation evoked release of ^{3}H-transmitters is a useful model to evaluate changes in catecholaminergic transmission induced by AMPH, as well as by the naturally occuring structurally related trace amines.

The influence of m-tyramine (TYR), p-TYR and β-phenylethylamine (PEA) on dopaminergic neurotransmission, as measured by their effects on the electrically-evoked release of ^{3}H-ACh from rat striatal slices was studied in parallel with AMPH. In addition, the stimulation of the α_2-adrenoceptor, which inhibits the electrically-evoked release of ^{3}H-5HT in hippocampal slices was employed as a model to study the influence of β-PEA on noradrenergic neurotransmission. Finally, since the possible role of β-PEA as a transmitter has received considerable attention (Loo et al., 1978), the possibility that ^{3}H-β-PEA can be taken up and released in a calcium-dependent manner from rat striatal slices, was also explored.

RESULTS

Facilitation of dopaminergic neurotransmission by β-PEA : similarities with AMPH, but not with m-TYR and p-TYR

When the release of ^{3}H-ACh from rat striatal slices is elicited at a low frequency of stimulation (1 Hz, 2 msec, 2 min), endogenously released DA does not activate DA receptors in the cholinergic interneuron, unless dopaminergic neurotransmission is enhanced by a DA releasing agent or by inhibition of the neuronal reuptake of DA (Baud et al., 1985). In this preparation, in the absence of monoamine oxidase (MAO) inhibition, exposure to AMPH inhibits ^{3}H-ACh release (Cantrill et al., 1983b ; Baud et al., 1985). Yet, under the same experimental conditions of intact MAO activity, m-TYR or β-PEA fail to modify ^{3}H-ACh release elicited by electrical stimulation (Cantrill et al., 1983a ; Arbilla et al., 1984).

When MAO of types A and B are inhibited by pargyline (Fuentes and Neff, 1975), in order to protect β-PEA and TYR from MAO activity (Yang and Neff, 1975), we observed that similarly to AMPH, m-TYR, p-TYR and β-PEA inhibit in a concentration dependent manner the electrically-evoked release of ^{3}H-ACh (Fig. 1). This inhibitory effect by the four amines on the release of ^{3}H-ACh closely parallels with their potency at enhancing the spontaneous outflow of ^{3}H-DA from rat striatal slices labelled with ^{3}H-DA (Fig. 1).

It is well established in behavioural tests that β-PEA, like AMPH can affect dopaminergic neurotransmission by an action on the reserpine resistant pool of DA (Moore, 1978 ; Fuxe and Ungerstedt, 1970 ; Scheel-Kruger, 1971), probably by promoting a latent dopaminergic neurotransmission, which does not involve the vesicular storage pool of DA (Cantrill et al., 1983b ; Langer and Arbilla, 1984). The inhibitory action of AMPH and β-PEA on the "in vitro" release of ^{3}H-ACh is not affected by reserpine (RES) pretreatment (Arbilla et al., 1984 ; Langer et al., 1984), providing additional support for the "in vivo" observations.

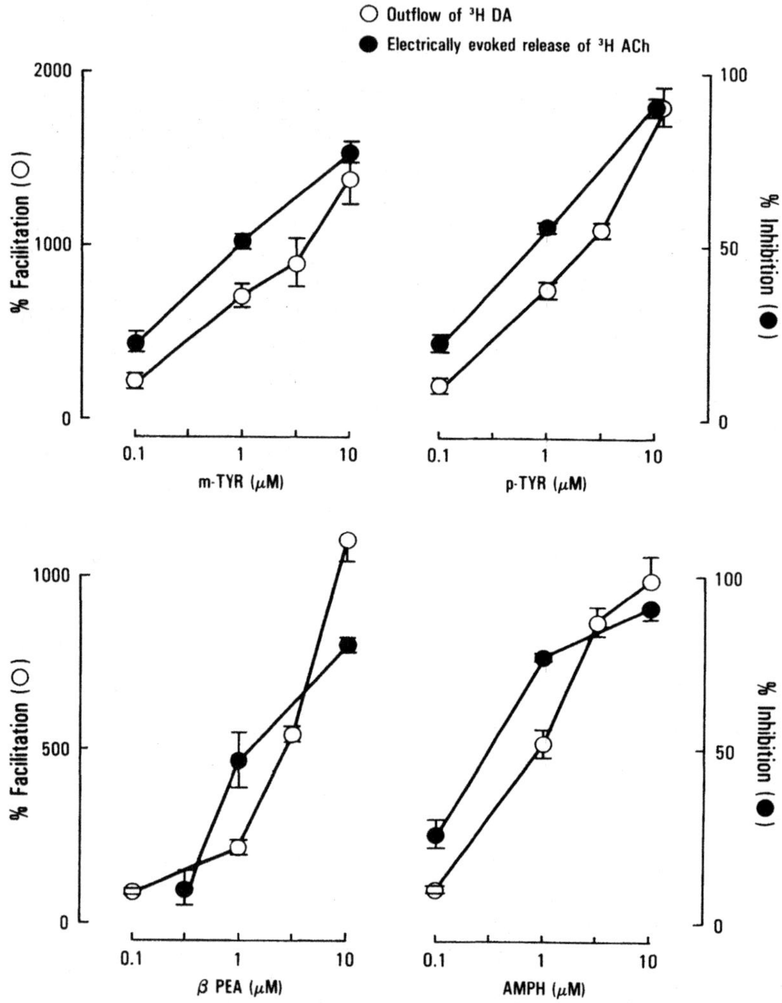

Figure 1 : Influence of m-TYR, p-TYR, AMPH and β-PEA on the spontaneous outflow of ^{3}H-DA and on the electrically-evoked release of ^{3}H-ACh from perfused rat striatal slices in the presence of pargyline 10 μM. Ordinate left (open circles) : percentage of facilitation by drugs of the spontaneous outflow of ^{3}H-DA measured as described by Arbilla et al. (1985). Ordinate right (full circles) : percentage of the inhibition of the electrically-evoked release of ^{3}H-ACh determined as described by Baud et al. (1985). Shown are mean values ± S.E.M. from 4 to 7 experiments.

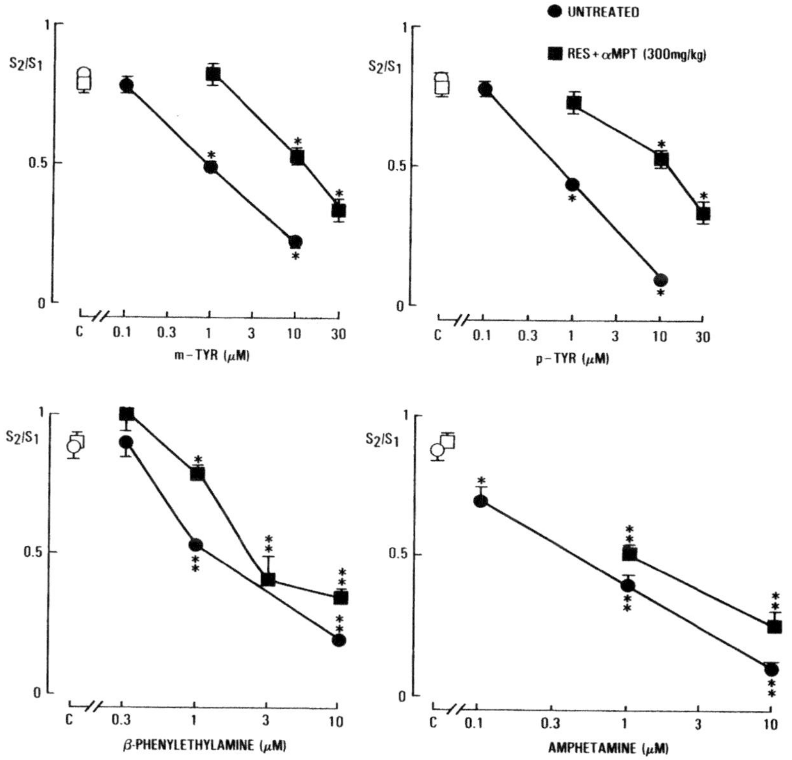

Figure 2 : Effects of depletion of endogenous DA on the inhibition by m-TYR, p-TYR, AMPH and β-PEA of the electrically-evoked release of ^{3}H-ACh from rat striatal slices.

Pargyline 10 μM was present in the Krebs medium. Two periods of electrical stimulation (S_1 and S_2) at 1 Hz, 24 mA for 2 min were applied to elicit ^{3}H-ACh release (Baud et al., 1985). S_2/S_1 represents the ratio between the two periods of stimulation under control conditions and after 20 min exposure before S_2 to the different drugs concentrations. RES (5 mg/kg s.c.) was injected 24 hours before the experiment. α-MpT (300 mg/kg i.p.) was injected 2 hours before the experiment and was present in the Krebs medium at 100 μM. Shown are mean values ± S.E.M. from 3 to 9 experiments. * $p < 0.05$ and ** $p < 0.001$ vs control (C) group (shown as open symbols).

In striatal slices from rats preteated with RES (5 mg/kg s.c., 24 h), combined with α-MPT (300 mg/kg i.p., 2 h + 100 μM in the Krebs medium) to ensure maximal depletion of DA stores and synthesis, the inhibitory action of m-TYR and p-TYR on ^{3}H-ACh release was markedly reduced, while AMPH and β-PEA retained their activity under these conditions (Figure 2 ; Langer et al., 1984 ; Arbilla et al., 1984). This inhibitory activity of β-PEA and AMPH under conditions of severe depletion of DA are mediated through the activation of DA receptors of D_2 subtype, because the inhibition is antagonized by S-sulpiride (Arbilla et al., 1984 ; Langer et al., 1984).

It is unlikely that the actions of AMPH and β-PEA on the cholinergic interneuron are mediated through direct activation of dopaminergic or specific receptors for phenylethylamines ; in fact, both amines are inactive after 6-OHDA induced denervation of DA neurons (Langer et al., 1984). Furthermore, the available data demonstrate that the integrity and the presence of dopaminergic nerve endings in the striatum is essential for AMPH and β-PEA to exert their inhibitory action on the stimulation-evoked release of ^{3}H-ACh.

<u>Electrically-stimulated release of ^{3}H-β-PEA but not of ^{3}H-AMPH from rat striatal slices</u>

In perfused rat striatal slices, labelled with either ^{3}H-DA, ^{3}H-β-PEA or ^{3}H-(+)AMPH, electrical stimulation (2 min, 3 Hz, 24 mA, 2 msec) releases ^{3}H-DA and ^{3}H-β-PEA, but not ^{3}H-(+)AMPH (Table 1). The release of ^{3}H-β-PEA is comparable to that of ^{3}H-DA (Table 1), and appears to be originated from vesicular stores, since it is abolished by the absence of calcium and does not occur in reserpinized tissues (Niddam et al., 1985). In addition, the release of ^{3}H-β-PEA might be associated with the release of DA, because it is facilitated by blockade of DA autoreceptors by S-sulpiride (Niddam et al., 1985). These results might indicate the existence of an exocytotic release of ^{3}H-β-PEA, probably from dopaminergic nerve terminals. However, since β-PEA does not modify by itself the electrically-evoked release of ^{3}H-ACh after 6-OHDA lesions (Langer et al.,1984), it appears that the cholinergic interneuron does not possess recognition sites for released β-PEA.

^{3}H-amine	n	S_1	S_2/S_1	Tissue content (nCi)[a]
^{3}H- DA	15	5.68 ± 0.42	0.88 ± 0.02	22.79 ± 2.74
^{3}H-β-PEA	37	5.06 ± 0.36	0.88 ± 0.06	6.11 ± 0.57
^{3}H-AMPH	12	0.13 ± 0.09	0.02 ± 0.02	2.71 ± 0.10

Table 1 : Depolarization induced release of ^{3}H-β-PEA and ^{3}H-DA, but not of ^{3}H-AMPH from perfused rat striatal slices.

S_2/S_1 represents the ratio between two consecutive periods of electrical stimulation, S_1 represents the percentage of total tissue radioactivity released by the first period of electrical stimulation and S_2 the second one applied 44 min after S_1. (a) radioactivity retained at the end of the superfusion. For details, see Niddam et al., 1985. Values are mean ± S.E.M. of n experiments per group.

Facilitation of noradrenergic transmission by β-PEA and AMPH

It is widely accepted that inhibitory α_2-adrenoceptors modulate the depolarization induced release of ^{3}H-5HT from rat hypothalamic (Langer and Moret, 1982 ; Galzin et al., 1983) and hippocampal slices (Frankhuyzen and Mulder, 1982).

In rat hippocampal slices, AMPH 1 µM, which facilitates noradrenaline release in brain slices by about 250 % (Kamal et al., 1983), inhibits significantly the electrically-evoked release of ^{3}H-5HT, without modifying the spontaneous outflow of radioactivity (Table 2).

Under the same experimental conditions, but with selective inhibition of MAO B by deprenyl 1 µM (Fuentes and Neff, 1975), exposure to β-PEA inhibits ^{3}H-5HT release (Table 2) by activation of α_2-adrenoceptors, since this effect is antagonized by idazoxan 1 µM (Table 2). The involvement of endogenously released noradrenaline in the effects of β-PEA on ^{3}H-5HT release is also supported by the lack of effects of β-PEA after inhibition of tyrosine hydroxylase activity with α-MpT (Table 2). From our experimental data, it appears that α_2-adrenoceptors in the serotoninergic nerve terminal are activated by noradrenaline, only when its release is promoted by the exposure to certain indirectly acting amines like β-PEA and AMPH. In fact, under our experimental conditions, blockade of α_2-adrenoceptors does not on its own facilitate ^{3}H-5HT release, neither in the hippocampus (S. Benkirane, S. Arbilla and S.Z. Langer, unpublished observations), nor in the hypothalamus (Galzin et al., 1983). The lack of effects of β-PEA on ^{3}H-5HT release after α-MpT or following the selective blockade of α_2-adrenoceptors by idazoxan indicates that β-PEA, at the concentration tested, does not activate 5HT receptors in the serotoninergic nerve terminals, and therefore excludes in our model a serotoninergic autoreceptor agonist activity of β-PEA.

CONCLUSIONS

In our experimental model of dopaminergic neurotransmission, when MAO is inhibited, m-TYR, p-TYR, β-PEA and AMPH exert an inhibitory effect on ^{3}H-ACh release, which parallels with their DA releasing action. In contrast to m-TYR and p-TYR, β-PEA has AMPH-like properties, since both are able to enhance a latent dopaminergic neurotransmission from a newly synthetised and reserpine-resistant pool of DA. It appears that β-PEA and AMPH are able to release newly synthesized DA in preference to the stored transmitter. However, it is necessary to inhibit MAO to demonstrate for β-PEA an activity similar to that of AMPH. The effects of β-PEA and AMPH in the striatal cholinergic interneuron appear to be mediated through the activation of DA receptors and do not involve interactions with specific AMPH or β-PEA receptors.

In the experimental model of noradrenergic neurotransmission, AMPH and β-PEA inhibit ^{3}H-5HT release. The effect of β-PEA is mediated by indirect activation of α_2-adrenoceptors by released noradrenaline and does not involve specific β-PEA or serotoninergic receptors.

In spite of the structural similarities between ^{3}H-β-PEA and ^{3}H-AMPH, only the first, which represents the naturally occuring analogue of AMPH, can be released by exocytosis. Striatal cholinergic interneurons and hippocampal serotoninergic nerve terminals do not seem to possess specific receptors for β-PEA. However, the release of ^{3}H-β-PEA, by electrical stimulation, suggests a possible neurotransmitter or neuromodulator role for this trace amine.

	µM	n	S_1	S_2/S_1	Sp_2/Sp_1
Normal Krebs	-	9	1.43 ± 0.15	0.99 ± 0.05	0.72 ± 0.04
+ AMPH (S_2)	1	7	1.59 ± 0.10	0.74 ± 0.04*	0.74 ± 0.02
Deprenyl (S_1 + S_2)	1	8	1.22 ± 0.09	0.83 ± 0.03	0.71 ± 0.06
+ β-PEA (S_2)	3	11	1.12 ± 0.10	0.53 ± 0.03*	0.78 ± 0.02
+ β-PEA (S_2)	10	9	1.12 ± 0.10	0.39 ± 0.06*	1.02 ± 0.09*
+ Idazoxan (S_1 + S_2)	1	9	1.21 ± 0.10	0.99 ± 0.04	0.63 ± 0.02
+ β-PEA (S_2)	3	10	1.53 ± 0.02	0.90 ± 0.05	0.80 ± 0.03
Deprenyl	1				
+ α-MpT (S_1 + S_2)	100	7	1.22 ± 0.12	0.91 ± 0.10	0.74 ± 0.02
+ β-PEA (S_2)	3	4	1.11 ± 0.20	0.89 ± 0.13	0.90 ± 0.06
+ β-PEA (S_2)	10	8	1.28 ± 0.19	1.03 ± 0.15	0.99 ± 0.06*

Table 2 : Inhibitory effects of β-PEA and AMPH on the electrically-evoked release of ^{3}H-5HT from perfused slices of rat hippocampus. S_1 corresponds to the percentage of total tissue radioactivity released by a first period of electrical stimulation (3 Hz, 2 min, 24 mA) and S_2 to an identical second one applied 44 min later. Sp_2/Sp_1 is the ratio between the spontaneous outflow of radioactivity obtained during the 4-min preceding S_2 (Sp_2) and the corresponding fraction released before S_1 (Sp_1). When indicated, deprenyl 1 µM was present in the Krebs medium. In experiments in the presence of α-MpT, the drug was injected (300 mg/kg i.p.) 2 hours before the experiment and added to the medium at the concentration of 100 µM. Other drugs were added 20 min before either S_1 or S_2. Values are the mean ± S.E.M. of n experiments per group. * $p < 0.05$ when compared with the corresponding control group.

REFERENCES

ANTELMAN, S.M., EDWARDS, D.J. and LIN, M. (1977) Phenylethylamine : evidence for a direct, postsynaptic dopamine-receptor stimulating action. Brain Research, 127 : 317-322.

ARBILLA, S., BAUD, P., CANTRILL, R. and LANGER, S.Z. (1984) Endogenous phenolamines : do they have a direct postsynaptic action on dopamine receptors in the central nervous system ? Br. J. Pharmacol., 82 : 852P

ARBILLA, S., NOWAK, J.Z. and LANGER, S.Z. (1985) Rapid desensitization of presynaptic dopamine autoreceptors during exposure to exogenous dopamine. Brain Res., in press.

BAUD, P., ARBILLA, S., and LANGER, S.Z. (1985) Inhibition of the electrically-evoked release of ^{3}H-acetylcholine in rat striatal slices : an experimental model for drugs that enhance dopaminergic neurotransmission. J. Neurochem., 44 : 331-337.

BRAESTRUP, C. and RANDRUP, A. (1978) Stereotypes behaviour in rats induced by phenylethylamine, dependence on dopamine and noradrenaline, and possible relation to psychoses. In : Noncatecholic phenylethylamines. A.D. Mosnaim and M.E. Wolf eds, pp. 245-269.

CANTRILL, R., ARBILLA, S. and LANGER, S.Z. (1983a) Inhibition by (+)-amphetamine of the electrically-evoked release of ^{3}H-acetylcholine from slices of the rat striatum : involvement of dopamine receptors. Eur. J. Pharmacol., 87 : 167-168.

CANTRILL, R., ARBILLA, S., ZIVKOVIC, B. and LANGER, S.Z. (1983b) Amphetamine enhances latent dopaminergic neurotransmission in the rat striatum : effects on ^{3}H-acetylcholine release. Naunyn-Schmiedeberg's Arch. Pharmacol., 322 : 322-324.

FRANKHUYZEN, A.L. and MULDER, A.H. (1982) Pharmacological characterization of presynaptic α-adrenoceptors modulating [^{3}H]noradrenaline and [^{3}H]5-hydroxytryptamine release from slices of the hippocampus of the rat. Eur. J. Pharmacol., 81 : 97-106.

FUENTES, J.A. and NEFF, N.H. (1975) Selective monoamine oxidase inhibitor drugs as aids in evaluating the role of type A and B enzymes. Neuropharmacology, 14 : 819-825.

FUXE, K. and UNGERSTEDT, U. (1970) Histochemical, biochemical and functional studies on central monoaminergic neurons after acute and chronic amphetamine administration. In : Amphetamines and related compounds, E. Costa and S. Garattini, eds, Raven Press, New York, pp. 257-288.

GALZIN, A.M., MORET, C. and LANGER, S.Z. (1983) Evidence that exogenous by not endogenous norepinephrine activates the presynaptic $alpha_2$-adrenoceptors on serotonergic nerve endings in rat hypothalamus. J. Pharmacol. Exp. Ther., 228 : 220-226.

HAUGER, R.L., SKOLNICK, P.H. and PAUL, S.M. (1982) Specific [^{3}H]β-phenylethylamine binding sites in rat brain. Eur. J. Pharmacol., 83 : 147-148.

KAMAL, L.A., ARBILLA, S., GALZIN, A.M. and LANGER, S.Z. (1983) Amphetamine inhibits the electrically-evoked release of [^{3}H]dopamine from slices of the rabbit caudate. J. Pharmacol. Exp. Ther., 227 : 446-458.

LANGER, S.Z. (1981) : Presynaptic regulation of the release of catecholamines. Pharmacol. Rev., 32 : 337-362.

LANGER, S.Z. and ARBILLA, S. (1981) Presynaptic receptors and modulation of the release of noradrenaline, dopamine and GABA. Posgraduate Medical Journal 57, suppl. 1 : 18-29.

LANGER, S.Z. and ARBILLA, S. (1984) The amphetamine paradox in dopaminergic neurotransmission. Trends in Pharmacological Sciences, 9 : 387-390.

LANGER, S.Z., ARBILLA, S. and KAMAL, L. (1980) Autoregulation of noradrenaline and dopamine release through presynaptic receptors. In : Neurotransmitters and their receptors, U.Z. Littauer, Y. Dudai, I. Silman, V.I. Teichberg and Z. Vogel eds, John Wiley and Sons Ltd.

LANGER, S.Z. ARBILLA, S. BAUD, P., NIDDAM, R. and SCATTON, B. (1984) Dopamine receptors modulating the release of several neurotransmitters in the peripheral and central nervous system. Clin. Neuropharmac., 7, suppl. 1 : 798-799.

LANGER, S.Z. and MORET, C. (1982) Citalopram antagonizes the stimulation by lysergic acid diethylamide of presynaptic inhibitory serotonin autoreceptors in the rat hypothalamus. J. Pharmacol. Exp. Ther, 222 : 220-226.

LEHMANN, J. and LANGER, S.Z. (1983a) Dopamine receptors modulating ^{3}H-acetylcholine release in slices of the

cat caudate : effects of (-)N-(2-chloroethyl)-norapomorphine. Eur. J. Pharmacol., 90 : 393-400.
LEHMANN, J., SMITH, P.V. and LANGER, S.Z. (1983b) Stereoisomers of apomorphine differ in affinity and intrinsic activity at presynaptic dopaminergic receptors modulating ^{3}H-dopamine and ^{3}H-acetylcholine release in slices of cat caudate. Eur. J. Pharmacol., 88 : 81-88.
LOO, Y.H., JERVIS, G.A. and HORNING, M.G. (1978) Possible role of 2-phenylethylamine in the pathology of the central nervous system. In : Non-catecholic phenylethylamines ; Part I : Phenylethylamine : biological mechanisms and clinical aspects, eds A.D. Mosnaim and M.E. Wolf (Marcel Dekker, New York), pp. 419-445.
MOGILNICKA, E. and BRAESTRUP, C. (1976) Noradrenergic influence on the stereotypes behaviour induced by amphetamine, phenylethylamine and apomorphine. J. Pharmac. Pharmacol., 28 : 253-255.
MOORE, K.E. (1978) The actions of amphetamine on neurotransmitters : a brief review. Biol. Psychiat., 12, 451-462.
NIDDAM, R., ARBILLA, S., BAUD, P. and LANGER, S.Z. (1985) ^{3}H-β-phenylethylamine but not ^{3}H-amphetamine is released by electrical stimulation from perfused rat striatal slices. Eur. J. Pharmacol., 110, 121-124.
PAUL, S.M., HULIHAN-GIBLIN, B. SKOLNICK, P. (1982) (+)-Amphetamine binding to rat hypothalamus : relation to anorexic potency of phenylethylamines. Science, 218 : 487-490.
SCHEEL-KRUGER, J. (1971) : Comparative studies of various amphetamine analogues demonstrating different interactions with the metabolism of the catecholamines in the brain. Eur. J. Pharmacol., 14 : 47-59.
SLOTIVER, R.S., CONNOR, J.D. and DRUST, E.G. (1980) Serotonergic properties of β-phenylethylamine in rats. Neuropharmacology, 19 : 1071-1074.
YANG, H.Y.T. and NEFF, N.H. (1973) β-Phenylethylamine : a specific substrate for type B monoamine oxidase of brain. J. Pharmacol. Exp. Ther., 187, 365-371.

FURTHER INVESTIGATION OF THE METABOLISM OF TWO TRACE AMINES, B-PHENYLETHYLAMINE AND p-TYRAMINE BY RAT AORTA SEMICARBAZIDE-SENSITIVE AMINE OXIDASE

C. GUFFROY, T. BOUCHER and M. STROLIN BENEDETTI.

Laboratoires Fournier, Centre de Recherche de Daix - 21121 Fontaine les Dijon (France)

INTRODUCTION

A number of rat tissues, particularly those of the cardiovascular system (Guffroy & Strolin Benedetti, 1984 a ; Lyles, 1984), contain an amine oxidase with a different substrate and inhibitor specificity from the flavin-dependent mitochondrial monoamine oxidase (MAO-A and B). The enzyme is resistant to MAO inhibiting concentrations (e.g. 1mM) of the acetylenic drugs, clorgyline and deprenyl, but is sensitive to carbonyl reagents such as semicarbazide which at similar concentrations produce little or no inhibition of MAO. Consequently, pyridoxal phosphate has been proposed as a possible cofactor for the semicarbazide-sensitive amine oxidase (SSAO). SSAO activity may conveniently be measured with low micromolar concentrations of the exogenous amine, benzylamine (K_m around 5 µM).

The physiological role of SSAO is far from being elucidated and merits further investigation. It is intriguing to speculate that benzylamine itself might be an endogenous substrate, as benzylamine has not so far been detected in man or rats. Hippuric acid, however, the glycine conjugate of benzoic acid, is found in urine and although it has been shown to arise mainly by benzoic acid of dietary origin, presence of hippuric acid in urine of fasting humans has been reported, suggesting that a small amount of hippuric acid is produced endogenously (Armstrong et al., 1955)

Recent studies have indicated that not only benzylamine but other MAO substrates such as 2-phenylethylamine (β-PEA) and p-tyramine (p-Tyr) are also metabolized by SSAO (Lyles, 1984 ; Urdin & Fuentes, 1983 ; Barrand & Callingham, 1982 ; Clarke et al., 1982 ; Guffroy & Strolin Benedetti, 1984 b). A considerable number of biochemical studies have indicated that these amines belong to the endogenous amines which occur in trace amounts in the central nervous system (Jones, 1983).

β-PEA and p-Tyr are present in rat blood and brain at about 10 nM concentrations (Boulton, 1979) . However, in studies with β-PEA and p-Tyr as substrates of SSAO, the concentrations used were never inferior to 10 μM (Tipton et al., 1984 ; Clarke et al., 1982). Moreover no information exists on the K_m and V_{max} values of this enzyme towards these two substrates, in the cardiovascular tissue. K_m values for the deamination of p-Tyr and β-PEA by SSAO and MAO A in homogenates of brown adipose tissue and brown adipocytes of the rat are given by Barrand and Fox (1984) : interestingly, the K_m values for the deamination of these two substrates by SSAO are lower than those for the deamination by MAO A. In a recent abstract Hasan and Tipton (1984) reported that comparison of the K_m and V_{max} values of rat intestinal MAO and SSAO indicates that the SSAO would play an important role in the metabolism of low concentrations of p-Tyr.
The purpose of the present work was :
- to study the oxidative deamination of these amines at low concentration (1μM) by MAO and SSAO in rat aorta, following the decrease of their oxidation as a function of increasing concentrations of clorgyline or semicarbazide ;
- to determine the kinetic parameters of the deamination of these amines by MAO and SSAO of rat aorta ;
- to investigate the "in vivo" contribution of SSAO to the metabolism of a small i.v. dose of ^{14}C-β-PEA.

MATERIAL AND METHODS

[7-^{14}C] p-tyramine hydrochloride was obtained from the

radiochemical centre, Amersham, U.K.; β-phenethylamine-[ethyl-1 ^{14}C] hydrochloride was obtained from New England Nuclear, Boston, MA, USA; clorgyline hydrochloride was a gift from May and Baker Ltd, Dagenham, U.K.; semicarbazide hydrochloride was obtained from Merck, Darmstadt, F.R. Germany; tranylcypromine hydrochloride was obtained from Sigma, St. Louis, MO, USA. All other reagents were standard laboratory reagents of analytical grade whenever possible.
Male Wistar rats weighing 220-280 g were supplied by IFFA CREDO, L'Arbresle, France.

"In vitro" experiments

Thoracic aorta MAO and SSAO activities were determined at pH 7.8 with fresh tissues, as previously described (Guffroy & Strolin Benedetti, 1984 a) . Preincubation time with clorgyline or semicarbazide was always 20 minutes at 37°C. The substrate concentration ranges used for the determination of the kinetic constants were :
- for p-Tyr : 16.66 to 100 μM (SSAO, MAO and total activity),
- for β-PEA : 4 to 64 μM (SSAO) and 50 to 1000 μM (MAO and total activity). Final protein concentration in the assay tube, determined by the method of Lowry et al.(1951), was 0.85 ± 0.03 mg/ml.

"In vivo" experiments

Rats received an i.v. injection of 0.154 mg/kg ^{14}C-β-PEA (0.2 mg/kg of β-PEA hydrochloride) in NaCl 0.9% (3 ml/kg). Injected radioactivity was 10 or 20 μCi/kg. One hour before the i.v. injection, rats were pretreated p.o. with tranylcypromine (5 mg/kg) and/or semicarbazide (100 mg/kg) (in aqueous solution, 5 ml/kg, dose expressed as base). Controls were treated with H_2O. Two to 60 minutes after the i.v. injection, rats were killed by decapitation and blood was collected on ice in 50 μl of a mixture of heparine (1000 units/ml), tranylcypromine (1.10^{-2}M) and semicarbazide (1.10^{-1}M). Five ml of 0.4 N perchloric acid were added to 1ml of plasma. After centrifugation at 15 000 gx10min, 1ml of the supernatant was taken for a total radioactivity count and 2ml for extraction of unchanged β-PEA as described by Fuller and Roush (1972) except that benzene was replaced by toluene. Under these conditions, β-PEA was extracted with a very high efficiency (92%) and p-Tyr was extracted with less than 1% efficiency. In a first experiment rats were pretreated with semicarbazide and compared to the controls. In a second

experiment, rats pretreated with semicarbazide + tranylcypromine were compared to rats pretreated with tranylcypromine. In a third experiment rats were pretreated with semicarbazide and/or tranylcypromine and compared to the controls. In this last experiment rats were killed 20 minutes after the i.v. injection of β-PEA ; blood was collected as in the two previous experiments, but brain was also rapidly removed, frozen in liquid nitrogen and stocked at -20°C until used.
Brain unchanged β-PEA and total radioactivity were measured as described by Strolin Benedetti et al., 1982.
Radioactivity due to brain and plasma metabolites of β-PEA was the difference between total radioactivity and radioactivity due to unchanged β-PEA.
Statistical analysis : as the hypothesis of equality of variance, tested by Bartlett's test (Snedecor & Cochran, 1957) cannot be assumed, a non-parametrical method (Kruskall and Wallis Test) (Schwartz, 1969) was used to compare several groups.

RESULTS AND DISCUSSION

"In vitro" experiments

In rat aorta oxidative deamination of 1μM β-PEA is carried out practically by SSAO (95%) (Fig. 1) whereas that of 1μM p-Tyr is carried out by SSAO (60%) and MAO-A (40%) (Fig. 1).

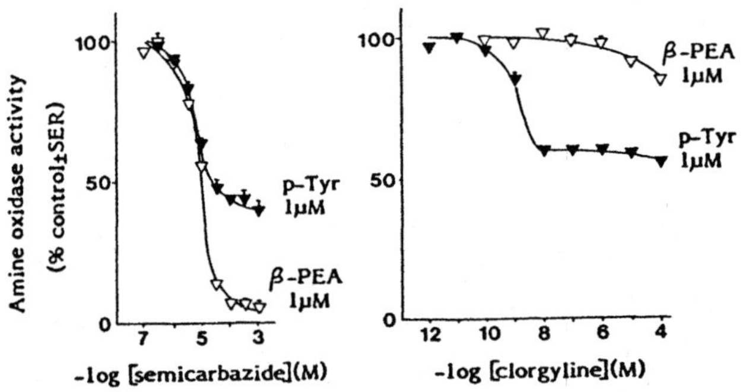

Fig. 1 : Oxidative deamination of β-PEA and p-Tyr in rat aorta homogenate as a function of increasing concentrations of a selective MAO (clorgyline) and SSAO (semicarbazide) inhibitor. Each point represents the mean of duplicate determinations with 4 or 5 homogenates, each of 3 aortas.

The K_m values for p-Tyr deamination in rat aorta by SSAO and MAO are 52 and 74 μM, respectively ; the corresponding V_{max} values are 0.50 and 0.48 nmoles.mg protein^{-1}.min^{-1}(Fig. 2). A previous work (Guffroy & Strolin Benedetti, 1984 b) where oxidative deamination of 100 μM p-Tyr in rat aorta was studied as a function of increasing concentrations of clorgyline, l-deprenyl and semicarbazide, allows us to affirm that the K_m and V_{max} values obtained in the present work for p-Tyr by MAO are those corresponding to the A form.

The K_m and V_{max} values for β-PEA deamination in rat aorta by SSAO are 14.7μM and 0.50 nmoles.mg protein^{-1}.min^{-1} (Fig.3 b). The deamination of β-PEA (between 50 to 1000 μM) in presence of semicarbazide (1.10^{-3}M) gives rise to a Lineweaver-Burk plot showing two components (Fig. 3 c) with approximate K_m values of 132 and 1420 μM and corresponding V_{max} values of 0.34 and 1.54 nmoles.mg protein^{-1}.min^{-1}. From the oxidative deamination of β-PEA as a function of increasing concentrations of clorgyline (Fig. 1) it would appear that the low-K_m enzyme component is that corresponding to MAO-B. K_m values for the

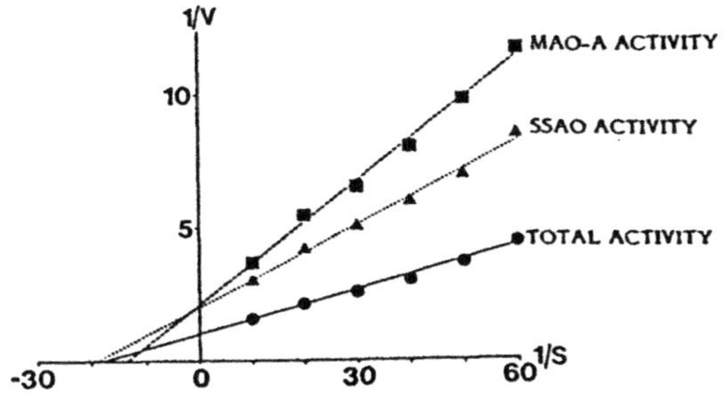

Fig. 2 : Lineweaver-Burk plots of p-Tyr deamination in rat aorta homogenate with and without clorgyline and semicarbazide. Each point represents the mean of triplicate determinations with 3 different homogenates, each of 3 aortas.
Abscissa : reciprocal of p-Tyr concentration (mM); ordinate : reciprocal of velocity (V) (nmoles.mg protein^{-1}.min^{-1})

●—● Total activity :
Km=57 μM ; V_{max}=0.103 nmoles.mg prot^{-1}.min^{-1}
▲···▲ SSAO activity (with clorgyline 1.10^{-4} M) :
Km=52 μM ;V_{max}=0.50 nmoles.mg prot^{-1}.min^{-1}
■···■ MAO-A activity (with semicarbazide 1.10^{-3} M) :
Km=74 μM ;V_{max}=0.48 nmoles.mg prot^{-1}.min^{-1}

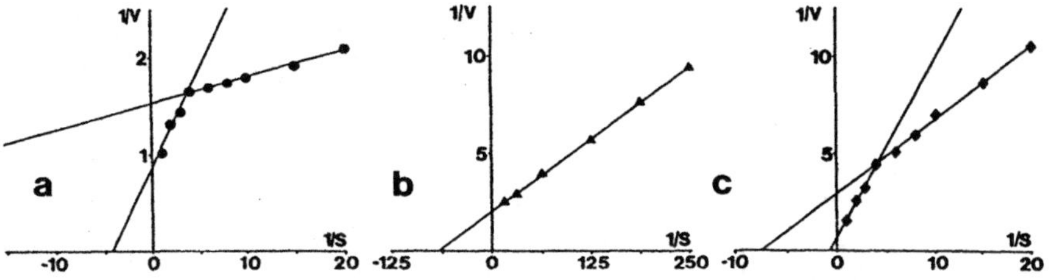

Fig. 3 : Lineweaver-Burk plots of β-PEA deamination in rat aorta homogenate with and without clorgyline and semicarbazide. Each point represents the mean of triplicate determinations with 4 different homogenates, each of 3 aortas.
Abscissa : reciprocal of β-PEA concentration (mM) ; ordinate : reciprocal of velocity (V) (nmoles.mg protein^{-1}.min^{-1})
Fig. 3 a : without inhibitor Km=18 and 242 µM ; V_{max}=0.65 and 1.19 nmoles.mg prot^{-1}.min^{-1}, respectively
Fig. 3 b : with clorgyline (1.10^{-4} M) : Km=14.7 µM ; V_{max}=0.50 nmoles.mg prot^{-1}.min^{-1}
Fig. 3 c : with semicarbazide (1.10^{-3} M) : Km=132 and 1420 µM ; V_{max}=0.34 and 1.54 nmoles.mg prot^{-1}min^{-1}, respectively

deamination of β-PEA and p-Tyr in rat aorta by SSAO are very similar to those found for these substrates in rat brown adipose tissue (Barrand & Callingham, 1982) (10µM and 40µM for β-PEA and p-Tyr, respectively).

Fig. 4 shows the dependence of the relative proportion of the activity that will be due to the SSAO on the concentration of each substrate. Values of Fig. 4 have been obtained using the following equation :

$$\% \text{ of total activity oxidized by SSAO} = \frac{v_{SSAO} \times 100}{v_{SSAO} + v_{MAO\text{-}A} + v_{MAO\text{-}B}}$$

where $$v = \frac{V_{max}}{1 + K_m/[S]}$$

v	= initial velocity
$[S]$	= substrate concentration
V_{max}	= maximum velocity
K_m	= Michaelis constant

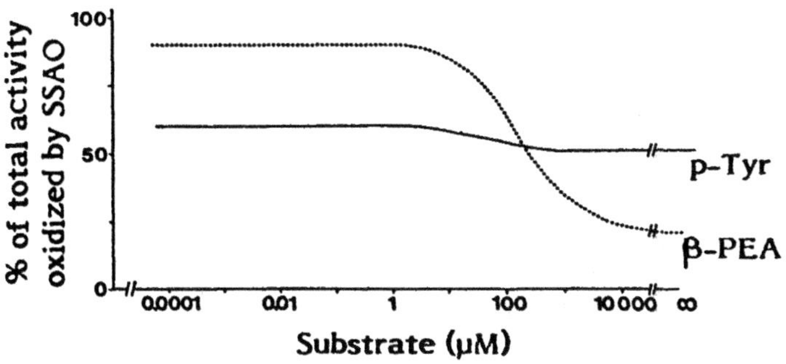

Fig. 4 : Percentage of the total amine oxidase activity due to the SSAO for β-PEA and p-Tyr in rat aorta homogenate. The curves are calculated from the appropriate K_m and V_{max} values of SSAO and MAO (shown in Fig. 2, 3 b and 3 c)

From Fig. 4 it appears that the relative contribution of SSAO to the deamination of p-Tyr is practically independent of the concentration of substrate used (about 60% until 10μM and between 60 and 50% at higher concentrations).
Results obtained from our calculations are in agreement with the experimental data of Clarke et al. (1982) who have studied the in vitro inhibition of p-Tyr deamination by clorgyline in homogenates from rat aorta from 10μM to 1mM.
On the contrary, the relative contribution of SSAO to the deamination of β-PEA is much more dependent upon the concentration of the substrate used (about 85% until 10μM, 65% at 100μM and between 65 and 20% at higher concentrations).
A previous work in Wistar Kyoto rat aorta on oxidative deamination of 100 μM β-PEA as a function of increasing concentrations of clorgyline (Guffroy & Strolin Benedetti, 1984 a) has given a higher contribution of SSAO compared with the present calculations ; however, data on oxidative deamination of 100 μM β-PEA in aorta of the same animals as those used in the present work have given a contribution of SSAO of about 75%.

"In vivo" experiments

For the "in vivo" experiments, β-PEA has been chosen rather than p-Tyr because the relative contribution of aorta SSAO to the deamination of endogenous concentrations of β-PEA is higher than the relative contribution of this enzyme to the deamination of p-Tyr (Fig.4).

Pretreatment of rats with semicarbazide increases and decreases respectively the plasma levels of unchanged β-PEA and of its metabolites during the first 10 minutes after injection of the amine (Fig. 5). It was previously demonstrated that the dose of semicarbazide used in this study completely inhibited SSAO activity in rat aorta for 24 hrs with no inhibition of MAO (Strolin Benedetti et al., 1984).

Pretreatment of rats with tranylcypromine has a much stronger and persistent effect on the increase of the plasma levels of unchanged β-PEA as well as on the decrease of the levels of β-PEA plasma metabolites (Fig.5).

When rats were pretreated simultaneously with tranylcypromine and semicarbazide, plasma levels of unchanged β-PEA were significantly higher than those measured in rats pretreated with tranylcypromine alone at least until 60 minutes after β-PEA injection ; in parallel, a significant decrease (up to 30 minutes) of β-PEA plasma metabolites was observed in the rats pretreated with the two compounds compared with those pretreated only with tranylcypromine (Fig.5).

In spite of a total MAO and SSAO inhibition the injected β-PEA is rapidly metabolized ; this is not surprising because other enzyme systems can metabolize this amine (Wu & Boulton, 1975 ; Jonsson et al., 1975 ; Boulton et al., 1974 ; Shannon et al., 1982 ; McQuade & Juorio, 1982) and their contribution might be enhanced when MAO and SSAO are inhibited.

Results from experiments presented in Fig. 5 are confirmed by those of an additional experiment (Table 1) where not only plasma levels of unchanged β-PEA and its metabolites but also the corresponding brain concentrations have been measured, 20 minutes after injection of β-PEA. In spite of the fact that rat brain contains negligible amount of SSAO if any (Guffroy & Strolin Benedetti, 1984 b ; Cao Danh et al., 1984), pretreatment of animals with semicarbazide increases slightly but significantly the brain content of unchanged β-PEA. The blockade of SSAO at the periphery, as well as the capacity of β-PEA to penetrate the blood brain barrier, can explain this result. Pretreatment of animals with tranylcypromine increases dramatically the brain content of unchanged β-PEA ; in parallel, brain β-PEA metabolites are strongly decreased.

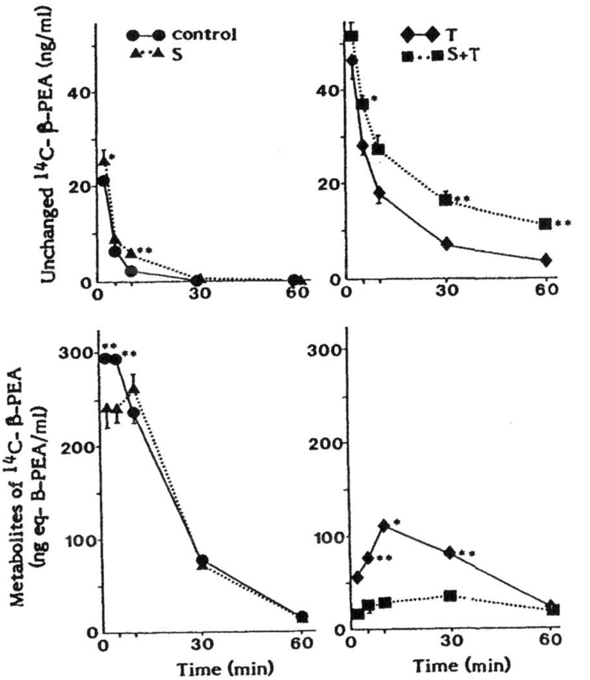

Fig. 5 : Kinetics of plasma concentrations of β-PEA and its metabolites after i.v. injection of 0.154 mg/kg ^{14}C-β-PEA, with and without pretreatment with semicarbazide (S) (100 mg/kg p.o.) and/or tranylcypromine (T) (5 mg/kg p.o.) one hour before

Statistical analysis : comparison, at each time, between control and rats pretreated with semicarbazide, or between rats pretreated with tranylcypromine and rats pretreated with tranylcypromine and semicarbazide. Kruskall and Wallis Test : * $p<0.05$, ** $p<0.01$

Pretreatment of rats with tranylcypromine and semicarbazide has the same effect as tranylcypromine alone on the brain content of unchanged β-PEA ; it is difficult to explain the effect of the combined pretreatment on the brain content of β-PEA metabolites without additional experiments.

Table 1 : Plasma and brain concentrations of β-PEA and its metabolites 20 minutes after i.v. injection of 0.154 mg/kg ^{14}C- β-PEA, with and without pretreatment with semicarbazide (100 mg/kg p.o.) and/or tranylcypromine (5 mg/kg p.o.) one hour before.

	Plasma concentrations ng/ml (n=7)		Brain content . ng in total brain (n=13)	
Pretreatment	β-PEA	Metabolites	β-PEA	Metabolites
Control	0.71±0.13	141±6	0.78±0.09	45±3
Semicarbazide	1.27±0.27	141±12	1.04±0.05 *	66±5
Tranylcypromine	11.5±0.8 **	109±3 **	223±4 **	13.8±0.5 **
Semicarbazide and tranylcypromine	18.3±1.5 **++	63±12 *+	236±25 **	56±12 +

Results for the metabolites are given in ng eq. β-PEA

Statistical analysis : Kruskall and Wallis Test ; compared to control * p<0.05 , ** p<0.01 ; compared to tranylcypromine : + p<0.05 ++p<0.01

A number of MAO are known to cause marked increases of β-PEA levels in brain after acute administration. Baker et al. (1982) found an important increase in the concentration of endogenous β-PEA in brain after administration of a high dose of pargyline (60 mg/kg i.p.), increase which was not further augmented by the injection of iproniazid (65 mg/kg i.p.), 60 minutes later. In contrast, the effect of pargyline on β-PEA brain concentrations was dramatically increased by injection of phenelzine (50 mg/kg i.p.) 60 minutes later. It is known that phenelzine is not only an inhibitor of MAO but also of SSAO and that it inhibits substantially the latter at doses inferior to those used by Baker et al.(Andree & Clarke, 1982 ; Lyles, 1984 ; Clarke et al., 1982). Therefore, the results of the present work, as well as the results from Baker et al. might indicate a contribution of SSAO to the metabolism of β-PEA "in vivo", although these authors suggested the possibility of a conversion of phenelzine to β-PEA.

In conclusion, the present work strongly suggests that one of the physiological roles of SSAO is to contribute to the metabolism of endogenous trace amines like β-PEA and p-Tyr, as well as to the detoxication of these amines when they are ingested with food. This must be confirmed, however, by parallel work in man.

Acknowledgements
We would like to thank H. Kadjan for his skilful technical assistance and M.C. Martinez for the preparation of the manuscript.

References

Andree T.H. and Clarke D.E. (1982) Characteristics and specificity of phenelzine and benserazide as inhibitors of benzylamine oxidase and monoamine oxidase. Biochem. Pharmacol. **31**, 825-830.

Armstrong M.D., Chao F.C., Parker V.J. and Wall P.E. (1955) Endogenous formation of hippuric acid. P.S.B.E.M. **90**, 675-679.

Baker G.B., LeGatt D.F. and Coutts R.T. (1982) Effects of acute and chronic administration of phenelzine on 2-phenylethylamine levels in rat brain. Proc. West Pharmacol. Soc. **25**, 417-420.

Barrand M.A. and Callingham B.A. (1982) Monoamine oxidase activities in brown adipose tissue of the rat : some properties and subcellular distribution. Biochem. Pharmacol. **31**, 2177-2184.

Barrand M.A. and Fox S.A. (1984) Amine oxidase activities in brown adipose tissue of the rat : identification of semicarbazide-sensitive (clorgyline-resistant)activity at the fat cell membrane. J. Pharm. Pharmacol. **36**, 652-658.

Boulton A.A. (1979) Trace amines in the central nervous system. International Review of Biochemistry, Physiological and Pharmacological Biochemistry **26**, 179-206.

Boulton A.A., Dyck L.E. and Durden D.A. (1974) Hydroxylation of β-phenylethylamine in the rat. Life Sci. **15**, 1673-1683.

Cao Danh H., Strolin Benedetti M., Dostert P. and Mousset A. (1984) Age-related changes in benzylamine oxidase activity in rat tissues. J. Pharm. Pharmacol. **36**, 592-596.

Clarke D.E., Lyles G.A. and Callingham B.A. (1982) A comparison of cardiac and vascular clorgyline-resistant amine oxidase and monoamine oxidase. Biochem. Pharmacol. **31**, 27-35.

Fuller R.W. and Roush B.W. (1972) Substrate-selective and tissue selective inhibition of monoamine oxidase. Arch. Int. Pharmacodyn. **198**, 270-276.

Guffroy C. and Strolin Benedetti M. (1984 a) Monoamine oxidase and semicarbazide-sensitive amine oxidase in spontaneously hypertensive and in normotensive control rats. Life Sci. **34**, 535-545.

Guffroy C. and Strolin Benedetti M. (1984 b) Monoamine oxidase and semicarbazide-sensitive amine oxidase in spontaneously hypertensive and normotensive control rats, in Monoamine Oxidase and Disease (Tipton K.F., Dostert P. and Strolin Benedetti M., eds) pp 624-625. Academic Press, London.

Hasan F. and Tipton K.F. (1984) The activities of intestinal amine oxidases in the metabolism of exogenous amines. J. Pharm. Pharmacol. **36**, Workshop suppl., 20W.

Jones R.S.G. (1983) Trace biogenic amines : a possible functional role in the CNS. TIPS. 426-429

Jonsson J., Lindeke B. and Cho A.K. (1975) Oxidation of phenethylamine yielding tyramine by rat liver microsomes. Acta Pharmacol. Toxicol. **37**, 352-360.

Lowry O.H., Rosebroogh N.J., Farr A.L. and Randall R.J. (1951) Protein measurement with the Folin phenol reagent. J. Biol. Chem. **193**, 265-275.

Lyles G.A. (1984) The interaction of semicarbazide-sensitive amine oxidase with MAO inhibitors, in Monoamine Oxidase and Disease (Tipton K.F., Dostert P. and Strolin Benedetti M., eds), pp 547-556, Academic Press, London.

McQuade P.S. and Juorio A.V. (1982) The effects of the administration of β-phenylethylamine on tyramine metabolism. Eur. J. Pharmacol. **83**, 277-282.

Schwartz D. (1969) Méthodes statistiques à l'usage des médecins et des biologistes, 3ème ed., pp 255-256.Flammarion Médecine-Sciences, Paris.

Shannon H.E., Cone E.J. and Yousefnejad D. (1982) Physiologic effects and plasma kinetics of β-phenylethylamine and its N-methyl homolog in the dog. J. Pharm. Exp. Ther. **223**, 190-196.

Snedecor G.W. and Cochran W.G. (1957) Méthodes statistiques, 6ème ed., pp 330-333. Associations de Coordination Technique Agricole, Paris.

Strolin Benedetti M., Dostert P., Boucher T. and Guffroy C. (1982) A new reversible selective type B monoamine oxidase inhibitor, in Monoamine Oxidase Basic and clinical Frontiers (Kamijo K., Usdin E. and Nagatsu T., eds), pp 209, Excerpta Medica, Amsterdam.

Strolin Benedetti M., Boucher T., Guffroy C. and Dostert P. (1984) MD 220661, an "in vivo" substrate of semicarbazide-sensitive amine oxidase.J. Pharm. Pharmacol. **36**, Workshop suppl., pp 29W.

Tipton K.F., O'Carroll A.M. and Hasan F. (1984) Enzymological and pharmacological aspects of monoamine oxidase, Vol.2 ,in IUPHAR 9th international Congress of Pharmacology (Paton W., Mitchell J. and Turner P., eds), pp. 179-185. The Macmillan Press Ltd, London.

Urdin M.D.C. and Fuentes J.A. (1983) Enzyme activity of an amine oxidase resistant to pargyline in the cardiovascular system of hypertensive rats. Biochem. Pharmacol. **32**, 2345-2347.

Wu P.H. and Boulton A.A. (1975) Metabolism, distribution, and disappearance of injected β-phenylethylamine in the rat. Can. J. Biochem. **53**, 42-50.

IN VIVO DISCRIMINATIVE STIMULUS (CUE) PROPERTIES IN RATS OF BETA-PHENYLETHYLAMINE

A.J. Goudie and D. Reid

Psychology Department, Liverpool University

P.O. Box 147, Liverpool L69 3BX, England.

INTRODUCTION

In recent years the drug discrimination (DD) bioassay has proved of value in analysing drug actions (Colpaert and Slangen, 1982). This procedure is a whole animal assay for analysis in vivo of actions of drugs and endogenous agents, thus it may be useful in analysing functions of endogenous trace amines. Colpaert et al. (1980) suggested that the cue properties of cocaine in the DD procedure are mediated by PEA, since the potencies of MAO inhibitors in generalising to cocaine depended on their relative potencies in inhibiting MAO Type B, for which PEA is a preferred substrate. Huang and Ho (1974) reported that rats discriminating amphetamine generalized to PEA after pretreatment with iproniazid. These data suggest that the cue properties of stimulant drugs may be mediated by endogenous PEA. We have therefore been involved in recent years with:- (I) Analysis of the actions of PEA as a "training" drug and, (II) Analysing PEA's actions in animals discriminating the stimulant dl-cathinone.

CHARACTERIZATION OF THE CUE PROPERTIES OF PEA

In an initial study, we trained rats to discriminate

TABLE 1

Dose/effect curve for the PEA cue

Behavioural Measure	Dose of PEA (mg/kg) 0	10	20	30	40
Percentage of trained rats selecting PEA lever	0	50	67	100	100
Response rate (% of non drug baseline)	101	82	76	70	24

PEA from saline in a two lever food rewarded FR10 DD procedure with the drug administered 15 mins prior to operant sessions (cf. Colpaert et al. 1980). The training dose of PEA was 40 mg/kg i.p. Subjects had difficulty learning this discrimination; of 12 rats in the study, only 6 reached a training criterion of 10 consecutive sessions of correct lever selection after as many as 75 training sessions. The trained subjects showed dose-related generalization to PEA itself, doses of PEA that produced generalization had profound effects on operant responding (Table 1). The cue properties of PEA in these animals were also time-related (Table 2). PEA was virtually undetectable 60 mins post injection. Thus, after extensive training, PEA produced a dose-related cue of very short duration, as would be expected on the basis of PEA's known short half-life. The difficulty experienced in training rats to discriminate PEA was probably due to its rapid metabolism, since rats may find it difficult to discriminate rapidly changing levels of this agent.

In the PEA trained rats, generalization tests were conducted with cocaine and amphetamine. In the amphetamine tests (Table 3), at the lowest dose tested all subjects selected the saline lever. At higher doses, no

TABLE 2

Time-effect curve for the PEA cue

	Time post injection (mins)				
	15	30	60	120	240
Percentage of trained rats selecting PEA lever	100	67	17	17	17

more than 50% drug lever selection was ever seen, even at the highest doses that could be tested which suppressed responding to 35% of the non-drug baseline. It is important to note that only some individual subjects showed generalization between PEA and amphetamine (Ss 1,2 and 4). A very similar pattern of data was seen in cocaine tests (Goudie, 1982). These data are suggestive of similarity between the PEA cue and that of stimulants, but the PEA cue <u>may</u> differ significantly in some respects or in some animals from the cue induced by stimulant drugs.

Since the studies described above failed to detect any agents which produced more than 50% generalization to PEA, we examined the cue properties of some potential PEA metabolites to determine whether these produced more complete generalization. The metabolites studied were N-Methyl PEA (NMPEA), Phenylethanolamine (PEOH) and N-Methylphenylethanolamine (NMPEOH). All these compounds are endogenous trace amines which have amphetamine-like actions (Shannon and De Gregorio, 1982). PEOH is synthesised by the action of dopamine beta hydroxylase on PEA, and NMPEA and NMPEOH are produced by the action of N-methyltransferases on PEA and PEOH (Blau, 1978). In this study we also assessed the ability of $(\alpha, \alpha\ d_2)$ deuterated PEA to generalize to PEA. We also assessed the specificity of the PEA cue by examining the extent to which fenfluramine, a non-stimulant substituted phenylethylamine, would generalize to PEA at doses known to have cue properties (Goudie, 1977). Subjects were trained to

TABLE 3

Results of generalization tests with d-amphetamine

	Amphetamine Dose (mg/kg)						Threshold for generalization (mg/kg)
	.0625	0.125	0.25	0.5	1.0	1.5	
Subject 1	S	D	D	D	D	D	0.125
2	S	S	D	D	D	D	0.25
3	S	S	S	S	S	S	NA
4	S	S	D	D	D	D	0.25
5	S	S	S	S	S	S	NA
6	S	S	S	S	S	S	NA
Percent of trained rats selecting PEA lever	0	16.7	50	50	50	50	
Response rate (% of non drug baseline)	99	112	110	90	46	35	

S = Saline lever selection, D = Drug lever selection
\- = No response, NA = No generalization seen at any dose

discriminate PEA from saline at 30 mg/kg i.p. (t = minus 15 mins) using a standard two lever FR10 operant DD procedure. Subjects were trained to a criterion of 10 consecutive correct lever selections. In this specific study, we succeeded in training 12 out of a total of 24 rats to discriminate PEA to criterion within 50 training sessions. The compounds shown in Figure 1 induced at least 50% generalization, ED_{50} values (mg/kg) being:- deuterated PEA

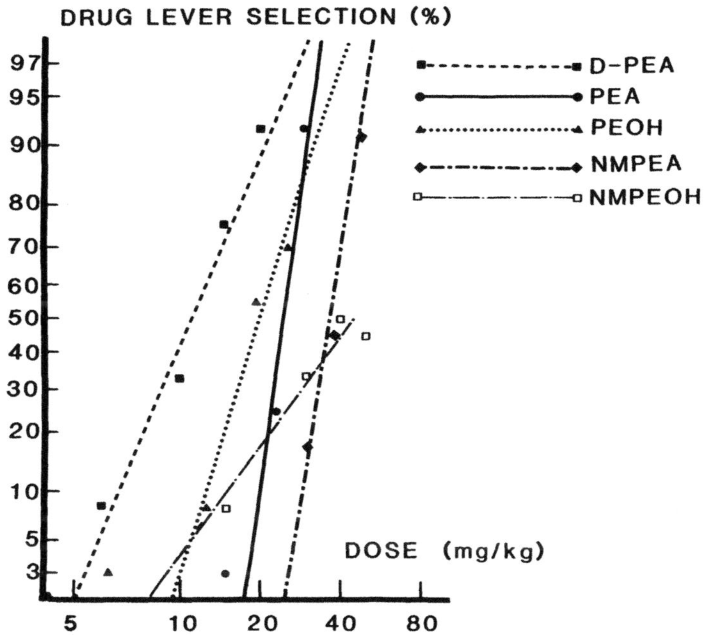

Figure One: Log-probit plot for agents showing at least 50% drug lever selection. Drug lever selection (%) is plotted against dose. Straight lines are calculated lines of best fit. Observed data points are also plotted for each drug.

(11.0), PEOH (19.7), PEA (24.7), NMPEA (36.1) and NMPEOH (46.7). All these compounds except NMPEOH caused greater than 50% generalization. Fenfluramine failed to generalize to PEA. At doses up to 3 mg/kg i.p. the highest level of drug lever selection was 16.7%. Thus, the PEA cue in the DD procedure shows pharmacological specificity. Deuterated PEA was twice as potent as PEA in inducing PEA lever selection, due presumably to resistance of deuterated PEA to metabolism by MAO (Blake et al. 1974). NMPEA and PEOH are the first reported compounds to show complete

TABLE 4

Compounds showing dose-related generalization to cathinone

Drug	Dose (mg/kg)	Response rate (% of non-drug baseline)	Percent cathinone Lever selection	ED_{50}*
Cathinone	0	104	0	
	0.5	88	29	0.83
	1	92	57	
	2	98	86	
Amphetamine	0.25	103	14	
	0.5	112	71	0.43
	1	76	86	
Cocaine	1.25	124	0	
	2.5	102	12.5	4.52
	5	66	50	
	10	55	100	
Pipradrol	1.75	105	14	
	3.5	65	71	3.51
	7	88	71	
	14	21	80	
Methylphenidate	0.5	109	0	
	1	99	25	2.21
	2	77	25	
	4	55	87.5	

* ED_{50} values were determined by probit analysis.

generalization to PEA. NMPEOH produced at most only 50% partial generalization.

TABLE 5

Compounds that failed to generalize to cathinone

Drug	Dose (mg/kg)	Response rate (% of non-drug baseline)	Percent cathinone lever selection
Fentanyl	0.1	98	0
	0.2	50	0
Fenfluramine	1	113	0
	3	69	0
Chlordiazepoxide	5	72	12.5
	7.5	96	12.5
	10	86	12.5
α-demethyl-cathinone	2	99	0
	4	101	0
	8	81	12.5

CHARACTERIZATION OF THE CUE INDUCED BY THE STIMULANT DRUG CATHINONE AND ITS RELATION TO THE PEA CUE

In the studies described above rats were trained to discriminate PEA and the PEA cue was analysed by examining generalization to PEA. In other studies we trained rats to discriminate cathinone, and assessed the extent to which PEA and deuterated PEA generalized to the cathinone cue. Cathinone is an amphetamine congener that shares many of its pharmacological actions (Kalix, 1984). We trained 8 rats to discriminate 2.0 mg/kg i.p. of dl-cathinone (t = minus 15 mins) versus saline to a criterion of ten consecutive correct lever selections. In generalization tests cathinone itself and all the compounds shown in Table 4 showed complete dose-related generalization to cathinone. The compounds shown in Table 5 failed to generalize to cathinone. These compounds, most of which

TABLE 6

Individual patterns of lever selection shown by subjects in tests with PEA and Deuterated PEA

Subject No.	PEA Doses (mg/kg) 7.5	15	30	40	Deuterated PEA Doses(mg/kg) 7.5	15	30
1	S	S	D	-	S	S	S
2	S	S	D	S	S	S	S
3	S	S	D	D	S	D	D
4	S	S	S	S	S	S	S
5	S	D	D	D	S	D	D
6	S	S	S	D	S	D	D
7	S	S	S	D	S	S	S
8	S	S	D	S	S	S	-
% Cathinone Lever Selection	0	12.5	62.5	57.1	0	37.5	42.9

S = Saline lever selection, D = Drug lever selection, - = No selection.

have cue properties at the doses studied come from a variety of drug classes. Thus the cathinone cue shows a high degree of specificity for stimulants.

The results of generalization tests conducted with PEA and deuterated PEA are shown in Table 6. At the lowest doses of PEA and deuterated PEA that were tested 0% cathinone lever selection was seen, at higher doses of both agents dose-related increases in cathinone lever selection were observed, although this never reached a level higher than 62.5%. Higher doses of PEA and deuterated PEA could not be tested due to their rate suppressant effects. Individual patterns of lever selection showed that some subjects (e.g. S5) selected the drug lever when treated with both PEA and deuterated PEA, whilst others (e.g. S4) consistently failed to select the drug lever with either PEA or deuterated PEA. Thus PEA and deuterated PEA showed partial generalization and this appeared to be due to some subjects consistently showing generalization whilst others did not at any dose. These data represent the converse pattern to that shown in Table 3 which indicates that the partial generalization to PEA observed with amphetamine was due to consistent generalization in some subjects but not in others.

PEA is α-demethylamphetamine; it is therefore of interest that α-demethylcathinone, a non stimulant congener of cathinone, failed to generalize to the cathinone cue (Table 5). Thus α-methyl groups of both cathinone and amphetamine are important structural determinants of their cue properties. Removal of the methyl group from either drug substantially reduced the ability of the drug to produce the cue of its parent compound.

CONCLUSIONS

These data go a considerable way towards characterizing the PEA cue and analyzing its relationship to that of stimulant drugs. PEA produces a pharmacologically specific cue of short duration. Relative to other drugs, it is difficult to train rats to discriminate PEA. Since deuterated PEA generalizes completely with PEA, and since it is longer acting, it is possible that deuterated PEA will be a useful tool in DD studies since the short half-life of PEA probably prevents animals detecting its presence or absence. Of all compounds so far studied only PEOH and NMPEA produced complete generalization to PEA. These compounds may be functionally important metabolites of PEA; although they may also have important actions of their own. Stimulants produce incomplete generalization to PEA. Conversely, PEA produces incomplete generalization to the cue induced by stimulants. The reason for such partial generalization probably resides in differential responses of individual subjects (cf. Colpaert and Janssen, 1984). The failure of some subjects treated with amphetamine at high doses to generalize to PEA (Table 3), and of some subjects treated with PEA at high doses to generalize to cathinone (Table 6) might be due simply to the fact that individual animals differ in sensitivity to drug cues (Colpaert and Janssen, 1982). Thus, if it had been possible to test higher doses of either the stimulants or of PEA, complete generalization might have been seen between these agents. However, this interpretation of the data is implausible. Amphetamine has effective cue properties in rats at doses as low as 0.16 mg/kg (Colpaert et al. 1976), yet some of our subjects failed to generalize to PEA when treated with amphetamine at 1.5 mg/kg, a dose which had marked rate suppressant effects (Table 3). We suggest that these data indicate that PEA has a complex, multidimensional cue in DD studies (cf. Koek and Slangen, 1984). One dimension of this cue may be related to that induced by stimulants whilst other(s) is/are not. Other evidence indicates clearly that one drug can induce different types of cues at different doses (Emmett-Oglesby et al. 1983), our hypothesis is therefore by no means implausible. One implication of this hypothesis is that the conceptualisation of PEA as an "endogenous amphetamine" (e.g. Sandler and Reynolds, 1976) may be somewhat tenuous. Whilst PEA undoubtedly has many actions in common with amphetamine, it also differs from it in potentially significant ways.

ACKNOWLEDGEMENTS

This research was supported by the Liverpool University Research Grants Scheme. Mrs. D. Foulds is thanked for her most conscientious typing.

REFERENCES

Blake M.I., Crespi H.K. and Katz J.J. (1975) Studies with deuterated drugs. J. Pharmaceutical Sci. 64, 367-391.

Blau, K. (1978) The significance of phenylethylamine in phenylketonuria and related disorders. In: Noncatecholic Phenylethylamines. Part 1 (Mosnaim

A.D. and Wolf M.E., eds.) pp. 385-396, Marcel Dekker, New York.

Colpaert F.C., Niemegeers C.J.E. and Janssen P.A.J. (1980) Evidence that a preferred substrate for Type B monoamine oxidase mediates stimulus properties of MAO inhibitors: A possible role for B-phenylethylamine in the cocaine cue. Pharmacol. Biochem. Behav. 13, 513-517.

Colpaert F.C., Kuyps J.J.M.D., Niemegeers C.J.E. and Janssen P.A.J. (1976) Discriminative stimulus properties of a low dl-Amphetamine dose. Arch. int. Pharmacodyn. Therap. 223, 34-42.

Colpaert F.C. and Janssen P.A.J. (1982) Factors regulating drug cue sensitivity: Limits of discriminability and the role of a progressively decreasing training dose in cocaine-saline discrimination. Neuropharmacology 21, 1187-1194.

Colpaert F.C. and Slangen J.L. (1982) Drug Discrimination: Applications in C.N.S. Pharmacology Elsevier Biomedical Press, Amsterdam.

Colpaert F.C. and Janssen, P.A.J. (1984) Agonist and antagonist effects of prototype opiate drugs in rats discriminating fentamyl from saline: Characteristics of partial generalization. J. Pharmac. exp. Therap. 230, 193-199.

Emmett-Oglesby M., Wurst M. and Lal H. (1983) Discriminative stimulus properties of a small dose of cocaine. Neuropharmacology 22, 97-101.

Goudie, A.J. (1977) Discriminative stimulus properties of fenfluramine in an operant task. Psychopharmacology 53, 97-102.

Goudie, A.J. (1982) Discriminative stimulus properties in an operant task of beta-phenylethylamine. In: Drug Discrimination: Applications in C.N.S. Pharmacology (Colpaert F.C. and Slangen J.L. eds.) pp. 165-180, Elsevier Biomedical Press, Amsterdam.

Huang J.T. and Ho B.T. (1974) The effects of pretreatment with iproniazid on the behavioral activities of B-phenylethylamine in rat. Psychopharmacologia 35, 77-81.

Kalix P. (1984) The pharmacology of Khat. Gen. Pharmacol. 15, 179-187.

Koek W. and Slangen J.L. (1984) External stimulus control in a "drug-discrimination" procedure: Drug effects and inter-animal variation. Psychopharmacology, 82, 168-173.

Sandler M. and Reynolds G.P. (1976) Does phenylethylamine cause schizophrenia? Lancet, 1, 70-71.

Shannon H.E. and De Gregorio C.M. (1982) Self-administration of the endogenous trace amines B-phenylethylamine, N-Methyl Phenylethylamine and Phenylethanolamine in dogs. J. Pharmac. exp Therap. 222, 52-60.

EFFECT OF PSYCHOACTIVE DRUGS ON BEHAVIOR INDUCED BY 2-PHENYLETHYLAMINE IN RATS

R. Ortmann, M. Schaub and P.C. Waldmeier

Research Department, Pharmaceuticals Division CIBA-GEIGY Ltd., CH-4002 Basel, Switzerland

INTRODUCTION

2-Phenylethylamine (PEA) combines 2 features which are important for the behavioral pharmacologist interested in the characterization of monoamine oxidase (MAO) inhibitors in rats: It is degraded rather specifically by the B-form of MAO (Yang and Neff 1974) and produces stereotyped behavior in rodents. Several authors have shown that the intensity of stereotyped behavior is greatly increased by inhibitors of the B-type of MAO, while MAO-A inhibitors were much less active (Braestrup et. al 1975,1978; Moja et al. 1976). We tried to establish a behavioral test system in vivo which would allow us to determine the inhibitory specificity of a given compound for the A- or B-type of MAO. The potency of various MAO-inhibitors to potentiate stereotyped behavior induced by PEA was compared with their effect on L-5-hydroxytryptophan (L-HTP)-induced excitation in rats. The latter test has been shown to be specific for MAO-A inhibitors (Ortmann et al. 1980). In addition we compared the behavioral effects of the test compounds in the PEA test with their ability to increase PEA levels of rat brain and to inhibit MAO in an ex vivo assay.

MATERIAL AND METHODS

Albino rats (Tif:RAIf) weighing 160-200g were used. The scoring of the behavioral effect was done as described previously for the PEA syndrome (Ortmann et al. 1984) and L-HTP-induced excitation (Ortmann et al. 1980) respectively. For the PEA test the presence or absence of forepaw padding ("piano playing") and continuous sniffing was observed every 5 minutes during a 30 sec interval. The number of rats showing continuous forepaw padding together with continuous sniffing (PEA-syndrome) for at least 15 seconds during one or more observation period during the test of 20 minutes was determined. Results were expressed as the number of animals per group showing the PEA-syndrome during at least one observation period. Statistical analysis of the behavioral data of the PEA test was done using Fischer's exact test for 2 x 2 table (Fisher 1958). Treated groups (N=6-12) were compared with pooled controls (N=126). Test compounds were injected as the salts at the indicated times before 30 mg/kg i.p. PEA or 100 mg/kg i.p. L-HTP in a volume of 2 ml/kg.

PEA level in whole rat brain were determined by GCMS as described by Lauber and Waldmeier (1984). MAO activity in rat brain was determined ex vivo as described by Ortmann et al. (1984a).

2-phenylethylamine-HCl was purchased from Sigma, St.Louis, USA, harmaline-HCl from Fluka AG, Switzerland, pargyline from Aldrich Comp. Belgium. These drugs were generously donated from the following companies: MD 780236 (Delalande, France; Strolin-Benedetti and Row 1983), tranylcypromine-sulfate (SKF, USA), citalopram-HBr (Lundbeck, Danmark). All other drugs were synthezised in the chemical laboratories of CIBA-GEIGY Basel or Summit, N.J.. CGP 6085 A (Waldmeier et al.1979) and brofaremin (CGP 11305 A; Waldmeier and Baumann 1983) are experimental drugs of Ciba-Geigy AG, Basel.

RESULTS

Dose response curves of PEA showed that up to 30 mg/kg i.p. the animals showed exophtalmus, locomotor hyperactivity, tachypnoea, salivation, stereotyped up- and down movements of the head ("head bobbing") and weak sniffing which lasted for 20-30 minutes. Only after 60 and 80 mg/kg i.p. PEA did the rat show bursts of both continuous sniffing and simultaneous forepaw padding ("piano playing") at the bottom or the wall of the cage without locomotor activity. The combined occurence of this behavioral signs was defined as the "PEA syndrome".

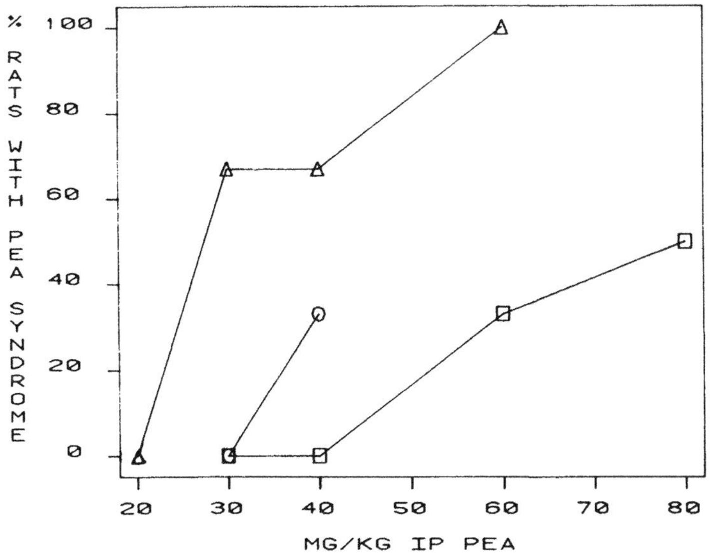

Figure 1: Stereotyped behavior induced by PEA in the presence of 10 mg/kg s.c. deprenyl ( ), 0,3mg/kg s.c. clorgyline ( ) and without pretreatment ( ). Deprenyl and clorgyline were given 2 hrs before the indicated doses of PEA.

Dose response curves of PEA in the presence of clorgyline and deprenyl showed that the MAO-B inhibitor deprenyl was much more potent to induce the PEA-syndrome at the 30 mg/kg i.p. dose of PEA than the MAO-A inhibitor clorgyline (Figure 1). For the quantitative evaluation of the tested compounds we therefore used this dose of PEA, because it seemed to offer the best separation of the behavioral effects of MAO-A and MAO-B inhibitors. None of the control animals injected only with 30 mg/kg i.p. PEA was ever observed to show the PEA syndrome as defined under our experimental conditions.

Effect of Amine Depletion and 5,7-DHT Lesions

α-methyl-p-tyrosine (α-MPT) has been shown to prevent stereotyped behavior induced by PEA, while reserpine was found to intensify it (Fuxe et al 1967; Braestrup and Randrup 1978). We confirmed this finding by demonstrating that the ability of 60 mg/kg i.p. PEA to induce the PEA

Table 1: Lack of effect of 5,7-DHT-lesions on the ability of PEA to induce the PEA syndrome

PEA (mg/kg i.p.)	Type of Lesion	% animals showing PEA syndrome	N
30	SHAM	0	6
30	5,7-DHT	0	6
60	SHAM	100	6
60	5,7-DHT	100	6

5,7-DHT or sham lesions were made 24 days before the experiment as described previously (Ortmann et al. 1981). The number of animals showing the PEA syndrome was evaluated as described in Methods.

syndrome was abolished by pretreatment with 250 mg/kg i.p. α-MPT, while reserpine slightly enhanced it (Data not shown). The finding that icv injections of 200μg 5,7-DHT did not affect the behavioral effect of PEA to induce the PEA syndrome (Table 1) suggests that the release of 5-HT does not play a crucial role in the mediation of the behavioral effects of PEA.

Effect of different MAO-Inhibitors

The potencies of various MAO-B inhibitors to induce the PEA syndrome correlated well with their ability to inhibit MAO-B in the ex vivo assay (Table 2) and to increase PEA level in whole rat brain (Table 3). In general only those doses of the inhibitors which inhibited the B-form of MAO by more than about 75% were found to induce the PEA syndrome and to increase PEA levels in the brain. Caroxazone was the only compound which potently induced the PEA syndrome and increased PEA levels with threshold doses between 10 and 30 mg/kg p.o., but was inactive in the ex vivo MAO assay. Since caroxazone has been shown to be a potent inhibitor of MAO-B in vitro (Moretti et al. 1981), these results indicate that the described behavioral test can pick up MAO-B inhibitor which are apparently inactive in the ex vivo assay.

Specific inhibitors of MAO-A like brofaremine had either no effect at all or only induced the PEA syndrome in rather high doses (Table 4). The combination of inactive doses of a MAO- A inhibitor like brofaremine or clorgyline with an inactive dose of deprenyl (1mg/kg s.c.) did, however, potently induce the PEA syndrome (Ortmann et al 1984a). These results suggest that MAO-A either contributes to the degradation of PEA when the B-form of the enzyme is inhibited or that the increase of synaptic dopamine levels due to the inhibition of MAO-A potentiates the behavioral effect of sub-threshold doses of MAO-B inhibitors.

Table 2: The effect of various MAO-B inhibitors on PEA-induced stereotypies and on MAO activity in rat brain homogenate.

DRUG	(MG/KG)	NUMBER OF RATS WITH SYNDROME	P	% INHIBITION MAO-B
DEPRENYL	1SC	0/12	NS	74[$]
	3SC	2/6	0.0017	75[$]
	10SC	8/12	<0.0001	86[$]
HARMALINE	30IP			
+PARGYLINE	20IP	6/12	<0.0001	75[$]
PARGYLINE	3 SC	1/6	0.0455	78[$]
	10SC	6/6	<0.0001	92[$]
TRANYLCY-PROMIN	0.3 SC	0/6	NS	49[$]
	1.0 SC	2/6	0.0017	77[$]
SU 11739	0.03SC	0/6	NS	26[$]
	0.1 SC	1/6	0.0455	87[$]
	0.5 SC	5/6	<0.0001	
MD 780236	10 PO	2/6	0.0017	
	25 PO	5/6	<0.0001	
CAROXAZONE	2.5 PO	0/6	NS	
	5 PO	1/12	0.087	
	10 PO	8/12	<0.0001	
	100 PO			24

Drugs were given 2 or 24([$]) hrs before the experiment.

TABLE 3: Comparison of the minimal effective dose of MAO-B inhibitors to increase PEA levels in rat brain and to induce the PEA syndrome.

DRUG	MG/KG	PEA-LEVELS (NG/G) CONTROLS	TREATED	PEA-SYNDROME (D.e.m.)
DEPRENYL	3 IP	3.5±1.7	7.0±0.7	3SC /10IP
PARGYLINE	3 SC	8.7±1.2	22±1.8	3 SC
MD 780236	10 IP	1.9±0.2	8.3±1.7	10PO /10IP
CAROXAZONE	30 PO	2.0±0.3	68.5±21	10 PO
TRANYLCY-PROMINE	1 SC	8.7±1.2	72.5±12.7	1 SC
SU 11739	0.1 SC	4.8±0.15	9.1±0.8	0.1 SC

For the determination of PEA levels drugs were given 4 hours, for the behavioral tests 2 hours before sacrifice or testing respectively. The doses given for the determinations of PEA levels were those increasing the mean concentrations (ng/g) of the group by at least 100%. The minimal effective doses (d.e.m.) given for the behavioral tests were the lowest doses inducing a significant effect ($P<0.05$).

Table 4 depicts the potencies of various MAO-inhibitors to potentiate the behavioral excitation induced by 100 mg/kg i.p. L-5-HTP in the rat and to induce the PEA syndrome. It can be seen that those compounds which inhibit preferentially the A-form of MAO are more active in potentiating the behavioral effects of L-5-HTP, while the opposite is true for compounds showing preference for the B-form in an ex vivo assay. Caroxazone does not seem to be a specific MAO-B inhibitor in vivo, because it was active in both tests in the same dose range.

Effects of Other Psychotropic Drugs

10 mg/kg i.p. oxaprotiline and maprotiline, two specific inhibitors of the noradrenaline uptake

system as well as 30 mg/kg i.p. citalopram and CGP 6085 A, specific 5-HT uptake inhibitors were unable to induce the PEA syndrome. CGP 6085 A was also inactive in the presence of 1 mg/kg s.c. deprenyl.

Table 4: Comparison of the ability of MAO-inhibitors to induce the PEA-syndrome and to potentiate the behavioral effect of L-5-HTP.

DRUG	MG/KG	L-5- HTP ED_{50}	PEA SYNDROME DEM	MAO-Inhibition Potency ex vivo
MOCLOBEMIDE	PO	1	100	A>B
CLORGYLINE	SC	0.1	10	A>B
BROFAREMINE	PO	4	>100	A>B
AMIFLAMINE	PO	1	50	A>B
CIMOXATONE	PO	2	>50	A>B
DEPRENYL	SC	10	3	B>A
SU 11739	SC	0.3	0.1	B>A
CAROXAZONE	PO	5	10	

ED_{50} values of the drugs to potentiate L-5-HTP-induced excitation in rats and the minimal effective doses (DEM) to induce the PEA syndrome are given. The relative potency to inhibit the A- and B-form of MAO respectively in an ex vivo assay is indicated qualitatively in the last column.

DISCUSSION

The main finding of the present study was the fact that those drugs induced the PEA syndrome most potently which also strongly inhibited the B-form of MAO ex vivo and elevated endogenous levels of PEA of rat brain. We therefore conclude that the induction of the PEA syndrome under our experimental conditions is a characteristic property of strong MAO-B inhibitors and can be used for the evaluation of these drugs in vivo. There is, however, no complete specificity of the test for MAO-B inhibitors as indicated by the finding that e.g. the MAO-A inhibitor clorgyline, also induced the syndrome at a dose which completely inhibited MAO-A but inhibited MAO-B by only 30% (Ortmann et al. 1984a). MAO-A inhibitors did not elevate endogenous PEA levels in rat brain in the doses used in our study (Lauber and Waldmeier 1984)and therefore do not seem to deaminate PEA under normal conditions. PEA levels are, however, much higher after the i.p. injections during the behavioral testing. The A-form of MAO may contribute to the degradation of PEA under these conditions of unphysiological high PEA-levels. Inhibitors of MAO-A may thus potentiate the behavioral effect of PEA, if there is also some concomitant inhibition of MAO-B.

Because of the lack of complete specificity of the behavioral test for MAO-B inhibitors we tried to improve the qualitative assessment of the MAO-inhibitory specificity of a test compound by determining the potency of the compounds to potentiate the behavioral effects of both L-5-HTP and PEA. The ED_{50} of specific MAO-A inhibitors to potentiate L-5-HTP induced excitation in the rat was 25 -100 time smaller than the smallest dose which induced the PEA syndrome. Deprenyl and SU 11739, which inhibit MAO-B more potently than the A-form were more potent in the PEA test than in the L-5-HTP test. The evaluation of MAO inhibitors in both behavioral tests thus allows a rather fast and accurate assessment of the inhibitory specificity of MAO inhibitors under in vivo conditions.

When injected alone none of the MAO-inhibitors induced the PEA syndrome, indicating that endogenous PEA levels are not high enough to reach overt behavioral significance even after almost complete blockade of MAO-B. This is also demonstrated by the finding that caroxazone has no effect on locomotor activity up to doses which dramatically increased central PEA levels. For the assessment of the behavioral effects of PEA, injections of rather high doses are therefore necessary to counterbalance the high turnover rate of PEA.

From experiments with 5-HT antagonists a possible serotonergic component of the behavioral effects of PEA has been suggested (Sloviter et al. 1980, Dourish 1981). Pretreatment of rats with p-chlorophenylalanine or 5,7-DHT did, however, not prevent the induction of a "serotonin behavioral syndrome" in rats by PEA suggesting that a release of 5-HT is not involved in PEA's behavioral effect (Sloviter et al. 1980). We confirmed the finding that lesions by icv injections of 5,7-DHT do not change the ability of PEA to induce the stereotyped behavior as measured as the "PEA-syndrome". The same 5,7-DHT lesions have been shown to induce behavioral supersensitity to L-5-HTP and 5-HT agonists (Ortmann et al. 1981, 1984b). Because we did not observe a supersensitivity of the lesioned rats towards PEA, the results suggest that a postsynaptic activity of PEA at 5-HT receptor sites is not involved in the mediation of PEA's behavioral effects.

REFERENCES

Braestrup C., Andersen H. and Randrup A. (1975) The monoamine oxidase inhibitor deprenyl potentiates phenylethylamine behaviour in rats without inhibition of catecholamine metabolite formation. Eur. J. Pharmacol. **34**,181-187

Braestrup C.and Randrup A. (1978) Stereotyped behavior in rats induced by phenylethylamine, dependence on dopamine and noradrenaline, and possible relation to psychoses? in Noncatecholic Phenylethylamines, eds. A.Mosnaim, M.E.Wolf; Marcel Dekker pp 245-269

Dourish C.T. (1981) Behavioural effects of acute and chronic β-phenylethylamine administration in the rat: Evidence for the involvement of 5-HT. Neuropharmacol. **20**, 1067-1072

Fischer R.A. (1958) Statistical methods for research workers. 13th edition, Oliver and Boyd Ltd, Edinburgh, p 96

Fuxe K.,Gobecker H. and Jonssen J. (1967) The effect of β-phenylethylamine on central and peripheral monoamine-containing neurones. Eur. J. Pharmacol. **2**, 202-207

Lauber J. and Waldmeier P.C. (1984) Determination of 2-phenylethylamine in rat brain after MAO inhibitors, and in human CSF and urine by capillary GC and chemical ionization MS. J Neural Tansm. **60**, 247-264

Moja E.A., Stoff D.M., Gillin J.C. and Wyatt R.J. (1976) Dose-response effects of β-phenylethylamine on stereotyped behavior in pargyline-pretreated rats. Biol. Psychiatry **11**, 731-742

Moretti A., Caccia C., Calderini G., Menozzi M. and Amico A. (1981) Studies on the mechanism of action of caroxazone, a new antidepressant drug. Biochem Pharmacol **30**, 2728-2731

Ortmann,R., Waldmeier,P.C., Radeke,E., Felner,A. and Delini-Stula,A (1980) The effects of 5-HT uptake- and MAO-inhibitors on L-5-HTP-induced excitation in rats. Naunyn-Schmiedeberg's Arch. Pharmacol. **311**, 185-192

Ortmann,R., Martin, S. and Waldmeier,P.C. (1981) Supersensitivity to L-5-HTP after hydroxytryptamine injections in desmethylimipramine- and nomifensine-pretreated rats: Behavioral evidence for postsynaptic supersensitivity. Psychopharmacol. **74**,109-114

Ortmann R., Schaub M., Felner A., Lauber J., Christen P. and Waldmeier P.C. (1984a) Phenylethylamine-induced stereotypies in the rat: a behavioral test system for assessment of MAO-B inhibitors. Psychopharmacol. **84**, 22-27

Ortmann R. (1984b) The 5-HT syndrome in rats as tool for the screening of psychoactive drugs. Drug Dev. Res. **4**, 593-606

Sloviter R.S., Connor J.D., Drust E.G. (1980) Serotonergic properties of β-phenylethylamine in rats. Neuropharmacol **19**, 1071-1074

Strolin-Benedetti M. and Dow J.(1983) A MAO-B inhibitor, MD 780236, metabolized essentially by the A form of the enzyme in the rat. J. Pharm. Pharmacol **35**, 238-245

Waldmeier,P.C., Baumann,P.A., Maître,L. (1979) CGP 6085 A a new, specific, inhibitor of serotonin uptake: Neurochemical characterization and comparison with other serotonin-uptake blockers. J. Pharm. Exp. Therap. **211**,42-49

Waldmeier, P.C., Baumann, P.A.(1983) Effects of CGP 11305 A, a new reversible and selective inhibitor of MAO A, on biogenic amine levels and metabolism in the rat brain. Naunyn-Schmiedeb. Arch. Pharmacol. **324**,20-26

Yang H.-Y.T. and Neff N.H. (1974) The Monoamine oxidases of brain: Selective inhibition with drugs and the consequences for the metabolism of the biogenic amines. J. Pharm. Exp. Therap. **189**, 733-740

POTENTIATION OF THE BIOCHEMICAL EFFECTS OF β-PHENYLETHYLAMINE AND TRYPTAMINE BY DEUTERIUM SUBSTITUTION

L.E. Dyck, B.A. Davis, D.A. Durden and A.A. Boulton
Psychiatric Research Division
Cancer and Medical Research Building
University of Saskatchewan
Saskatoon, Saskatchewan
Canada S7N OWO

INTRODUCTION

The stable isotope deuterium (D) can exhibit a primary isotope effect when it is substituted for hydrogen (H) in certain carbon-H bonds (Blake et al., 1975; Baillie, 1981; Foster, 1984). Deuterium substitution in the α,α-positions of the alkyl side chain of several biogenic amines, including β-phenylethylamine (PE), significantly decreases their rate of oxidative deamination by monoamine oxidase (MAO) (Belleau and Moran, 1963; Yu et al., 1981 and 1982). Oxidative deamination by MAO is the major pathway whereby both PE and tryptamine are catabolised; consequently, one would predict that the α-deuterated analogs would be more potent drugs <u>in vivo</u> because they would be catabolised significantly more slowly than the unsubstituted amines. This prediction has been borne out. Belleau et al. (1961) showed that α, α-deuterotryptamine produced greater physiological effects in the cat than unsubstituted tryptamine. Similarly, the effects of α,α,β,β-deutero PE on rat behaviour are greater than the effects of the unsubstituted PE (Dourish et al., 1983). The purpose of the following study was to demonstrate that the enhanced potency of deutero-PE and deutero-tryptamine compared to the unsubstituted amines was due to the greater resistance of the deutero-amines to the actions of MAO <u>in vivo</u>. To accomplish this, rats were injected with an equimolar mixture of the deuterated and the non-deuterated amine, and the amounts of the amines present in brain, liver and plasma were measured at various times.

MATERIALS AND METHODS

Male Wistar rats (body weight 180-250 g) were injected i.p. with (a) an equimolar mixture of PE and either $\alpha,\alpha,\beta,\beta$-$D_4$PE; α,α,-D_2PE or β,β-D_2PE (160 μmol kg^{-1} of each) <u>or</u> (b) an equimolar mixture of tryptamine and either α,α-D_2-tryptamine or β,β-D_2-tryptamine (160 μmol kg^{-1} of each). In some experiments, the rats were pretreated with pargyline (50 mg kg^{-1}) or iproniazid (40 mg kg^{-1}) 4h before injection with the respective amines. The rats were killed at various times after injection by cervical dyslocation, and the brain and part of the liver dissected out. After cutting the hepatic vein, blood was collected into heparinized tubes and centrifuged to obtain plasma. The brain, liver and plasma were homogenised in 0.04 N perchloric acid containing known amounts of the internal standard (D_9PE or D_4 tryptamine). The amines in the samples were dansylated and successively separated unidimensionally on two silica gel thin layer plates. The amines (PE and D_4PE, PE and D_2PE or tryptamine and D_2 tryptamine) were quantitated simultaneously using a mass spectrometric technique described previously (Dyck et al., in press).

RESULTS

Studies comparing PE and various deuterated PE's

After injecting rats with an equimolar mixture of $\alpha,\alpha,\beta,\beta$-$D_4$PE and PE, the amounts of D_4PE and PE present in the brain, liver and plasma 10 min after injection were very large; thereafter, the amounts of D_4PE and PE decreased rapidly (see Fig. 1). It is important to note, however, that at all times more D_4PE than PE was present in brain, liver and plasma though an equimolar mixture had been injected into the rats. While the ratios between the amounts of D_4PE and PE in the brain, plasma and in the liver were very similar at 10 and 20 min following injection, the D_4PE to PE ratios in the brain at the later time periods were larger than those observed in the plasma and liver (which were still similar to each other). It is also interesting to note that the concentration of D_4PE and PE in the brain were greater than those in the plasma at all times.

When rats were injected with pargyline before receiving an injection of D_4PE and PE, the ratios of the amounts

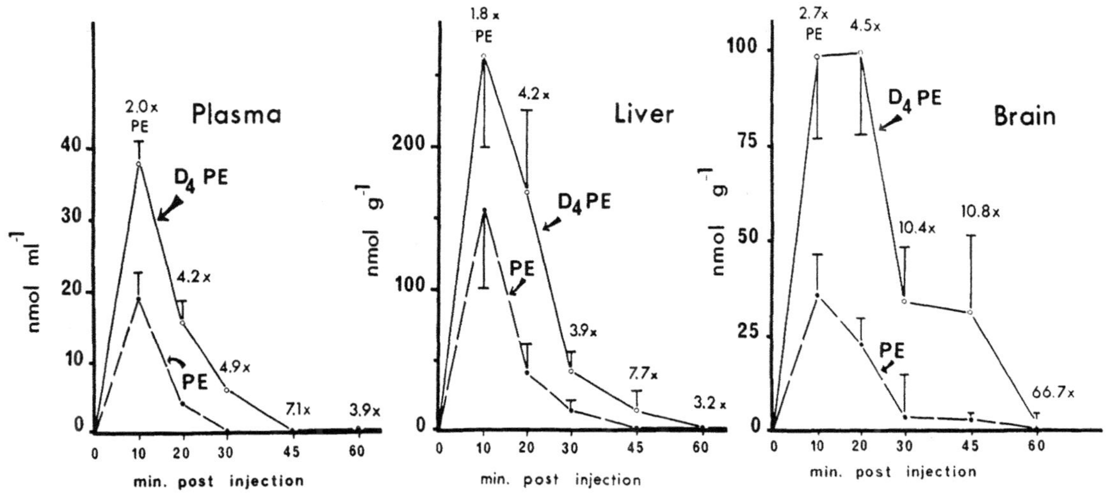

Fig. 1. Amounts (mean ± S.E.M.) of D_4PE and PE in plasma, liver and brain at various times after the i.p. injection of 160 µmol kg^{-1} each of D_4PE and PE. Numbers above the D_4PE values indicate the ratios of the D_4PE to DE values at a particular time period, for eg., at the 10 min time period in the plasma the D_4PE concentration was 2.0 x larger than the PE concentration.

of D_4PE and PE in whole brain were attenuated compared to those observed in the absence of pargyline (see Fig. 2).

In the next series of experiments, substituting the α-hydrogens of PE with deuterium was compared to substituting the β-hydrogens with deuterium. When equimolar amounts of α,α-D_2PE and PE were injected simultaneously, significantly greater amounts of α,α-D_2PE than PE were present in the brain and plasma 30 min after injection (see Table 1). In contrast, when β,β-D_2PE was injected along with PE, the levels of β,β-D_2PE and PE were equal in all three samples. Interestingly, the levels of PE in brain after an injection of PE and D_2PE was significantly greater (14.4 times) when the PE was injected with α,α-D_2PE than when it was administered with β,β-D_2PE. The amounts of PE in the liver and in the plasma after an injection of PE and D_2PE were similar whether α,α-D_2PE or β,β-D_2PE were injected.

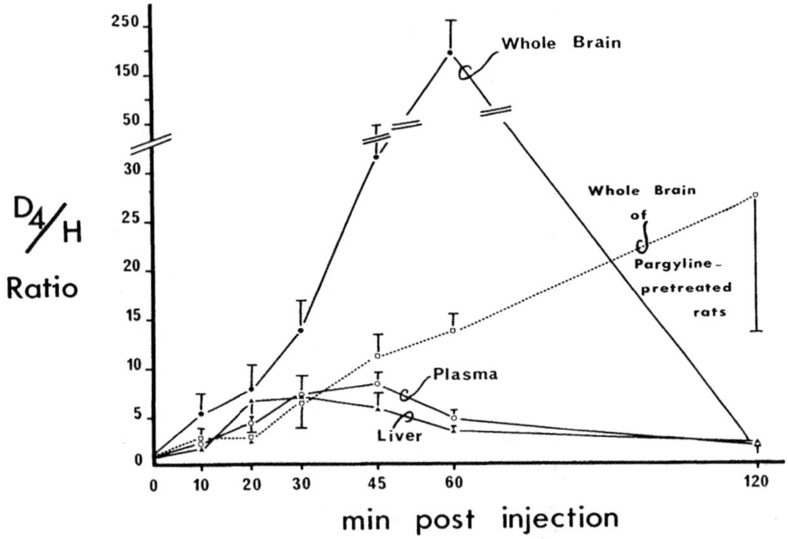

Fig. 2. Ratios (mean ± S.E.M.) of the amounts of D_4PE and PE in whole brain, liver, plasma at various times after the i.p. injection of 160 μmol kg^{-1} each of D_4PE and PE. When the rats were pretreated with pargyline (50 mg kg^{-1}, 4h), 32 μmol kg^{-1} each of D_4PE and PE were injected.

We have attempted to show that the effects of deuterium substitution in vivo were due to the greater resistance of D_4PE and D_2-tryptamine to actions of MAO and aldehyde dehydrogenase by pretreating the aminals with an MAO inhibitor. This affects the isotope effects of deuterated PE and tryptamine in two ways. First, it has a direct effect by inhibiting MAO. Secondly, this affects aldehyde dehydrogenase by inhibiting the production of its substrate, phenylacetaldehyde or indoleacetaldehyde. As expected, the D_4PE to PE ratios, and the D_2 tryptamine to tryptamine ratios, were reduced in the group pretreated with an MAO inhibitor. Similarly, our data on the effects of α,α- versus β,β-deuterium substitution support the suggestion that the greater resistance of the deuterated amines to oxidative deamination is responsible for the observed effects. Only those amines with α,α-deuterium substitution were enriched in brain, liver and plasma compared to the non-deuterated amines. Since the α-hydrogens but not the β-hydrogens are removed during deamination of PE and tryptamine by MAO and by aldehyde dehydrogenase, these data support the conclusion that appropriate deuterium substitution protects PE and tryptamine from such degradation.

Deuterium substitution can also alter the physicochemical properties of a compound, which might affect its transport into the brain by altering its lipophilicity. The dissociation constant of an amine group is altered slightly by the presence of deuterium rather than hydrogen on carbon atoms neighbouring the amine group. The basicity of amphetamine (Vree et al., 1971) and phenylethylamine (Durden et al., this meeting) are increased slightly by α-deuteration. This tends to reduce the lipophilicity of the deuterated compound so that less deuterated PE and deuterated tryptamine than PE and tryptamine would be transported into the brain. Thus, this secondary effect of deuterium substitution if anything would tend to counteract the observed enrichment of the deuterated amines in the brain compared to the non-deuterated amines.

The observation that deuterium substitution affected PE and tryptamine metabolism _in vivo_ was not too surprising in view of the previous _in vitro_ data; however, the unexpected finding was the magnitude of the effect, especially with deuterated PE. _In vitro_ the isotope effect, $(V/K)_H/(V/K)_D$ of the α,α-substitution was 2.1 for PE (Yu et al., 1981). Interestingly, the D_4PE to PE ratios (which are a measure of the isotope effect) found in brain, liver and plasma were approximately 2, ten minutes after injection; however, at later times, much larger ratios (10 or more) were observed. In the whole animal, then, a markedly greater effect of deuterium substitution occurred. When D_4PE and PE (or D_2T and T) are injected i.p. into a rat, they are subject to a first-pass effect; i.e., they are subjected to the action of hepatic drug metabolising enzymes (primarily MAO in this case) which may degrade large amounts of the injected amines before they reach the circulatory system.

PE and tryptamine can apparently cross the blood-brain barrier to enter the brain proper, though tryptamine is taken up to a smaller extent (Oldendorf, 1971). Though the plasma levels of deuterated tryptamine and tryptamine were greater than those of deuterated PE and PE, much smaller amounts of the tryptamine were present in the brain. These findings are compatible with the observations that rodents do not exhibit much of a central behavioural response to an i.p. injection of tryptamine alone, but do respond significantly when they are pretreated with an MAO inhibitor (Vogel and Cohen, 1977; Marsden and Curzon, 1978; Luscombe et al., 1983). On the other hand, rodents do exhibit a central behavioural response to i.p. administration of PE without pretreatment with an MAO inhibitor (Randrup and Munkvad, 1970).

In conclusion, then, this study shows quite convincingly that appropriate deuterium substitution can significantly protect PE and tryptamine from oxidative deamination in vivo. The effect with PE was quite marked and demonstrates that much greater deuterium isotope effects can occur in the whole animal than expected from in vitro observations. Such large effects may be useful in developing more potent centrally-acting drugs. In order for potentiation to occur several conditions would probably have to be met: first, appropriate deuterium substitution must be able to hinder substantially the major degradative pathway of the drug; second, the drug must be subject to a first-pass effect, and finally, the drug must be degraded in the brain in the same manner as the periphery.

ACKNOWLEDGEMENTS

We thank Saskatchewan Health and the MRC for financial support, and M. Mizuno and R. Mag-Atas for technical assistance. A patent is pending regarding some aspects of this work.

REFERENCES

1. Baillie T.A. (1981). The use of stable isotopes in pharmacological research. Pharmacol. Rev. 33, 81-132.
2. Blake M.I., Crespi H.L. and Katz J.J. (1975). Studies with deuterated drugs. J. Pharm. Sci. 64, 367-391.
3. Belleau B., Burba J., Pindell M. and Reiffenstein J. (1961). Effects of deuterium substitution in sympathomimetic amines on adrenergic receptors. Science 133, 102-104.
4. Belleau B., Fang M., Burba J. and Moran J. (1960). The absolute optical specificity of monoamine oxidase. J. Am. Chem. Soc. 82, 5752-5754.
5. Belleau B. and Moran J. (1963). Deuterium isotope effects in relation to the chemical mechanism of monoamine oxidase. Proc. Ann. New York Acad. Sci. 107, 822-839.
6. Dourish C.T., Greenshaw A.J. and Boulton A.A. (1983). Deuterium substitution enhances the effects of β-phenylethylamine on spontaneous motor activity in the rat. Pharmacol. Biochem. Behav. 19, 471-475.
7. Dyck L.E., Durden D.A. and Boulton A.A. (in press). Formation of β-phenylethylamine from the antidepressant, β-phenylethylhydrazine. Biochem. Pharmacol.
8. Foster A.B. (1984). Deuterium isotope effects in studies of drug metabolism. TIPS 5, 524-527.
9. Luscombe G., Jenner P. and Marsden C.D. (1983). Alterations in brain 5-HT and tryptamine content during indoleamine-induced myoclonus in guinea pigs. Biochem. Pharmacol. 32, 1857-1864.
10. Marsden C.A. and Curzon G. (1978). The contribution of tryptamine to the behavioural effects of L-tryptohan in tranylcypromine-treated rats. Psychopharmacol. 57, 71-76.
11. Randrup A. and Munkvad I. (1970). Biochemical, anatomical and psychological investigations of stereotyped behavior induced by amphetamine, in Amphetamines and

Related Compounds (eds. E. Costa and S. Garratini) pp. 695-713, Raven Press, New York.
12. Olendorf W.H. (1971). Brain uptake of radiolabelled amino acids, amines and hexases after arterial injection. Amer. J. Physiol. 221, 1629-1639.
13. Vogel W.H. and Cohen I. (1977). Effects of tryptamine on the conditioned avoidance response in rats. Brain Res. 122, 162-164.
14. Vree T.B., Gorgels J.P.M.C., Muskens A.T.J.M. and van Rossum J.M. (1971). Deuterium isotope effects in the metabolism of N-alkylsubstituted amphetamines in man. Clin. Chim. Acta 34, 333-344.
15. Yu P.H., Bailey B., Durden D. and Boulton A.A. (submitted). Stereospecific deuterium substitution of α-carbon position of dopamine and its effect on oxidative deamination catalysed by MAO-A and MAO-B from different tissues. J. Neurochem.
16. Yu P.H., Barclay S., Davis B.A. and Boulton A.A. (1981). Deuterium isotope effects on the enzymatic oxidative deamination of trace amines. Biochem. Pharmacol. 30, 3089-3094.
17. Yu P.H., Kazakoff C., Davis B.A. and Boulton A.A. (1982). Deuterium isotope effect on the enzymatic oxidation of dopamine and serotonin. Biochem. Pharmacol. 31, 3697-3698.

POSSIBLE PATHWAYS FOR SOME TRACE AMINE-CONTAINING NEURONS

A. V. Juorio, A. J. Greenshaw and A. A. Boulton
Psychiatric Research Division
Cancer and Medical Research Building
University of Saskatchewan
Saskatoon, Saskatchewan
Canada S7N OWO

The trace amines are present in discrete areas of the mammalian brain and possess a high metabolic rate as determined by their accumulation after monoamine oxidase inhibition (see review, Juorio 1984). In the rat brain the highest concentration of β-phenylethylamine (PE) is present in the hypothalamus and the striatum and a similar regional distribution was found for phenylacetic acid (Table 1). The highest concentrations of p-tyramine (p-TA), m-tyramine (m-TA) and tryptamine, or their corresponding acid metabolites, were found in the striatum (Table 1). By comparison with the striatal concentrations of dopamine (DA), or 5-hydroxytryptamine (5-HT) (about 1,000-10,000 ng g^{-1}), the concentrations of PE, p-TA, m-TA or tryptamine are very small and range between 1.3 and 11.1 ng g^{-1} (Table 1).

The iontophoretic administration of small amounts of the trace amines results in no change in neuronal firing rate; these small amounts, however, potentiate changes in cell firing rate induced by iontophoresing some of the classical transmitters such as DA, NA or 5-HT (see review, Jones, 1984). Such findings suggest that the trace amines possess a function in neurotransmission; the precise anatomical localization of possible trace aminergic pathways, however, remains to be fully investigated. The strategy adopted in order to identify some of these pathways has been to lesion brain nuclei and measure the trace amine concentrations in their known terminal projection sites.

Table 1

The rat brain regional distribution of some trace amines and trace acids

Regions ng g^{-1}	PE	PAA	*p*-TA	*p*-HPAA	*m*-TA	*m*-HPAA	*p*-OA	T	IAA
Whole brain	1.8[1]	31.8[3]	2.0[4]	10.6[5]	0.3[6]	2.3[5]	n.d.	0.5[7]	11.6[8]
Hypothalamus	2.6[9]	60.1[3]	1.2[2]	4.5[5]	0.2[2]	1.2[5]	3.4[10]	0.9[7]	23.7[8]
Striatum	1.3[9]	64.6[3]	11.1[2]	28.3[5]	3.0[2]	5.3[5]	n.d[10]	2.9[7]	63.6[8]
Brain Stem	1.6[2]	33.1[3]	0.4[2]	8.6[5]	0.1[2]	1.8[5]	1.1[10]	0.2[7]	11.7[8]
Cerebellum	2.1[2]	31.3[3]	0.1[2]	8.1[5]	0.01[2]	1.2[5]	n.d.	0.3[7]	10.1[8]
Rest	1.4[2]	27.6[3]	0.5[2]	5.3[5]	0.2[2]	1.7[5]	n.d.	0.3[7]	-

The superscripts indicate the source of reference. [1]Durden *et al.*, 1973; [2]Philips unpublished, cited by Boulton, 1976; [3]Durden and Boulton, 1982; [4]Philips *et al.*, 1974a; [5]Durden and Boulton, 1981; [6]Philips *et al.*, 1975; [7]Philips *et al.*, 1974b; [8]Warsh *et al.*, 1977; [9]Greenshaw *et al.*, 1985.; [10]Danielson *et al.*, 1977; n.d., not determined.

β-Phenylethylamine

The finding that PE concentrations and PE recognition sites are unevenly distributed in the mammalian brain (see review, Juorio, 1984; Hauger, Skolnick and Paul, 1982) and that the caudate nucleus is one of the regions with the highest levels suggested that PE-containing neuron terminals may exist in the caudate nucleus (Table 1) in a manner similar to the presence of dopamine-containing neurons that project from the nigra to the caudate nucleus (Andén, Carlsson, Dahlström, Fuxe, Hillarp and Larsson, 1964). To investigate this possibility we have assessed the effects of lesions of the substantia nigra pars compacta (SNC) on the striatal concentration of PE (Greenshaw and Juorio, 1985). Injections of 6-hydroxydopamine (6-OHDA) (8 μg) were placed unilaterally in the SNC of male Wistar rats and the caudate nucleus concentrations of PE, DA, 3,4-dihydroxyphenylacetic acid (DOPAC), and homovanillic acid (HVA) were estimated.

On the seventh post-lesion day rats were pretreated with l-deprenyl (2 mg kg^{-1}) and killed 2 hours later. The results show that 7 days following the 6-OHDA lesion, the PE concentrations in the caudate nucleus concentrations were significantly reduced (to 55% of the respective intact contralateral side) (Table 2). In addition, the lesioned

Table 2

The effect of unilateral 6-hydroxydopamine (8 μg) lesions of the substantia nigra pars compacta (left side) on the striatal concentration of PE and DA with deprenyl-treated (2 mg kg^{-1}, 2 hours) rats

Type of Treatment	PE ng g^{-1}	DA ng g^{-1}	DOPAC ng g^{-1}	HVA ng g^{-1}
Sham controls				
right side	11.7	10,220	760	490
left side	11.4	9,260	750	480
Lesions				
right side	14.9	10,360	710	470
left side	8.2*	3,060*	340*	210*

Values are in ng g^{-1} of fresh tissue. *Significantly different from control side. Results are from Greenshaw and Juorio, 1985.

side exhibited the expected significant reduction in caudate nucleus DA, DOPAC and HVA (to 30%, 48% and 45% of their respective levels in the intact contralateral side) (Table 2).

Sham controls injected unilaterally in the SNC with the 0.1% ascorbic acid vehicle showed no changes in either PE, DA, DOPAC or HVA (Table 2).

The results indicate either the existence of PE-containing neurons in the nigro-striatal pathway or PE-containing neurons whose axons pass through the region in and/or around the SNC or that PE is located in dopamine-containing neurons and is affected in a similar way.

Tyramines

The p- and m-isomers of TA are normal constituents of the brain and attain their highest concentrations in the caudate nucleus (Table 1); the concentrations of both amines are reduced by reserpine, tetrabenazine or intraventricularly administered 6-OHDA (Table 3), suggesting an intraneuronal location in storage granules (Boulton, Juorio, Philips and Wu, 1977; Juorio, 1979). The location of the cell bodies that give rise to terminals containing either p- or m-TA is unknown. The richest source of amine innervation to the caudate is the dopamine-containing projection from the SNC and it seems possible that p- and m-TA are located either within dopaminergic neurons or in separate neurons arising in the SNC or passing through it.

The effects of unilateral electrolytic lesions of the SNC in male Wistar rats on concentrations of p-TA, m-TA, dopamine and HVA were studied at various survival times. At 12 and 24 hours following lesioning, the ipsilateral level of p-TA was significantly reduced compared to the contralateral side, whereas the concentrations of m-TA, DA and HVA were significantly increased (Table 3). Thus, in the short term, the lesion results in an increase in DA turnover, which is accompanied by an increase in m-TA levels and a decrease in p-TA levels (Juorio and Jones, 1981). At survival times of 2, 11 and 25 days, the ipsilateral concentrations of m-TA, DA and HVA were reduced along with p-TA (Table 3). These longer-term alterations in amine levels are most likely a consequence of the degen-

Table 3

The effect of intraventricular cerebral administration of 6-hydroxydopamine (6-OHDA) (250 µg/rat) or lesion of the substantia nigra compacta on the rat striatal _p_-TA, _m_-TA and DA

Treatment	Time days	_p_-TA ng/g	_m_-TA ng/g	DA ng/g
Vehicle	10	10.6[1]	2.1[1]	8440[1]
6-OHDA	10	3.6[1]*	0.9[1]*	3150[1]*
Control side	0.5	16.6[2]	5.9[2]	8850[2]
Operated side	0.5	6.2[2]*	8.1[2]*	14490[2]*
Control side	1	11.8[2]	3.4[2]	7120[2]
Operated side	1	6.0[2]*	5.2[2]	10260[2]*
Control side	2	12.9[2]	3.5[2]	7120[2]
Operated side	2	2.8[2]*	1.5[2]*	3730[2]*
Control side	11	14.8[2]	3.5[2]	8890[2]
Operated side	11	4.1[2]*	2.2[2]*	3130[2]*
Control side	25	13.9[2]	4.2[2]	10650[2]
Operated side	25	6.5[2]*	2.2[2]*	4140[2]*

The asterisks indicate that the results are statistically significantly different from corresponding controls and the superscript indicates the source of reference. [1]Boulton, Juorio, Philips and Wu, 1977; [2]Juorio and Jones, 1981.

eration of nigro-striatal axonal projections.

In an attempt to delineate the anatomical location of the nigrostriatal TA-containing neurons, electrolytic lesions were placed 1 mm dorsal to the SNC (Juorio and

Jones, 1981). Following this lesion the concentration of p-TA was slightly reduced; DA was unaffected, but the concentration of m-TA was profoundly increased, even more so than after the SNC lesion. This possibly indicates the existence of specific m-TA-containing cell bodies or axons located dorsal to the substantia nigra (Juorio and Jones, 1981).

Electrical stimulation of the SNC for 1 hour induced a substantial increase in DA turnover in the rat caudate nucleus as indicated by an increase in HVA; concurrently there was an increase in striatal m-TA and a substantial decrease in p-TA (Table 4) (Jones, Juorio and Boulton, 1983). Following pretreatment with a tyrosine hydroxylase inhibitor, the effect of stimulation of the SNC on m-TA was reversed and the decrease in p-TA was partially prevented (Table 5). The effect on m-TA suggested that α-methyl-p-tyrosine inhibits the hydroxylation of phenylalanine to m-tyrosine while that on p-TA reflects a reduction in the hydroxylation of p-tyrosine to DOPA with a tendency to increase p-TA (Jones, Juorio and Boulton, 1983).

Table 4

Effects of stimulation or lesion of the substantia nigra compacta on amine levels in the right or left caudate nucleus

Type of Treatment		p-TA $ng\ g^{-1}$	m-TA $ng\ g^{-1}$	DA $ng\ g^{-1}$
Urethane	right side	10.7	3.7	10,900
	left side	11.0	3.5	10,800
Stimulated	right side	12.2	4.3	13,300
	left side	6.3*	5.3*	10,600*
Lesion	right side	11.5	4.6	11,300
	left side	8.7*	7.9*	12,700*

Urethane anaesthesia (1.2 g kg^{-1}, 1 hour); values are in ng g^{-1} of fresh tissue. *Significantly different from control side. results are from Jones, Juorio and Boulton, 1983.

Table 5

Effects of stimulation or a lesion of the substantia nigra pars compacta on amine levels in the right or left caudate nucleus in rats pretreated with α-methyl-p-tyrosine (150 mg kg^{-1}, 1 hour and 20 min)

Type of Treatment		p-TA ng g^{-1}	m-TA ng g^{-1}	DA ng g^{-1}
Urethane	right side	17.7	1.8	7,620
	left side	17.8	1.8	7,150
Stimulated	right side	16.2	1.9	7,370
	left side	12.7*	1.3*	5,500*
Lesion	right side	20.6	2.9	6,920
	left side	27.0	3.6	7,310

Urethane administration (1.2 g kg^{-1}, 1 hour); values are in ng g^{-1} of fresh tissue. *Significantly different from control side. Results are from Jones, Juorio and Boulton, 1983

Recent experiments have shown that electrolytic lesions of the dorsal and median raphé nuclei induce an increase in striatal DA turnover, a reduction in p-TA and an increase in m-TA (Juorio and Greenshaw, this book), thus further confirming the reciprocal relationship between p- and m-TA concentrations and DA turnover (see review, Juorio, 1982). In addition, the fact that the p- and m-TA concentrations are unchanged at 7 days after raphé nuclei lesions indicates that striatal TA-containing terminals do not originate in or pass through the dorsal or median raphé nuclei.

Octopamines

Earlier experiments have shown that the administration of divided doses of 6-OHDA (200 mg kg^{-1}, 48 hours) produced marked reductions in heart, spleen and salivary gland octopamine (OA) concentration (Molinoff and Axelrod, 1969). In these experiments, OA was determined by a radioenzymatic method that does not resolve the p- and m-isomers. Nevertheless, it supported that, at least in some peripheral organs, p- and/or m-OA are associated with nerve terminals. The hypothalamus contains the highest concentration of brain p-OA (Table 1) but the location of the corresponding OA-containing cell bodies has as yet to be determined.

Tryptamine

The possible existence of tryptamine-containing neurons originating in the midbrain raphé is suggested by several reports of tryptamine-mediated responses to electrical stimulation of the raphé nuclei (Jones and Broadbent, 1982; Jones, 1982a,b; Cox, Davis, Juxon, Lee and Martin, 1981). To assess this hypothesis, the effects of electrolytic lesions of the median and dorsal raphé nuclei on striatal, hypothalamic and hippocampal concentrations of tryptamine, 5-HT and 5-HIAA were investigated. In addition, the rat striatal tryptophan concentrations were also determined. No changes in the concentrations of tryptamine were observed at 1-2 weeks after lesioning the dorsal and median raphé nuclei at which time the 5-hydroxyindoles were markedly reduced (Table 6); furthermore, no reductions were observed in tryptamine concentrations in the striatum, hypothalamus or hippocampus of rats pretreated with a monoamine oxidase inhibitor (Juorio and Greenshaw, 1985). The only change observed in these rats was a limited increase in striatal

Table 6

Effect of dorsal and median raphé nuclei (RN) lesions on rat striatal tryptamine, 5-hydroxytryptamine (5-HT) and 5-hydroxyindole acetic acid (5-HIAA) concentrations

	Tryptamine $ng\ g^{-1}$	5-HT $ng\ g^{-1}$	5-HIAA $ng\ g^{-1}$
Sham controls	0.35	600	620
RN lesions	0.42	260*	190*

The rats were killed 1-2 weeks after the lesions. Values are means in ng g^{-1} of fresh tissue. *Significantly different from controls. Results are from Juorio and Greenshaw, 1985.

Table 7

Effect of dorsal and median raphé nuclei lesions (RN) on striatal tryptamine, 5-hydroxytryptamine (5-HT) and 5-hydroxyindole acetic acid (5-HIAA) concentrations of pargyline-treated rats (200 mg kg^{-1}, 2 hours)

Type of Lesion	Time days	Tryptamine ng g^{-1}	5-HT ng g^{-1}	5-HIAA ng g^{-1}
Sham	0.5	81.8	-	-
RN lesion	0.5	95.7	-	-
Sham	1	69.4	-	-
RN lesion	1	87.6*	-	-
Sham	2	81.0	-	-
RN lesion	2	83.6	-	-
Sham	7	78.0	1.13	0.31
RN lesion	7	67.5	0.57*	0.15*
Sham	14	89.9	-	-
RN lesion	14	88.0	-	-

Values are means in ng g^{-1} of fresh tissue. *Significantly different from controls. Results are from Juorio and Greenshaw, 1985.

tryptamine and tryptophan observed at one day after lesioning (Table 7 and Juorio and Greenshaw, 1985). The results indicate that tryptamine concentrations are independent of the integrity of 5-HT-containing neurons of the midbrain raphé nuclei. Furthermore, if tryptamine-containing neurons exist, their possible terminal projections to the striatum, hypothalamus and hippocampus have their cell bodies located in regions outside the dorsal and median raphé nuclei.

Conclusion

The results of the lesion and stimulation studies described above may represent a useful link between the contrasting levels of the analysis of shared biosynthetic

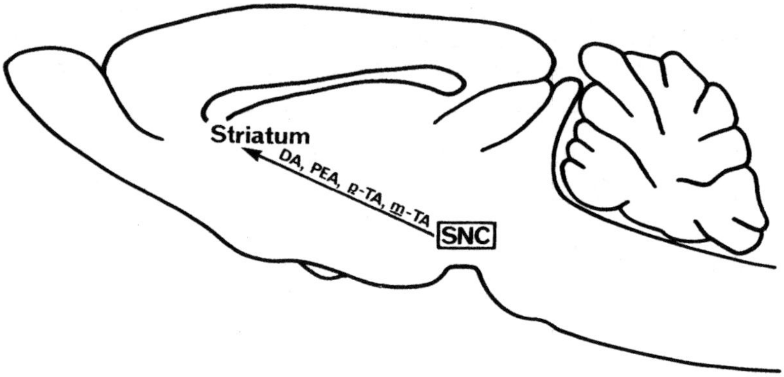

Fig. 1. Possible transmitters involved in projections from the SNC to the striatum.

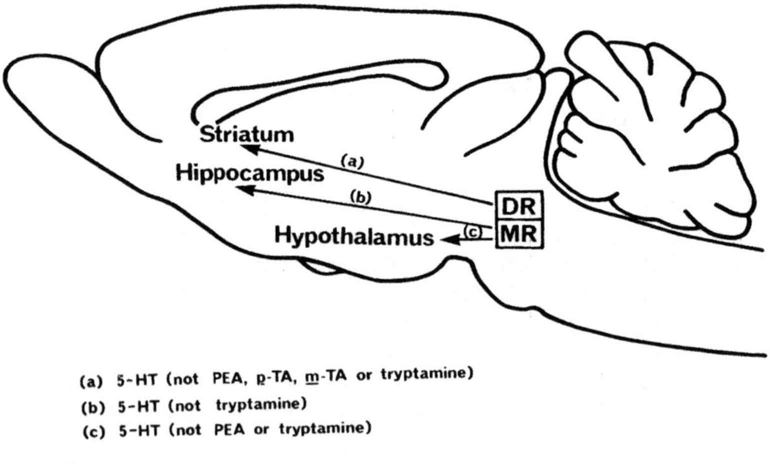

Fig. 2. Possible transmitter involved in projection from the RN to the striatum, hypothalamus or hippocampus.

pathways for amines and attempts to establish a functional role for trace-amine activity in the CNS. The current status of our knowledge concerning the anatomical distribution of possible trace-amine containing neurons in relation to those associated with dopamine and 5-HT is illustrated in Figures 1 and 2.

Further elucidation of the localisation of trace amines in terms of distinct neuroanatomical pathways should serve to provide a useful framework for the analysis of the involvement of trace amines in behavioural responses to lesions or to stimulation of specific areas of CNS.

ACKNOWLEDGEMENTS

We thank Saskatchewan Health for financial support. A.J. Greenshaw is the recipient of a Saskatchewan Health Research Board Fellowship.

REFERENCES

Andén N.-E., Carlsson A., Dahlström A., Fuxe K., Hillarp N.-A. and Larsson K. (1964) Demonstration and mapping out of nigro-neostriatal dopamine neurons. Life Sci. 3, 523-530.

Boulton A.A. (1976) Identification, distribution, metabolism and function of meta- and para-tyramine, phenylethylamine and tryptamine in brain. Adv. Biochem. Psychopharmacol. 15, 57-67.

Boulton A.A., Juorio A.V., Philips S.R. and Wu P.H. (1977) The effects of reserpine and 6-hydroxydopamine on the concentrations of some arylalkylamines in rat brain. Br. J. Pharmac. 59, 209-214.

Cox B., Davis A., Juxon V., Lee T.F. and Martin D. (1981) Core temperature changes following electrical stimulation of the midbrain raphé and intrahypothalamic tryptamine in the rat. Br. J. Pharmac. 74, 216.

Danielson T.J., Boulton A.A. and Robertson H.A. (1977) m-Octopamine, p-octopamine and phenylethylamine in mammalian brain: a sensitive, specific assay and effects of drugs. J. Neurochem. 29, 1131-1135.

Durden D.A. and Boulton A.A. (1981) Identification and distribution of m- and p-hydroxyphenylacetic acid in the brain of the rat. J. Neurochem. 36, 129-135.

Durden D.A. and Boulton A.A. (1982) Identification and distribution of phenylacetic acid in the brain of the rat. J. Neurochem. 38, 1532-1536.

Durden D.A., Philips S.R. and Boulton A.A. (1973) Identification and distribution of β-phenylethylamine in the rat. Can. J. Biochem. 51, 995-1002.

Greenshaw A.J. and Juorio A.V. (1985) β-Phenylethylamine and dopamine: concurrent depletion after lesion of the substantia nigra pars compacta. Soc. Neurosci. Abstr. 11 (in press).

Greenshaw A.J., Juorio A.V. and Boulton A.A. (1985) Behavioural and neurochemical effects of deprenyl and β-phenylethylamine in Wistar rats. Brain Res. Bull. (in press).

Hauger R.L., Skolnick P. and Paul S.P. (1982) Specific [^{3}H]-phenylethylamine binding sites in rat brain. Eur. J. Pharmacol. 83, 147-148.

Jones R.S.G. (1982a) Responses of cortical neurones to stimulation of the nucleus raphé medianus: A pharmacological analysis of the rate of indoleamines. Neuropharmacol. 21, 511-520.

Jones R.S.G. (1982b) Tryptamine modifies cortical neurone responses evoked by stimulation of nucleus raphé medianus. Brain Res. Bull. 8, 435-437.

Jones R.S.G. (1984) Electrophysiological studies of the possible role of trace amines in synaptic function. In, Neurobiology of the Trace Amines (A.A. Boulton, G.B. Baker, W.G. Dewhurst and M. Sandler, eds), Humana Press, New Jersey, U.S.A., pp. 205-223.

Jones R.S.G. and Broadbent, J. (1982) Further studies on the roles of indoleamines in the responses of cortical neurones to stimulation of nucleus raphé medianus: Effect of indoleamine precursor loading. Neuropharmacol. 21, 1273-1277.

Jones R.S.G., Juorio A.V. and Boulton A.A. (1983) Changes in levels of dopamine and tyramine in the rat caudate nucleus following alterations of impulse flow in the nigrostriatal pathway. J. Neurochem. 40, 396-401.

Juorio A.V. (1979) Drug-induced changes in the formation, storage and metabolism of tyramine in the mouse. Br. J. Pharmac. 66, 377-384.

Juorio A.V. (1982) A possible role for tyramines in brain function and some mental disorders. Gen. Pharmacol. 13, 181-183.

Juorio A.V. (1984) Drug-induced changes in the central metabolism of tyramine and other trace monoamines: their possible role in brain functions. In, Neurobiology of the Trace Amines (A.A. Boulton, G.B. Baker, W.G. Dewhurst and M. Sandler, eds) The Humana Press, New Jersey, pp 145-162.

Juorio A.V. and Greenshaw A.J. (1985) Tryptamine concentrations in areas of 5-hydroxytryptamine terminal innervation after electrolytic lesions of midbrain raphé nuclei. J. Neurochem. (in press).

Juorio A.V. and Jones R.S.G. (1981) The effect of mesencephalic lesions on tyramine and dopamine in the caudate nucleus of the rat. J. Neurochem. 36, 1898-1903.

Molinoff P. and Axelrod J. (1969) Octopamine: Normal occurrence in sympathetic nerves of rats. Science 164, 428-429.

Philips S.R., Davis B.A., Durden D.A. and Boulton A.A. (1975) Identification and distribution of m-tyramine in the rat. Can. J. Biochem. 53, 65-69.

Philips S.R., Durden D.A. and Boulton A.A. (1974a) Identification and distribution of p-tyramine in the rat. Can. J. Biochem. 52, 366-373.

Philips S.R., Durden D.A. and Boulton A.A. (1974b) Identification and distribution of tryptamine in the rat. Can. J. Biochem. 52, 447-451.

Warsh J.J., Chan P.W., Godse D.D., Coscina D.V. and Stancer H.C. (1977) Gas chromatography-mass fragmentographic determination of indole-3-acetic acid in rat brain. J. Neurochem. 29, 955-958.

$[^3H]$TRYPTAMINE RECEPTORS IN RAT BRAIN

P.L. Wood, L.L. Martin and C.A. Altar.

Neuroscience Research, Pharmaceutical Division,

Ciba-Geigy Corp., Summit, N.J. 07901.

Tryptamine concentrations in the rodent brain range from 16 fmol/mg protein (cerebellum) to 160 fmol/mg protein (striatum); concentrations which have lead to its designation as a trace amine (Philips et al., 1974; Philips et al., 1980; Juorio and Durden, 1984). However, despite its low levels, this amine has a rapid turnover as evidenced by 6- to 12-fold increases in regional tryptamine levels after inhibition of monoamine oxidase (Philips et al., 1980; Juorio and Durden, 1984). A clear functional role for these dynamic pools of tryptamine has not been defined but the recent descriptions of a unique tryptamine receptor in the rat brain have suggested a potential neurotransmitter or neuromodulator role.

Tryptamine Receptor Pharmacology

At the present time, very little is known about agonists and antagonists of tryptamine receptors. For this reason, most investigators interested in studying tryptamine receptor pharmacology have approached the problem by comparing the ability of serotonin receptor antagonists to block the *in vitro* or *in vivo* effects of tryptamine and serotonin. This approach is based on the assumption that tryptamine and serotonin receptors should have similar topographies and should be affected similarly by drugs since the two indolethylamines differ only by the presence or absence of a hydroxyl group. Indeed, tryptamine has been shown to bind to serotonin receptors *in vitro* (Martin and Sanders-Bush, 1982) and to activate serotonin receptors both in the central nervous system (Martin and Sanders-Bush, 1982) and in the periphery (Stollak and Furchgott, 1983). Clearly, specific tryptamine receptor agonists and antagonists must be developed to further study the roles of CNS tryptamine receptors.

In previous studies, metergoline, a serotonin receptor antagonist, has been found to be more potent in blocking the effects of tryptamine than in blocking the effects of serotonin. Tryptamine-mediated effects which have been shown to be selectively blocked by metergoline include tryptamine-induced rat forepaw clonus (Clineschmidt and Lotti, 1974), changes in body temperature in both rats (Cox et al., 1981) and mice (Quock and Weick, 1978) and changes in neuronal firing rates produced by iontophoretic administration of tryptamine (Jones, 1982). Metergoline, however, was shown in one study to be nonselective in antagonizing myoclonus in guinea pigs produced by either 5-hydroxytryptophan (a serotonin precursor) or tryptamine (Luscombe, et al., 1982). Triflupromazine also appears to possess some selectivity for tryptamine receptors (Cox et al, 1981) as does tetrahydro-B-carboline (THBC) which antagonizes tryptamine-mediated contraction of the rat tail artery (Hicks and Langer, 1983). Other serotonin antagonists do not possess such selectivity for tryptamine receptors. These agents include phenoxybenzamine (Winter and Gessner, 1968), methysergide, ketanserin, cyproheptadine, methiothepin, cinanserin, 2-bromolysergic acid diethylamide, and mianserin. Methysergide, however, has also been shown to be a selective antagonist of tryptamine-induced myoclonus in guinea pigs (Luscombe et al., 1982) and cyproheptadine appears to be selective in blocking the enhancement of the flexor reflex in chronic spinal dogs by tryptamine (Martin and Eades, 1970, 1972).

Summarizing the above data, tryptamine and serotonin receptors appear to possess some similar pharmacological characteristics since there is considerable overlap in their antagonism by drugs. Nevertheless, metergoline, triflupromazine and THBC appear to possess some selectivity for tryptamine receptors in the rat as does methysergide in the guinea pig and cyproheptadine in the dog. These drugs may prove to be useful in the study of tryptamine receptor function and pharmacology in these species. Further work needs to be done, however, to demonstrate whether species-related differences in the selectivity of drugs for serotonin and tryptamine receptors is related to differences in tryptamine or serotonin receptor pharmacology or to both.

Recently, high affinity binding sites for $[^3H]$tryptamine were demonstrated to exist in the rat brain (Kellar and Cascio, 1982). The binding kinetics, anatomical and subcellular distribution and pharmacological specificity of these binding sites are consistent with their being neurotransmitter receptors for tryptamine (Cascio and Kellar, 1983; Wood et al., 1984). Stereospecificity of $[^3H]$tryptamine binding sites has also been demonstrated using isomers of alpha-methyltryptamine as well as amphetamine (Cascio and Kellar, 1983; Wood et al., 1984). The sites have high affinity for tryptamines, B-carbolines and phenethylamines but very low affinity for agonists and antagonists of other neurotransmitter receptors or for various classes of therapeutic agents (Cascio and Kellar, 1983; Rommelspacher and Bruning, 1984; Wood et al., 1984). The low affinity of the serotonin antagonists for [^{3}H]tryptamine receptors was unexpected given their activity in blocking the effects of tryptamine *in vitro* and *in vivo*.

The structural requirements of [^{3}H] tryptamine binding sites has been studied quite extensively. Substitution of bulky and especially hydrophilic groups at the 5-position on tryptamine can drastically reduce affinity. For example, 5-hydroxytryptamine (serotonin) is several hundred-fold weaker in displacing [^{3}H] tryptamine than tryptamine itself. N-substitutions and alpha-substitutions also greatly reduce affinity of tryptamine (Cascio and Kellar, 1983; Wood et al., 1984).

Rommelspacher and Bruning (1984) and Wood et al. (1984) have investigated the structure-activity relationships of the B-carbolines. Similar to the effects of ring-substituents on the affinity of tryptamine analogs, substitution of B-carbolines at the 6-position by a methoxy group or especially by a hydroxy group causes a reduction in affinity. Substitution at the 1- or 3-position also reduces affinity. In addition, it appears that dihydro-B-carbolines are more potent than either the tetrahydro- or fully unsaturated B-carbolines. The high affinity of B-carbolines for [^{3}H]tryptamine binding sites is particularly interesting since B-carbolines are formed *in vivo* and therefore could act as endogenous modulators for tryptamine receptors (Rommelspacher and Bruning, 1984).

The structure-activity relationships of the phenethylamines have been characterized by Martin et al. (1984). Unlike the effect of indole substitutions, the order of potency of phenethylamine ring substitutions was methoxy > bromo > chloro > methyl > hydroxy > fluoro > hydrogen. Para substitutions were most effective. The optimum side chain length was four carbons. The conclusion of these studies was that the tryptamine receptor possesses a large lipophilic site which is probably necessary to accomodate the large indole ring of tryptamine. The long optimum chain length was unexpected, however, and its purpose is unclear at the present time.

Tryptamine Receptor Modulation

Treatments which chronically perturb the normal amount of information which flows through a synapse can induce a number of adaptive changes in neuronal function which are intended to blunt the overall effect of the treatment on neuronal transmission. One means by which neurons adapt to chronic changes in synaptic transmission is by altering either the affinity or the number of its receptors. This phenomenon occurs in a number of neurotransmitter systems and in response to a variety of treatments and has been the subject of numerous reports (Reviewed by Creese and Sibley, 1981). In order to determine whether [^{3}H]tryptamine binds to functional tryptamine receptors, we have begun to evaluate the effects of chronic drug treatments on [^{3}H]tryptamine binding. The assumption is that functional tryptamine receptors may adapt to treatments which chronically alter tryptaminergic function whereas mere chemical recognition sites should not.

Because of the lack of specific agonists and antagonists of tryptamine receptors, we chose two monoamine oxidase (MAO) inhibitors, clorgyline and deprenyl, for study. It was thought that MAO inhibitors should enhance tryptaminergic transmission by inhibiting the degradation of tryptamine by MAO in the brain. The binding of [^{3}H]tryptamine to membrane preparations of the rat frontal/parietal (F/P) cortex was determined by the method of Wood et al. (1984). The F/P cortex was removed from decapitated, male, Sprague-Dawley (Marland Farms, Hewlett, NJ) rats (200-225 g) and frozen until assay. The effects of chronic (11-day)

Table 1: Effects of chronic clorgyline and deprenyl on [^{3}H]tryptamine binding in rat F/P cortex.

Drug Treatment	[^{3}H]Tryptamine Bound (pmol/g tissue)
Vehicle	7.82 ± .39
Clorgyline, 1.0 mg/kg x 11 days	5.80 ± .25*
Deprenyl, 1.0 mg/kg x 11 days	8.88 ± .61

Rats were given daily injections of vehicle, clorgyline (1.0 mg/kg) or deprenyl (1.0 mg/kg) for 11 days and sacrificed 24 hours after the last injection. *$p < .05$ compared to vehicle.

treatments of clorgyline (1.0 mg/kg) and deprenyl (1.0 mg/kg) on [^{3}H]tryptamine binding are shown in Table 1. Chronic administration of clorgyline, but not deprenyl, significantly reduced the binding of [^{3}H]tryptamine. The effect of clorgyline was dependent on a chronic treatment regimen since specific binding was not found to be affected 2.0 hr following the acute administration of a single, large dose (11 mg/kg) of clorgyline (vehicle = 6.04 ± .73, treated = 5.62 ± .40 pmol [^{3}H]tryptamine/g tissue). Kinetic analysis (Table 2) showed that the reduction in [^{3}H]tryptamine binding after chronic clorgyline administration was due to a decrease in the number (Bmax) of the receptors with no change in their affinity (Kd). The reduction in binding could not be attributed to a direct interaction of clorgyline with [^{3}H] tryptamine binding sites since clorgyline had very low affinity for these sites in vitro (Table 3) and acute clorgyline administration had no effect on [^{3}H]tryptamine binding as noted above.

Table 2: [^{3}H]Tryptamine binding kinetics in rat F/P cortex after chronic clorgyline treatment.

Treatment	Kd (nM)	Bmax (pmol/g tissue)
Vehicle	1.59 ± .14	10.7 ± 1.0
Clorgyline	1.32 ± .06	8.21 ± .24 *

Rats were injected with clorgyline (1.0 mg/kg) or vehicle daily for 11 days and were sacrificed 24 hr after the last dose. Values represent the mean ± sem of 6 determinations.
*p < .05 compared to vehicle

Table 3: Affinity of MAO inhibitors for [^{3}H] tryptamine sites.

Compound	IC50 (nM)
Tryptamine	3.84 ± .29
Clorgyline	5550 ± 1580
Deprenyl	> 10,000

Data represent the mean ± sem of 3 determinations.

These data provide the first evidence that [^{3}H]tryptamine binding sites may be modulated by chronic drug treatments and support the notion that [^{3}H]tryptamine binds to functional tryptamine receptors. We believe that the reduction in [^{3}H]tryptamine binding is a neuronal adaptation to the accumulation of tryptamine in the synapse following inhibition of MAO. These data provide further evidence that brain tryptamine is degraded primarily by MAO type A since [^{3}H]tryptamine binding was affected by clorgyline, a selective MAO type A inhibitor, but not by deprenyl, a selective MAO type B inhibitor. Phillips and Boulton (1979) have previously demonstrated that brain tryptamine levels are increased by clorgyline but not by deprenyl.

Quantitative Autoradiography of [^{3}H]Tryptamine Binding

Our search for functional roles of tryptamine receptors may be accelerated by a localization of these sites throughout brain. [^{3}H]Tryptamine binding is relatively high in membrane homogenates of the hippocampus, caudate-putamen, and frontal and parietal neocortex (Kellar and Cascio, 1983; Wood et al, 1984). The techniques of quantitative receptor autoradiography (Penny et al, 1981) and computer-assisted image analysis (Altar et al, 1984) were combined to generate autoradiographs of [^{3}H]tryptamine (34.6 Ci/mmol; NEN) binding. We therefore determined the distribution and equilibrium and pharmacological properties of these sites in coronal and horizontal sections of rat brain.

The brains of male rats were frozen and sectioned (20 um) in either the coronal or horizontal planes. The sections were thaw-mounted onto gelatin-coated microscope slides. After storage at -20° C, the sections were thawed, dried, and incubated for 60 min at 4 ^{o}C in buffer (50 mM TRIS-HCl, 5 mM ascorbate, 20 uM pargyline; pH 7.7 at 25° C). Non-specific binding was determined with 10 uM tryptamine (Kellar and Cascio, 1982). To determine K_d and B_{max} values by equilibrium analysis, adjacent coronal sections from 5 regions of brain were incubated in 1, 2, 4, 8, 14, or 20 nM of [^{3}H]tryptamine. To determine the pharmacological specificity of sites at the level of the caudate-putamen, coronal sections were incubated in 3.0 nM [^{3}H]tryptamine and various concentrations of a displacer. The 3.0 nM concentration was chosen because it is close to previously reported affinity constants for [^{3}H]tryptamine binding to homogenized brain membranes (Cascio and Keller, 1983; Wood et al, 1984). The displacers tested included p-methoxyphenylpropylamine (PMPPA; synthesized by Dr. Dennis M. Roland, CIBA-GEIGY Corp., Summit, NJ, USA), harmaline, 5-methyltryptamine, norepinephrine (NE), serotonin (5-HT) (Sigma Chemical Co., St. Louis, MO, USA) or spiperone (Janssen Pharmaceutica, N. V., Beerse, Belgium).

The sections were rinsed six times for 1.0 sec each in buffer at 4° C and dried. The sections and [^{3}H]thymidine autoradiography standards were exposed to Ultrofilm for 3 weeks and the film subsequently developed with standard techniques.

The quantification of tryptamine binding site concentrations in numerous brain regions (Table 4) were accomplished with computer-assisted imaging techniques as recently described (Altar et al., 1984; 1985).

The capacity (B_{max}) and affinity (K_d) of tryptamine binding sites were calculated according to the best fit to a parabola by iterative, nonlinear regression analysis (Bliss and James, 1966). IC_{50} values and Hill coefficients (n_H) were calculated by logit-log analysis (Rodbard and Lewald, 1970).

Specific [^{3}H]tryptamine binding to coronal or horizontal sections was distributed heterogeneously and accounted for up to 58% of the total binding (Table 4). A 10-fold quantitative variation of specific binding occurred throughout coronal sections. [^{3}H]Tryptamine binding to 14 regions of coronal sections from five rostral-caudal levels was saturable and of high affinity. The range of binding capacities in these regions is only about 3-fold, yet is probably greater throughout brain. The low levels of specific binding in many other brain regions were so low as to preclude accurate calculation of B_{max} or K_d values. Perhaps of considerable significance is the narrow range (K_d = 3.6-8.9) of affinities of [3H]tryptamine for its high-affinity binding site throughout brain.

The greatest concentrations of sites were in the nucleus accumbens, olfactory tubercle, and claustrum (coronal sections) and subiculum/CA1 of hippocampus (horizontal sections). The lateral septum, cingulate cortex, frontal and

<u>Table 4</u>: Affinity and capacity of [^{3}H]tryptamine binding sites in rat brain as determined by quantitative autoradiography.

Brain Region	Kd (nM)	Bmax (fmol/mg protein)
Medial N. accumbens	4.9 ± 0.5	758 ± 35
Lateral N. accumbens	5.4 ± 1.0	701 ± 61
Claustrum	5.6 ± 0.6	661 ± 37
CA2/CA3	7.4 ± 1.6	601 ± 59
Lateral septum	4.3 ± 1.2	577 ± 84
Olf. tubercle	4.5 ± 0.6	535 ± 36
Dorsal raphe N.	8.9 ± 2.1	520 ± 70
Frontal cortex	3.6 ± 0.4	515 ± 82
Caudate-putamen	5.1 ± 1.6	441 ± 82

Consecutive sections were incubated in 1 of 6 concentrations of [^{3}H]tryptamine with or without 10 uM tryptamine to define non-specific binding. Values are expressed as mean ± SEM; N=6.

parietal cortex, caudte-putamen, and CA1/CA2 of hippocampus contained intermediate binding. Low concentrations of sites were observed in the raphe nuclei, the medial geniculate, globus pallidus, entorhinal cortex, and cerebellum.

The pharmacological specificity of $[^3H]$tryptamine binding was investigated in 7 regions of coronal sections obtained at the level of the caudate-putamen. 5-Methyltryptamine displaced specific binding with K_i values of 12.5-17.2 nM . The displacement was from an apparent single site or multiple sites with identical affinities for 5-methyltryptamine, since Hill coefficients were close to one in each of the 7 brain regions. Binding was also potently displaced by the phenylalkylamine, PMPPA (K_i = 3.9-9.3 nM) and by harmaline, a B-carboline (K_i = 11.8-25.9 nM). Hill coefficients for these displacers were less than one for most brain regions. Neither norepinephrine nor spiperone displaced $[^3H]$tryptamine, even at uM concentrations. Serotonin, at uM concentrations, displaced $[^3H]$tryptamine apparently from multiple sites. An important result of the present research is that the binding characteristics for $[^3H]$tryptamine as determined autoradiographically are very similar to those obtained with homogenates (Kellar and Cascio, 1982; Cascio and Kellar, 1983; Wood et al., 1984).

Summary

In summary, our data strongly support the presence of a high affinity binding site for $[^3H]$tryptamine in rat brain. Furthermore, this binding site is characterized as stereoselective and heterogenous in distribution with the greatest concentrations being observed in the N. accumbens. Theses sites are also downregulated following chronic injections of MAO type A inhibitors; suggesting that the binding sites we have labelled are functional tryptamine receptors.

REFERENCES

Altar, C.A., Walter, R.J., Neve, K.A. and Marshall, J.F. (1984) Computer-assisted video analysis of [^{3}H]spiropiridol binding autoradiographs. J. Neurosci. Meth. 10, 173-188.

Altar, C.A., Wasley, A.M. and Martin, L.L. (1985) Autoradiographic localization and pharmacology of unique [^{3}H]tryptamine binding sites in rat brain. Neurosci. (in press)

Bliss, C.I. and James, A.T. (1966) Fitting the rectangular hyperbola. Biometrics 22, 573-602.

Cascio, C. S. and Kellar, K. J. (1983) Characterization of 3H-tryptamine binding sites in brain. Eur. J. Pharmacol. 95, 31-39.

Clineschmidt, B. V. and Lotti, V. J. (1974) Indoleamine antagonists: relative potencies as as inhibitors of tryptamine and 5-hydroxytryptamine-evoked responses. Brit. J. Pharmacol. 50, 311-313.

Cox, B., Lee, T. F. and Martin, D. (1981) Different hypothalamic receptors mediate 5-HT and tryptamine-induced core temperature changes in the rat. Brit. J. Pharmac. 72, 477-482.

Creese, I. and Silbey, D.R. (1981) Receptor adaptations to centrally acting drugs. Ann. Rev. Pharmacol. 21, 357-391.

Hicks, P. E. and Langer, S. Z. (1983) Antagonism by tetrahydro-B-carboline of the vasoconstrictor responses to tryptamine in rat tail arteries. Eur. J. Pharmacol. 96, 145-149.

Jones, R. S. G. (1982) A comparison of the responses of cortical neurones to iontophoretically applied tryptamine and 5-hydroxytryptamine in the rat. Neuropharmacology 21, 209-214.

Kellar, K. J. and Cascio, C. S. (1982) [3H]-Tryptamine: High affinity binding sites in rat brain. Eur. J. Pharm. 78, 475-478.

Luscombe, G., Jenner, P. and Marsden, C. D. (1982) Tryptamine-induced myoclonus in guinea-pigs pretreated with a monoamine oxidase inhibitor indicates pre- and post-synaptic actions of tryptamine upon central indoleamine systems. Neuropharmacology 21, 1257-1265.

Martin, W. R. and Eades, C. G. (1970) The action of tryptamine on the dog spinal cord and its relationship to the agonist actions of LSD-like psychotogens. Psychopharmacology 57, 242-257.

Martin, W. R. and Eades, C. G. (1972) Cross tolerance to tryptamine in the LSD tolerant dog. Psychopharmacology 27, 93-98.

Martin, L. L., Roland, D. M., Neale, R. F. and Wood, P. L. (1984) ^{3}H-Tryptamine binding in rat brain: Effects of phenylalkylamine derivatives. Pharmacologist 26, 135.

Martin, L. L. and Sanders-Bush, E. (1982) Comparison of the pharmacological characteristics of $5HT_1$ and $5HT_2$ binding sites with those of serotonin autoreceptors which modulate serotonin release.Naunyn-Schmiedeberg's Arch. Pharmacol. 321, 165-170.

Penny, J.B., Pan, H.S., Young, A.B., Frey, K.A. and Dauth, G.W. (1981) Quantitative autoradiography of [^{3}H]muscimol binding in rat brain. Science 214, 1036-1038.

Phillips, S.R. and Boulton, A.A. (1979) The effect of monoamine oxidase inhibitors on some arylalkylamines in rat striatum. J. Neurochem. 33, 159-167.

Quock, R. M. and Weick, B. G. (1978) Tryptamine-induced drug effects insensitive to serotonergic antagonists: Evidence of specific tryptaminergic receptor stimulation? J. Pharm. Pharmac. 30, 280-283.

Rodbard, D. and Lewald, J.E. (1970) Computer analysis of radioligand assay and radioimmunoassay data. Acta Endocrinol. 147, 79-103.

Rommelspacher, H. and Bruning, G. (1984) Formation and function of tetrahydro-B-carbolines with special reference to their action on [3H]-tryptamine binding sites. In: Progress in Tryptophan and Serotonin Research, Schlossberger, H. G., Kochen, W., Linzen, B. and Steinhart, H. (eds.), Walter de Gruyter, Berlin.

Stollak, J. S. and Furchgott, R.F. (1983) Use of selective antagonists for determining the types of receptors mediating the actions of 5-hydroxytryptamine and tryptamine in the isolated aorta. J. Pharmacol. Exp. Ther. 224, 215-221.

Winter, J. C. and Gessner, P. K. (1968) Phenoxybenzamine-antagonism of tryptamines,their indene isosteres and 5-hydroxytryptamine in the rat stomach fundus preparation. J. Pharmacol. Exp. Ther. 162, 286-293.

Wood, P. L., Pilapil, C., LaFaille, F., Nair, N. P. V. and Glennon, R. A. (1984) Unique [3H]tryptamine binding sites in rat brain: distribution and pharmacology. Arch. Int. Pharm. Ther. 268, 194-201.

TRYPTAMINE, TYRAMINE AND HISTAMINE IMMUNOREACTIVITIES IN SPECIFIC NEURONES OF THE GASTROPOD CNS

N.N. OSBORNE

Nuffield Laboratory of Ophthalmology,
University of Oxford,
Walton Street, Oxford, U.K.

Previous reports based on spectrofluorometric, microspectrofluorometric, chromatographic and radiometric methods have established that the following monoamines are present in the snail CNS: dopamine, serotonin, histamine, octopamine, tyramine and tryptamine (see Gerschenfeld 1973; Leake and Walker 1980; Robertson and Juorio 1976). There are plenty of physiological and pharmacological data supporting the opinion that dopamine, serotonin, histamine and octopamine have functional roles in the gastropod CNS (see Gerschenfeld 1973; Leake and Walker 1980; Osborne 1978; Weinreich 1977; Saavedra et al. 1974; Robertson and Juorio 1976).

The evidence for dopamine and serotonin having specific transmitter functions is best documented, partly because they can be localised histochemically in specific neurones and the pathways of the various neurones mapped out. No histochemical methods have previously existed for localising histamine, tyramine and tryptamine, although biochemical data on hand-dissected neurones have shown that histamine is associated with specific gastropod neurones (Weinreich 1978). Such results must be viewed with caution because of contamination problems associated with the methodology (Osborne 1974, 1977).

Within the last fifteen years it has been recognised that it is in fact possible to make antibodies to small neuroactive molecules which are not antigenic themselves (Fillip and Schneider 1964, Ranadive and Sehon 1967) by coupling them to a carrier protein such as bovine serum albumin. Such an antibody to serotonin was subsequently raised and used only recently in histochemical studies to localise serotonergic pathways (Steinbusch et al. 1978; Osborne et al. 1982a). In the last few years, antibodies to dopamine, noradrenaline and GABA have been produced by similar means and used successfully to localise the substances in nervous tissues histochemically (Vieillemaringe et al. 1983; Geffard et al. 1985).

This new methodological approach to raising antibodies to small molecules which are not antigenic themselves makes it theoretically possible to localise the trace amines in nervous tissues and establish whether they are associated with neuronal or glial elements. We have therefore used a variety of procedures to produce immunogens to the following compounds: histamine, octopamine, tyramine, tryptamine, β-phenylethanolamine and phenylethylamine for subsequent antibody production. While a number of our experiments proved negative, we have recently obtained positive data suggesting that antibodies are produced to localise histamine, tyramine and tryptamine. These preliminary results are described in this report.

Preparation of Immunogens

(a) Histamine. The immunogen was prepared as described by Wilcox and Seybold (1982) by mixing together 0.1g histamine dihydrochloride with 0.325g methylate BSA in 50ml buffered phosphate, saline, giving a final histamine concentration of 2mg/ml. This emulsified histamine/methylated BSA solution was injected into rabbits.

(b) Tryptamine. 1mmole Tryptamine was dissolved in 1ml 0.1N HCl to which was added dropwise diazotised para-aminobenzoic acid, and it was stirred constantly until a precipitate was maximally formed. Following a centrifugation at 12000 x g the precipitate was washed five times with distilled water and then dried. 200mg of the dried precipitate in 10ml 10mM phosphate buffer pH 6.5 was mixed with 400mg thyroglobulin in 10ml phosphate buffer and 200 mg 1-ethyl-3-(3-dimethylaminepropyl) carbodiimide HCl in 10ml phosphate buffer. The mixture was stirred vigorously at room temperature for 4 hours and then dialysed for 3 days against running tape water. Following a centrifugation, the precipitate was lyophilised and stored at 4°C.

(c) Tyramine. 100nmole Thyroglobulin and 400nmole tyramine were dissolved in 1ml 0.1M phosphate buffer pH 7.5. To this solution was then added dropwise over 15 mins.

1.5ml 0.25% glutaraldehyde. After a vigorous stir the mixture was left overnight at 4°C. An aglutinous mass forms which is then homogenised and aliquoted.

Preparation of Antisera

Antibodies to various immunogens were raised in rabbits. Each animal received about 0.5ml. of an immunogen-containing solution (approximately 2mg/ml emulsified in about 0.5ml complete Freund's adjuvant) injected intracutaneously in different sites. Booster injections with incomplete Freund's adjuvant were given every four weeks.

Serum samples from animals were taken just before immunisation (preimmune serum) and 7 days after the booster injections.

Immunocytochemistry Methodology

Serum collected after the fourth injection was tested on the snail nervous system. Positive immunohistochemical data were obtained for histamine, tyramine and tryptamine from bleeds taken respectively after the 7th, 6th and 5th injection. For immunohistochemistry, nervous tissues were dissected and fixed in 4% paraformaldehyde made up in 0.1M phosphate buffer pH 7.2 for 1 hour and processed as described elsewhere (Osborne et al. 1982). In brief, ten-micrometer frozen sections were obtained at -20°C, incubated at room temperature with antiserum to histamine, tyramine or tryptamine (serum diluted 1:50 with phosphate buffer containing 0.2% Triton X-100) for two hours. After rinsing, sections were developed with fluorescein isothiocyanate-conjugated rabbit immunoglobulin anti-rat IgG for forty minutes at room temperature. The sections were then mounted and examined with a microscope equipped with epi-fluorescence optics.

Immunocytochemistry Results

In order to establish the specificity of the histamine, tyramine and tryptamine antisera, a variety of experiments was performed. First, each antiserum was rendered inactive by preabsorption with the respective immunogen. Second, the pre-immune serum in each case gave negative results. Third, sera from animals injected with conjugates to substances produced in the same way as histamine (carnosine conjugate), tryptamine (adrenaline-conjugate) and tyramine (serotonin conjugate) gave negative (carnosine) or different (adrenaline,serotonin) results.

(a) Histamine. Histamine-like immunoreactivity was observed in the cytoplasm of somata throughout the nervous system of *Helix aspersa* (Fig. 1a). In addition some immunoreactivity was associated with processes situated in the neuropile. Although the detailed distribution of histamine-positive cells was not analysed, they would appear to be few in number compared with the CCK or substance P-like positive neurones (Osborne et al. 1982b; Osborne and Dockray 1982) and distributed more or less in all ganglia.

While the experiments on *Helix* showed us that our antiserum was "staining" specific molecules, it did not prove that it was revealing the presence of histamine. Now Turner and Cottrell have reported the presence of histamine (1977), revealed by microchemical analysis, in an identified giant neurone situated in the visceral ganglion of the pond snail *Lymnaea stagnalis*. This animal was therefore analysed immunohistochemically. Figure 1b. is a section taken through the visceral ganglion of *Lymnaea* where it can be seen that histamine-like immunoreactivity is associated with a giant neurone. This cell is the largest in the

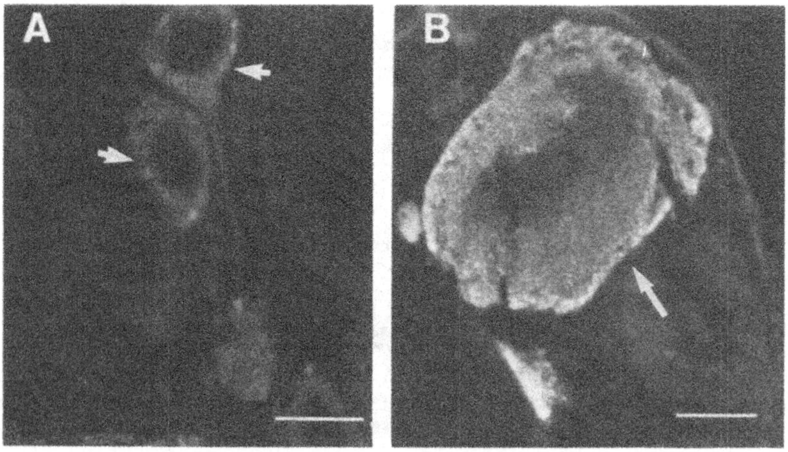

Fig.1: Sections taken through the nervous systems of *Helix* (A) and *Lymnaea* (B) and stained for the localisation of histamine immunoreactive neurones (arrows). Section shown in B is through the visceral ganglion. Scale bar = 50 µm.

visceral ganglion and is therefore the same neurone as that described by Turner and Cottrell (1977). We interpret our data as showing that our antiserum recognises the histamine molecule.

(b) Tryptamine. Neuronal cell bodies and cell processes containing tryptamine immunoreactivity were identified in the CNS of the snail (Fig. 2). The number of tryptamine-positive neurones was few and the cells were usually small. From a limited study (serial sections from three central ganglia) it was not possible to establish the precise positions of the neurones. It was estimated that 60-90 neurones stained for tryptamine inthe whole CNS and the immunoreactive cell bodies were found in all ganglia. An analysis of serial sections alternatively processed for localising tryptamine and 5-hydroxytryptamine (serotonin) in some instances revealed the co-occurrence of immunoreactivities in the same neurone but on other occasions neurones stained either for tryptamine or 5-hydroxytryptamine immunoreactivity (Fig. 3). These results showed that the antibody raised for localising tryptamine did not cross react and revealed 5-hydroxytryptamine.

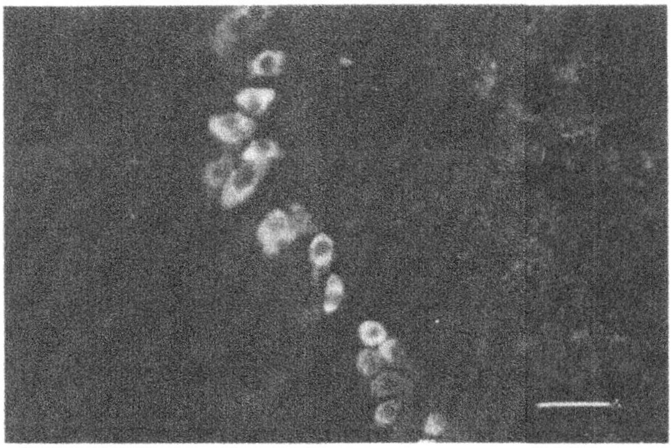

Fig.2: Section taken through the pedal ganglion of *Helix* showing a cluster of tryptamine immunoreactive cell bodies. It can be seen that the antibody stains specific neurones. Scale bar = 50 µm.

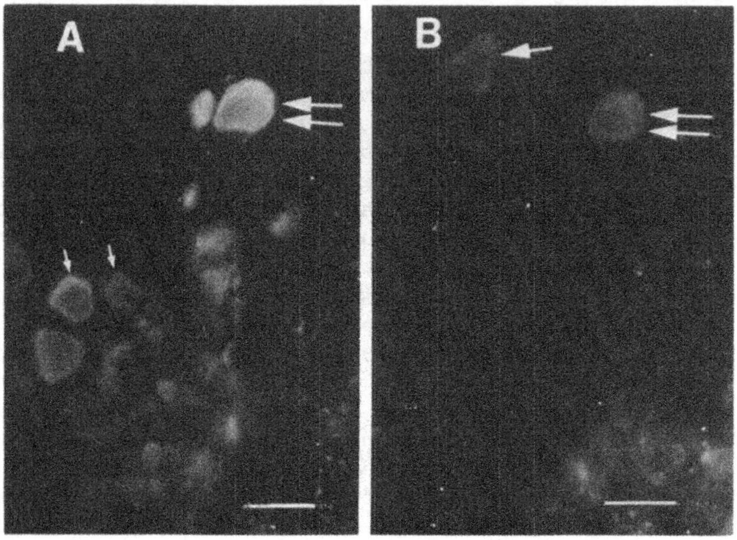

Fig. 3: Serial sections through the snail brain processed for the localisation of serotonin (A) and tryptamine (B) immunoreactivities. It can be seen that some cells stain for tryptamine (arrow) or serotonin (small arrow) or for both substances (double arrow).
Scale bar = 50 µm.

(c) Tyramine. Tyramine immunoreactivity was shown to be associated with specific neurones in the snail brain (see Fig. 4) and there was no evidence of the co-existence of tyramine with tryptamine or histamine. The tyramine immunoreactivity was generally weaker than either the tryptamine or histamine immunoreactivities, but this may be due to the titre of the antibodies, rather than reflecting the concentrations of the different amines. It was estimated that around 30-60 tyramine-positive neurones occur in the whole of the central ganglia and they appeared to be present in all ganglia and not obviously abundant in one ganglion.

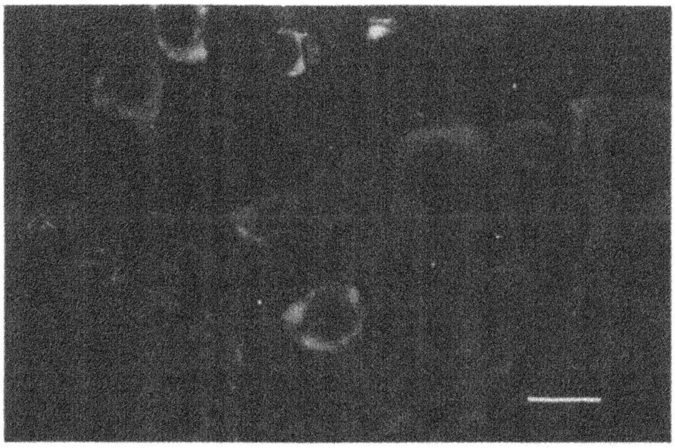

Fig. 4: A section taken through the snail brain and processed for the localisation of tyramine. It can be seen that positive immunoreactivity is associated with specific cell bodies.
Scale bar = 50 µm.

Characterisation of Antisera

The reactivities of the individual antisera to tryptamine, tyramine, histamine and some structurally related compounds were tested by coupling these compounds (0.1mmole) to nitrocellulose paper activated with polylysine and glutaraldehyde and incubating the paper with primary and secondary antibodies, thus simulating the immunohistochemistry of tissue sections. The neat antisera of each substance reacted positively with the corresponding amine, e.g. the tyramine antisera gave a positive reaction with tyramine. The reactions were not very intense, however, and diluted sera (1:10) did not produce positive reactions. From the limited experiments carried out we were unable to determine any obvious cross-reactivity between the antibodies with related substances. For example, tryptamine, tyramine and histamine antisera gave no positive reactions with 5-hydroxytryptamine, octopamine or histidine respectively.

Significance of Results

The present work provides histochemical evidence for the occurrence of histamine, tyramine and tryptamine in specific gastropod neurones. To our knowledge, this is the first description of the histochemical localisation of tyramine and tryptamine-like immunoreactivities in specific neurones. It is known that all three amines, viz. histamine, tyramine and tryptamine are present in the gastropod CNS (see Leake and Walker 1980; Robertson and Juorio 1976). The presented data show that each of the antisera "stains" specific molecules without proving that the individual amines are revealed. The finding that the histamine antiserum "stains" a neurone which has been shown biochemically to contain histamine suggests very strongly that this antiserum does recognise the histamine molecule. The characterisation experiments support this idea; histamine was incorporated into nitrocellulose paper and then shown to react with the antiserum. However, the results of the characterisation experiments for all the substances must be viewed with care, firstly because only the neat antisera gave positive data and secondly, because the antisera did not cross-react with substances of closely related structures. It might have been expected that the antisera would have some cross-reactivity with closely related substances.

The finding that histamine, tyramine and tryptamine-like immunoreactivities are in specific neurones supports the view that each of these substances may be a neurotransmitter (Weinreich 1978; Schwartz et al. 1980; Jones 1982; Robertson and Juorio 1976). Tyramine is also a precursor of octopamine and the evidence for octopamine being a transmitter in invertebrates is persuasive (Saavedra et al. 1974; Leake and Walker 1980). In studies on rat brain some positive staining was observed in the hypothalamus with the histamine antisera but negative results were obtained with the tyramine and tryptamine antisera. The significance of all these studies will only be understood after further investigations; detailed characterisation of the antisera is clearly necessary.

Acknowledgements: The financial support from the Stiftung Volkswagenwerk is gratefully acknowledged.

REFERENCES

Fillip G and Schneider H. (1964) Die Synthese der Serotoninazoproteins. Experimente zur Frage der Serotoninimmunität. Acta Allergol, 19, 216

Geffard M, Henrich-Rock A.M., Dulluc J and Sequela P. (1985) Antisera against small neurotransmitter-like molecules. Neurochem.Int. (in press)

Gerschenfeld H.M. (1973) Physiol.Rev. 53, Chemical transmission in invertebrate central nervous systems and neuromuscular junctions, 1-119

Jones R.S.G. (1982) Tryptamine: a modulator or neurotransmitter in mammalian brain? Progress in Neurobiol. 19, 117-140

Leake L.D. and Walker R.J. (1980) Invertebrate neuropharmacology. Blackie. Glasgow

Osborne N.N. (1974) Microchemical analysis of nervous tissue. Pergamon Press, Oxford

Osborne N.N. (1977) Do snail neurones contain more than one neurotransmitter? Nature 270, 622-623

Osborne N.N. (1978) Biochemistry of characterised neurones. Pergamon Press, Oxford

Osborne N.N. and Dockray G.J. (1982) Bombesin-like immunoreactivity in specific neurones of the snail *Helix aspersa* and an example of the co-existence of substance P and serotonin in an invertebrate neurone. Neurochem. Int. 4, 175-180

Osborne N.N., Nesselhut T, Nicholas D.A., Patel S and Cuello A.C. (1982a) Serotonin-containing neurones in vertebrate retinas. J. Neurochem. 39, 1519-1528

Osborne N.N., Cuello A.C. and Dockray G.J. (1982b) Substance P and cholecystokinin-like peptides in *Helix* neurones and cholecystokinin and serotonin in a giant neurone. Science 216, 409-411

Ranadive N.S. and Sehon A.H. (1967) Antigenicity of 5-hydroxyindole-3-acetic acid, a derivative of serotonin: II Antibodies specific to 5-hydroxyindole-3-acetic acid. Canadian J. Biochem. 45, 1701-1710

Robertson H.S. and Juorio A.V. (1976) Octopamine and some related noncatecholic amines in invertebrate nervous systems. Int. Rev. Neurobiol. 19, 173-224

Saavedra J.M., Brownstein M.J., Carpenter D.O. and Axelrod J. (1974) Co-existence of several putative neurotransmitters in single identified neurone of *Aplysia*. Science 185, 364-365

Schwartz J.C., Pollard H. and Quach T.T. (1980) Histamine as a neurotransmitter in mammalian brain neurochemical evidence. J. Neurochem. 35, 26-33

Steinbusch H.W.M., Verhofstad A.A.J. and Joosten H.W.J. (1978) Localization of serotonin in the central nervous system by immunohistochemistry: description of a specific and sensitive technique and some applications. Neuroscience 6, 557-618

Turner J.D. and Cottrell G.A. (1977) Properties of an identified histamine-containing neurone. Nature 267, 447-448

Vieillemaringe J., Duris P., Geffard M., Le Moal M., Delaage M., Bensch C. and Girardie J. (1984) Immunohistochemical localization of dopamine in the brain of the insect *locusta* in comparison with the catecholamine distribution determined by the histochemical technique. Cell Tissue Res. 237, 391-394

Weinreich D. (1977) Synaptic responses mediated by identified histamine-containing neurones. Nature 267, 854-856

Weinreich D. (1978) Histamine-containing neurones in *Aplysia*. In: Biochemistry of characterised neurones (ed. N.N. Osborne) Pergamon Press, Oxford pp. 153-175

Wilcox B.J. and Seybold V.S. (1982) Localisation of neuronal histamine in rat brain. Neuroscience Letters 29, 105-110

BEHAVIORAL AND PHARMACOLOGICAL CHARACTERISATION OF TRYPTAMINE-INDUCED EXCITATION SYNDROME IN RATS

A. Delini-Stula and E. Radeke

Research Department, Pharmaceuticals Division, CIBA-GEIGY Ltd., CH-4002 Basel, Switzerland

INTRODUCTION

Tryptamine is naturally present in the brain of several animal species as well as in the human brain. Te exact amounts are disputed and those reported in the literature, even for the same species, vary over the wide range of ng/g concentrations. But, the fact is that by comparison to the brain levels of monoamine neurotransmitters, which are in µg/g ranges, those of tryptamine are very low (for ref. see Jones, 1982). Tryptamine is normally formed by decarboxylation of L-tryptophan and metabolised by monoamineoxidase (MAO) into indolacetic acid (IAA). Although there are other claims (Housley and Tipton, 1974), tryptamine is considered to be a mixed substrate for MAO-A and MAO-B enzymes. For a long time not much credit has been given to the possibility that tryptamine plays a role in physiological brain functioning and rather little attention has been paid to this trace amine.
Recent development of new sensitive and specific assays for detection and quantification of tissue contents of tryptamine revived the interest in this trace amine. Moreover, cumulating evidence that sensitive receptor (or acceptor) sites, different from the 5-HT binding sites, exist in the brain (Wood et al. 1984; this symposium) suggests possible functional importance of tryptamine in the neurotransmission processes. We

found, therefore, of interest to study and characterise behavioral effects of tryptamine in rats in more depth.

MATERIAL AND METHODS

Male albino mice (Tif: MAGf, Tierfarm Sisseln), weighing 18 - 24 gr and rats (IVA:SDIV, (SIV 50) Ivanovas), weighing 140 - 160 gr were used. Animals were kept in standart cages with free access to food and water. Light conditions (12 h light/dark cycle) as well as the room temperature (21°C) were constant.

Rats were assigned to groups of 6 or 12 at random and injected with 75 mg/kg s.c pargyline 2 h prior to tryptamine (TRYP). For the testing of drug effects a dose of 15 mg/kg i.p. TRYP was used and drugs were given 30 (i.p.) or 60 (p.o.) min before injection. Reserpine was given 30 min and also 22 h beforehand.
Behavioral changes were observed during 2 h. Each symptom was arbitrarily rated form 0 - 3 points in 30 min intervals. The criteria for rating are summerized in Table 1. In addition, the presence of backward motion (BM) was recorded ("yes/no").

TABLE 1: CRITERIA FOR RATING THE TRYP - SYNDROME

SCORE	VOCALIZATION	TREMOR	IRRITABILITY	MOTOR AGITATION
0	NO VOCALIZATION	NO TREMOR	NO EXAGGERATED RESPONSE TO ANY STIMULATION	NO ABNORMAL ACTIVITY
1	RARE VOCALIZATION, OCCURING ONLY AFTER STIMULATION (TAPPING ON THE BACK)	FINE TREMOR HEAD APPEARING ONLY IF THE ANIMALS MOVES	TAPPING ON THE BACK PROVOKES JERKING MOVEMENTS	CONTINOUS SHAKING (RIGHT - LEFT) OF THE HEAD
2	OCCASIONAL SPONTANEOUS VOCALIZATION	TREMOR OF THE HEAD AND PROXIMAL BODY PART ALSO AT REST	SLIGHT TOUCH INDUCES JUMPS, SQEEZING AND INCREASED RESTLESSNESS	AS ABOVE AND INTERMITTENT AGITATED MOVEMENTS OF THE FOREPAWS
3	INTENSE FREQUENT VOCALIZATION	INTENSE TREMOR OF THE WHOLE BODY	NOISE OR APPROACH INCREASES ALL RESPONSES	INTENSE AND CONTINUOUS AGITATION AND PHASIC CIRCLING MOVEMENTS

For each dose at least 12 rats were used and several experiments were performed with the observer blind to the treatment. The results are expressed as mean cumulative score for each symptom and for the group. The percentage of animals displaying BM was calculated. The consistent reproduction of the results in different experiments was taken as criterion of the validity of the findings.
In order to compare the effectivness of different MAOI in eliciting behavioral changes after tryptamine and 5-HTP treatment in mice and rats, experiments were perfomed in which only the presence or absence of characteristic excitation signs was recorded. Standard doses of tryptamine and 5-HTP were used which in the combination with 75 mg/kg s.c. of pargyline produced typical effects.

Intracerebral administration

Pilot experiments were performed in order to assess behavioral changes produced by direct injection of tryptamine (in pargyline pretreated rats), N,N,-dimethyl - tryptamine (DMT) and N-methyl-tryptamine into the brain. For intraventricular administration a cannula was stereotaxically implanted into the left ventricle under general pentobarbital anaesthesia and fixed to the skull with the bone cement. Intracisternal administration was performed under light ether anaesthesia. Controls were injected with sterile physiological solution. The injection volume was 10 μl. Per dose 5 - 10 rats were used.

RESULTS

In doses of 15, 30 and 60 mg/kg i.p. tryptamine alone produced slight increase in activity of rats, but no other notable changes in general behavior. At a dose of 100 mg/kg animals showed only slight hypoactivity of about 2 h duration. In mice pretreated with MAO-A, but not MAO-B inhibitors, tryptamine induced intense, stereotyped-like head waving (shaking), which was easily discriminated from the typical 5-HTP head-twitch in mice pretreated with MAOI. The 5-HTP head-twitch developed after treatment with MAOI-A as well as after high doses of MAOI-B (Table 2,B). In the rat, in the presence of even high doses of selective either MAOI-A or MAOI-B, tryptamine failted to induce signs of characteristic behavioral excitation (Table 2, A). The full syndrome as described also by Marsden and Curzon (1978), developed only after a high dose of pargyline, as well as after the combination of clorgyline and deprenyl. Behavioral changes consisted of following signs: vocalisation, tremor, irritability motor agitation and backward movements (see also Table 1) which appeared within 20 - 30 min after tryptamine and were maximal between 1 and 2 h post drug. Backward motion usually preceeded other symptoms. The intensity of the excitation signs appeared to be dependent of the tryptamine dose (Fig. 1).

TABLE 2 DEVELOPMENT OF TRYPTAMINE AND 5-HTP EXCITATION SYNDROME BY DIFFERENT MAO-INHIBITORS

A

MAO-I	DOSE MG/KG	TRYP MG/KG (15 IP.)	5-HTP MG/KG (100 IP.)
PARGYLINE	10 s.c. 75 s.c.	0 +	+ +
DEPRENYL	10 s.c.	0	+
MD 780236	10 I.P. 25 I.P.	0 0	N.T. +
CLORGYLINE	10 s.c.	0	+
BROFAREMINE	10 I.P.	0	+
DEP. + CLORG.	10+10 s.c.	+	

B

MAO-INH.	DOSE MG/KG	TRYP MG/KG (100 IP.)	5-HTP MG/KG (75 IP.)
PARGYLINE	10 s.c.	0	+
DEPRENYL	25 s.c.	0	+
MD 780236	25 s.c.	0	+
CLORGYLINE	0.5 s.c.	+	+
BROFAREMINE	2.5 s.c.	+	+
CIMOXATONE	0.5 P.O.	+	+

N.T. = NOT TESTED

IN RATS THE PRESENCE OR ABSENCE OF 5-HTP SYNDROME (ORTMANN ET AL., 1984) AND IN MICE THE 5-HTP HEAD-TWITCH WERE RECORDED. IN MICE TRYP INDUCED ALSO HEAD-WAVING WHICH WAS, HOWEVER, MORE INTENSE AND MORE PRONOUNCED THAN AFTER 5-HTP.

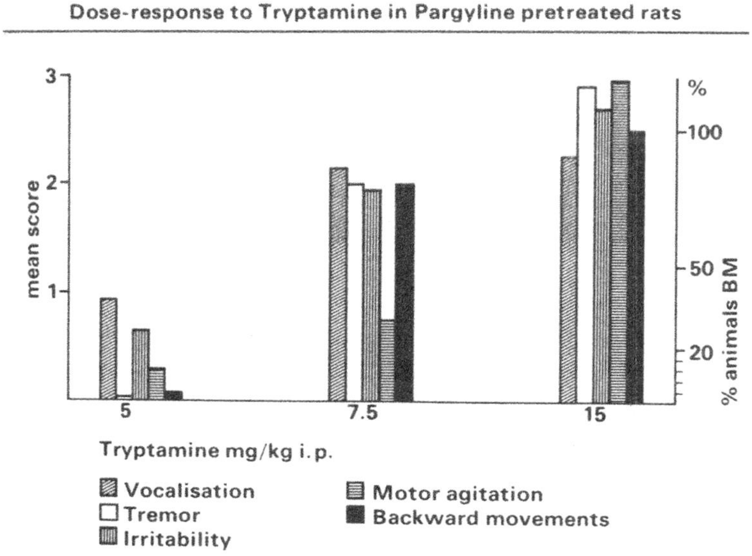

Fig. 1 For explanation see Methods

Full syndrome of marked intensity developed also after intraventricular injection of 100 µg of TRYP in pargyline pretreated rats (Table 3). Of interest is that about 90 min after injection animals showed, in addition, pronounced aggressiveness. Vocalisation and clear-cut irritability, but neither tremor nor motor agitation developed after i.c.v. administration 300 µg DMT. Backward locomotion, by contrast, was present at 100 µg and was apparently dose-dependent. N-methyl-tryptamine produced dose-related tremor, agitation and paw clonus, which ended up in convulsions and death of some animals.

TABLE 3 BEHAVIORAL CHANGES INDUCED BY DIRECT I.C.V. ADMINISTRATION OF TRYPTAMINE, DMT AND N-METHYL TRYPTAMINE

DRUG	DOSE µg	SPECIFIC SYMPTOMS V	T	I	A	BM %	OTHER	N
TRYPTAMINE*	100	++	++	++	++	100	AGGRESSION	5
DMT	100	0	0	0	(+)	40	SEDATION	10
	200	(+)	0	(+)	0	90	"	10
	300	+	0	++	0	100	"	5
N-METHYL ** TRYPTAMINE	100	0	+	0	(+)	0	CONVULSIVE FITS	10
	300	0	++	0	++	0		5

* RATS WERE PRETREATED WITH PARGYLINE 75 MG/KG SC 2H BEFOREHAND

** INJECTED INTRACISTERNALLY

There was no uniformity in the pattern of inhibitory effects of various 5-HT antagonists or neuroleptics on different behavioral signs induced by tryptamine. As illustrated in Table 4 and Fig. 2, tremor and backward locomotion appeared to be inhibited consistently and more or less dose-dependently by methergoline, methysergide, cyproheptadine and mianserin. Cinanserine and ketanserine, however, decreased tremor and irritability but failed to reduce the intensity of other symptoms to more than 50% at tested doses. Moreover, for the suppression of each of the particular signs, different dose of the same drug was sometimes necessary. Thus, for instance, backward motion was almost completely inhibited by 0.05 mg/kg of methergoline or 10 mg/kg of mianserin. By contrast, the same extent of inhibition of tremor was not achieved even at ten times higher doses of methergoline or with 30 mg/kg of mianserine, respectively.

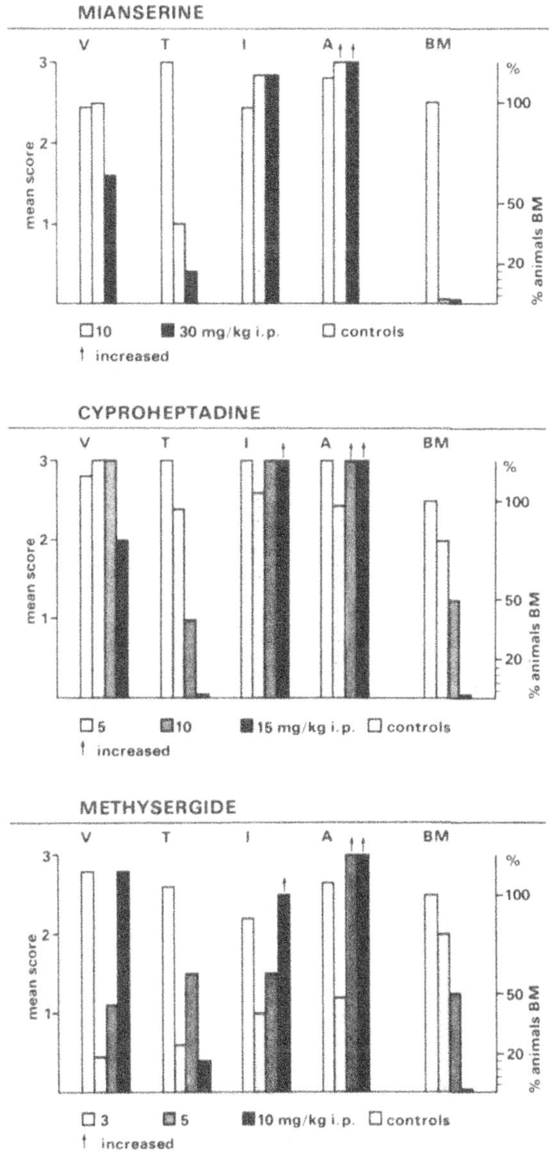

Fig. 2: Each score is the mean of n = 12. For other details see text and Fig. 1.

The most consistent and also clearly dose-dependent effects of neuroleptics, except in the case of clozapine, was the inhibition of agitated, restless movements. Other symptoms were variably affected (Fig 3, Table 4 and 5).

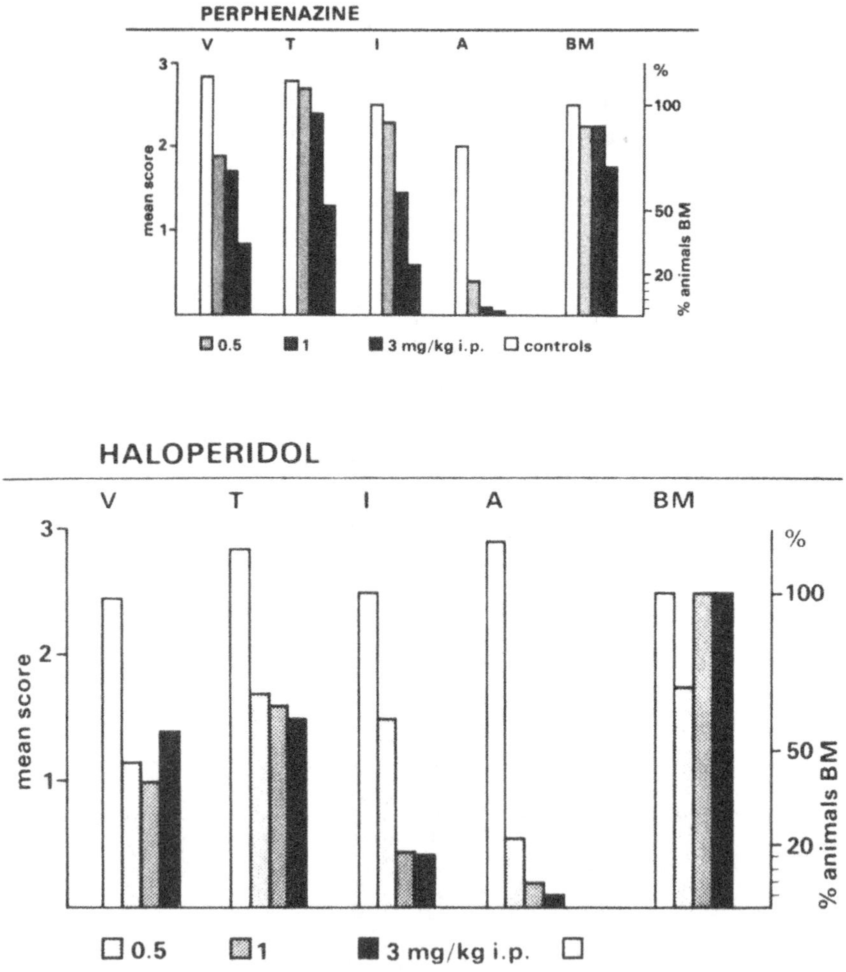

Fig. 3: For explanation see Fig. 2.

TABLE 4

RELATIVE POTENCIES OF SOME 5-HT ANTAGONISTS AND NEUROLEPTICS IN INHIBITING TRYPTAMINE-INDUCED SYMPTOMS IN THE RAT

DRUG	DOSE IN MG/KG (>50 % DECREASE IN MEAN SCORE)					ED_{50} μ MOL/KG ^3H- SPIPERONE BINDING (IN VIVO)[1]
	VOC	TREM	IRRIT	M.AGIT	BACK.MOV.	
METHERGOLINE	> 0.5	0.5[a]	> 0.5	0.1	< 0.05[a]	0.4
CYPROHEPTADINE	>15	10[a]	>15	>15	10[a]	0.06
KETANSERINE	> 1	1	1	> 1	> 1	-
CINANSERINE	10	10[a]	10	>10	>10	6.1
MIANSERIN	>10	10[a]	≫30	30(↑)	≪10[b]	1.0
THIORIDAZINE	10	1	10	1[a]	10	15.3
THIOPROPERAZINE	10	>10	10	< 1[a]	>10	≫47
PERPHENAZINE	3	> 3	3	< 0.5[a]	> 3	1.1
SPIPERONE	> 1	1	1	0.5[a]	1	6.8

a DOSE-DEPENDENT EFFECT, b 100 % INHIBITION, (↑) INCREASE
1 IN RAT CORTEX (FROM ORTMANN ET AL. 1982)

TABLE 5 COMPARISON OF RELATIVE 5-HT AND TRYPTAMINE ANTAGONISTS POTENCIES OF 5-HT-RECEPTOR BLOCKERS AND NEUROLEPTICS

DRUG	TRYP EFF. DOSE (>50 %↓) MG/KG		TRYP* ED_{50} MG/KG (PAW CLONUS)	5-HTP* ED_{50} MG/KG (HEAD-TWITCH)	5-HT[1] OEDEMA ED_{50} MG/KG
	T	BM			
METHERGOLINE	0.5	< 0.05	0.14	0.47	0.15
CYPROHEPTADINE	10	10	5	0.12	0.9
METHYSERGIDE	3-10	5	2.3	2.5	0.04
CINANSERINE	10	> 10	40	1.5	-
MIANSERIN	>10	≪ 10**	1.5	-	6.6
CHLORPROMAZINE	> 3	> 3	-	-	10
THIORIDAZINE	1	10	-	-	7.4
PERPHENAZINE	3	> 3	-	-	12.4
HALOPERIDOL	> 3	> 3	-	-	26.6

ALL DOSES I.P.

* FROM LUSCOMBE ET AL. 1982
1 5-HT-INDUCED PAW OEDEMA IN THE RAT WAS ESTIMATED AS DESCRIBED BY ORTMANN ET AL. 1982. (DATE ARE FROM OWN LABORATORY).

Table 5 illustrates relative potencies of 5-HT antagonists in inhibiting tryptamine and L-5-HTP induced effects in the rat.

Sometimes the inhibition was associated with marked general depression, or with signs of catalepsy (chlorpromazine, haloperidol, thioproperazine, perphenazine). By contrast to the other neuroleptics, clozapine produced particular type of effect. At doses of 1 - 3 mg/kg it reduced tremor in a dose-dependent manner, but rather increased the irritability. At 10 mg/kg almost all signs of tryptamine induced excitation were antagonized (Fig. 4): at the same time bursts of hyperactive, but in appearance normal type of locomotion appeared. It should be mentioned that such effects were sometimes observed after 5-HT antagonists, in particular after methergoline.

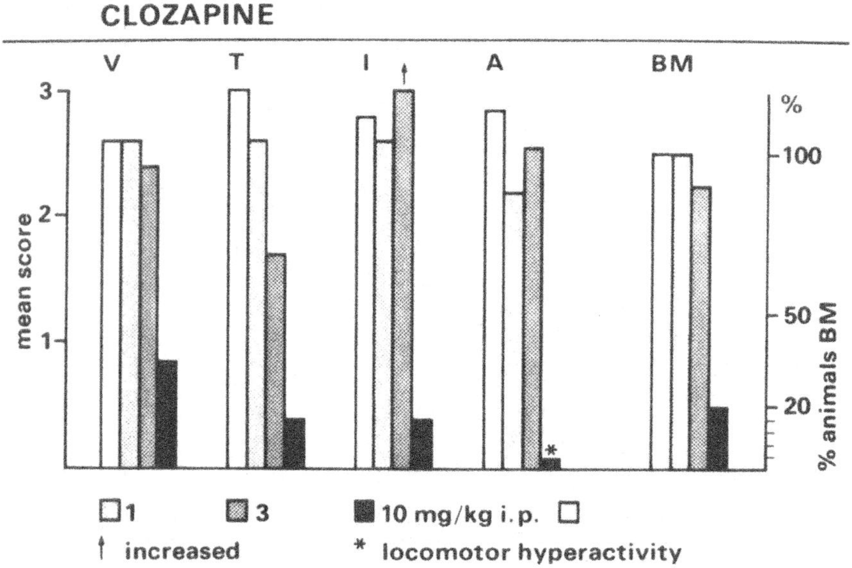

Fig 4: For explanation see Fig. 3

In naive rats on the other hand, clozapine at a dose of 10 mg/kg suppressed spontaneous locomotor activity to more than 50 % (unpublished data). Reserpine given 30 min and even more so if given 22 h prior to tryptamine, clearly enhanced all tryptamine-induced effects (data not shown).

None of the excitation signs induced by tryptamine was affected by phentolamine, an α-adrenergic blocking drug, or by pyrilamine, an antihistaminic. Scopolamine (0.3 - 5 mg/kg i.p.) increased irritability as well as vocalisation. Physostigmine (0.2 mg/kg i.v.) inhibited them to almost 100 %, but neostigmine was inactive (data not shown). Only at a high dose of 10 mg/kg i.p. diazepam consistently suppressed vocalisation and irritability. Interestingly enough, the same pattern of effects was produced by ß-receptor antagonists such as propranolol, oxprenolol and alprenolol but at doses far in excess of the ß-blocking doses (30 mg/kg i.p.)

DISCUSSION

Several observations from this study appear interesting. If it is generally recognized that tryptamine is a mixed substrate for the MAO-A and MAO-B enzymes, our results suggest that this applies for rats, but not for mice where tryptamine appears to be a preferential MAO-A substrate. It seems worth mentioning that recently Bieck et al. (1984) reported marked increase in urinary tryptamine excretion in volunteers treated with highly selective MAO-A inhibitor brofaremine and only negligible changes in its urinary output after deprenyl. This findings suggest that in man tryptamine is also a preferential substrate for MAO-A. In rats, however, particular tryptaminergic excitation could apparently develop if a) both forms of the MAO enzyme are almost completely inhibited and b) an excess of tryptamine is present. This assumption is supported by other observations. Marsden and Curzon (1978), for instance, have described that small increase in hyperactivity, produced by the combination of MAOI and L-tryptophan in rats, is greatly enhanced by either increasing tryptophan doses or by addition of tryptamine. The super-imposed behavioral excitation was obviously associated with large increases in the brain tryptamine content, since under such conditions the 5-HT levels were only minimally augmented. The finding that tryptamine, injected also directly into the brain of pargyline pretreated rats, produces intense and characteristic behavioral excitation further supports such observations. Interesting in this respect is the finding that some, but not all of the components of the syndrome could be induced by, although large, amounts of N,N-dimethyl-tryptamine (DMT) and N-methyl tryptamine alone, which in addition was obviously the most toxic and convulsive agent. It is noteworthy, that backwards motion, which typically occurred after DMT injection, is considered by some authors as animal counterpart of hallucinatory behavior in man. Indeed, Saavedra and Axelrod (1972) have suggested that in the presence of MAOI N-methylated tryptamine products are formed in the brain in mesurable concentrations; these could possibly exert a pathogenetic effect on brain functions. Although contested by some authors, the hypothetical role of N-methylated tryptamine metabolites in the pathogenesis of psychotic states is still of considerable attractiveness. DMT, for instance, has been identified in schizophrenic, psychotic patients treated with MAOI.

Tryptamine-induced aberrant behavioral signs were differently influenced by various pharmacological treatments. Particularly, it was evident that 5-HT antagonists influence some signs preferentially and dose-dependently, whereas for the inhibition of some others unproportionally high doses were necessary. Remarkable inhibition of backward movements seen after methergoline and mianserin treatment, by comparison to the antagonism of other effects, as well as less potent effectiveness of other 5-HT antagonists on this particular sign, may suggest the involvement of specific tryptaminergic and not 5-HT receptors in the mediation of backward motion. This is supported by other findings (Cox et al. 1981, Vargaftig et al. 1971) showing higher selectivity of methergoline and mianserin, but not cyproheptadine or cinanserine, towards tryptamine-induced effects. Likewise, Luscombe et al. (1982) found marked differences in the ability of methysergide and cyproheptadine to block tryptamine and 5-HT-induced myoclonus in guinea-pigs. It is noteworhty also, that some neuroleptics which, by comparison to the antidopaminergic potency show prevalent 5-HT receptor blocking action (chlorpromazine, thioridazine), or equal affinity for both 5-HT and DA-receptors (Ortmann et al. 1982), in general influence backward motion less than other symptoms. Distinctly different effect of clozapine could possibly be explained by its known strong 5-HT-antagonistic and, perhaps also, specific, antitryptaminergic property. Of interest is that clozapine, but not haloperidol, blocks the disruptive effect of DMT on conditioned avoidance behavior in rats (unpublished).

It has been suggested, and our results favour this assumption, that different neurochemical systems are imlicated in different components of tryptaminergic excitation. Tryptamine enhances the activity of the 5-HT system, although some responses of 5-HT (excitatory) on cortical neurons are antagonised (see Jones, 1982). Preferential and consistent inhibition of signs of motor agitation by neuroleptics strongly suggests the involvement of dopaminergic mechanisms in their elicitation. This may be associated with the fact that the levels of tryptamine after MAOI are particularly high in the striatum and hippocampus (Philips and Boulton, 1979). Further, modulation of cholinergic transmission by tryptamine has been demonstrated and our data show that physostigmin and scopolamine specifically affect some signs of tryptaminergic excitation and do this in opposite direction. It remains an open question by which mechanisms high doses of diazepam and ß-antagonists suppress some of the excitation signs. Nevertheless, it seems that the inhibition

they produce is not due to general depressant effect. α - antagonists and antihistaminics even at sedative doses do not inhibit any of the symptoms. This fact also suggests, in accordance with observations of other authors, that neither α - adrenergic nor histamin-receptors play a role in the mediation of tryptamine effects.
Finally, the potentiation of tryptamine effects in reserpine depleted rats should be commented. This finding indicates that a) tryptamine effects are not dependent on the release of monoamines and b) postsynaptic supersensitive response to tryptamine develops in depleted animals.
In conclusions, our findings confirm and extend the observations of other authors that excessive amounts of tryptamine produce aberrant behavioral responses. Pharmacological interactions support the notion that these responses result from the activation of different transmitter systems. Possibly they also relate to the activation of specific tryptaminergic receptors. The fact that some components of the tryptamine-induced excitation may be produced by direct administration of N-methylated tryptamine products in non MAOI pretreated rats, reinforces the assumption that under unphysiological conditions, such as abnormally low MAO levels, excessive formation of N-methylated metabolites may also represent an additional pathogenetic factor in eliciting disturbances in brain functions. It should be recalled, that exacerbations or provocation of psychotic state in depressive as well as in schizophrenics may occur under MAOI treatment. Whether or not tryptamine also plays a role in phsiological transmission processes, still remains an open question.

REFERENCES

Bieck, P.R., Nilsson, E., Schick, C., Waldmeier, P.C., and Lauber, J. (1984), Urinary excretion of Tryptamine in comparison to normetanephrine and beta-phenylethylamine in human volunteers after subchronic treatment with different monoamine oxidase inhibitors, in: Neurobiology of trace amines, eds. A.A. Boulton, G.B. Bauer, W.G. Dewhurst and M. Sandler, Humana Press, 1984, pp. 525 - 539

Cox, B., Lee, T.F. and Martin, D. (1981). Different hypothalamic receptors mediate 5-Hydroxytryptamine- and tryptamine-induced core temperature changes in the rat, Br. J. Pharmac. 72, 477 - 482

Housely, M.D. and Tipton, K.F., (1974), A kinetic evaluation of monoamine oxidase activity in rat liver mitochondrial outer membrane, Biochem. J. 139, 645 - 652

Jones, R.S.G., (1982) Tryptamine: a neuromodulator or neurotransmitter in mammalian brain? Prog. Neurobiol. 19, 117 / 139

Luscombe, G., Jenner, P. and C.D. Marsden (1982) Tryptamine-induced myoclonus in guinea-pigs pretreated with a monoamine oxidase inhibitor indicates pre- and post-synaptic actions of tryptamine upon central indoleamine systems, Neuropharmacol. 24, 1257 - 1265

Ortmann, R., Bischoff, S., Radeke, E., Buech, O. and Delini-Stula, A. (1982) Correlations between different measures of antiserotonin activity of drugs, Naunyn-Schmiedeberg's Arch. Pharmacol. 329, 265 - 270

Ortmann, R., (1984) The 5-HT syndrome in rats as tool for the screening of psychoactive drugs Drug. Dev. Res. 4, 593 - 606

Philips, S.R., and Boulton, A.A. (1979) The effect of monoamine oxidase inhibitors on some aryl-alkylamines in rat striatum. J. Neurochem. 33, 159 - 167

Saavedra, J.M. and Axelrod, J. (1972), Psychotomimetic N-methylated tryptamines: Formation in brain in vivo and in vitro, Sci. 175, 1365 - 1366

Vargaftig, B.B., Coignet, J.L., de Vos C.J., Grijsen, H. and Bonta, I.L. (1971) Mianserin hydrochloride: peripheral and central effects in relation to antagonism against 5-hydroxy-tryptamine and tryptamine, Eur. J. Pharmacol. 16, 336 - 346

Wood, P.L., Pilapil, C., La Faille, F., Noir, N.V.P. and Ennon, R.A. (1984) Unique 3-H-tryptamine binding sites in rat brain: distribution and pharmacology. Arch. Int. Pharmacology, Therap. 268, 194 - 201

BIOSYNTHESIS OF ß-CARBOLINES

Rudy Susilo and Hans Rommelspacher

Department of Neuropsychopharmacology,
Free University
Ulmenallee 30, D-1000 Berlin 19, F.R.G.

There is good evidence that ß-carbolines (BC's), potentially psychoactive compounds, are present endogenously in mammalian tissues (McIsaac, 1961; Honecker and Rommelspacher, 1978; Johnson et al., 1984). The biosynthesis of these compounds, however, is at present not established. BC's have been synthesized chemically by the Bischler-Napieralski cyclisation (Whaley and Govindachari, 1951a; route I of fig. 1) and by the Pictet Spengler condensation (Whaley and Govindachari, 1951b; route II and III of fig. 1).

Kveder and McIsaac (1961) detected trace amounts of 6-methoxyharmalan, a BC, in the urine of rats after administration of ^{14}C-labelled melatonin. This observation suggests a possible biosynthesis pathway of BC's via the Bischler-Napieralski reaction in mammals (route I).

The synthesis of tetrahydroharman from tryptamine and acetaldehyde under so-called physiological conditions (25°C, in buffer pH 5-6) was demonstrated by Hahn and Ludewig (1934; route II). This pathway is favoured by many investigators for the biosynthesis of BC's because aldehydes are very reactive agents, especially formaldehyde. Furthermore, several authors (Peura et al., 1980; Rommelspacher et al., 1980) showed that ethanol loading in rat and man increased the BC concentration. However, we found that harman concentration in the brain, indeed, increased after ethanol loading but there was no correlation between the increase of acetaldehyde in the blood and the rise of the harman concentration

(Rommelspacher et al., 1984).

Hahn et al. (1935) showed that tryptamine reacted under mild conditions easily with pyruvic acid yielding tetrahydroharman-1-carboxylic acid (1-CTHH, route III). They assumed that pyruvic acid condenses much more easily with tryptamine than do the aldehydes. Based on this consideration they suggested the 'pyruvic acid route' in which pyruvic acid serves as the carbonyl precursor of the BC biosynthesis. On the other hand Schöpf et al. (1940) revealed that aldehyde reacts approximately hundred times faster than pyruvic acid. The evidence that the condensation product of pyruvic acid and indoleethylamine occurs in mammals, was presented by Collins et al. (1982) who detected 6-hydroxy-1-CTHH in CSF of monkeys.

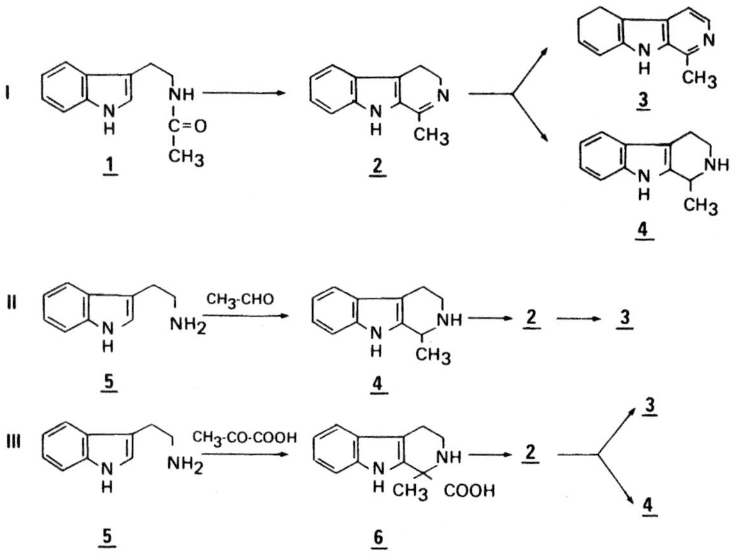

Fig.1 Routes of synthesis of BC's. I: N-acetyltryptamine (1) is cyclodehydrated to harmalan (2) according to the Bischler-Napieralski cyclisation. Harmalan is converted either to harman (3) or tetrahydroharman (4). Route II: Tryptamine condenses with aldehydes, e.g. acetaldehyde to yield tetrahydroharman (4) according to Pictet Spengler. Route III: Tryptamine condenses with α-keto-acids, e.g. pyruvate yielding 1-carboxy tetrahydroharman (6), again according to Pictet Spengler. This compound is converted to harmalan (2).

Thus, evidence is presented for the existence of the end-products of all three pathways in vivo. Since there are only a few investigations of route I and route III, it is necessary to examine the possible role of these cyclisation reactions for the biosynthesis of BC's in mammals. The main interest of the present study was to investigate these two possible pathways for the BC formation. Furthermore, we have reinvestigated the Hahn experiments (1934, 1935) to examine whether the condensation reactions occur in vitro with approximately physiological tryptamine concentrations or not.

Materials and Methods

Chemicals

^{3}H N-acetyltryptamine (NAT) was synthesized from G- ^{3}H tryptamine hydrochloride (1.4 Ci/mmol, Amersham) and acetanhydride in ethanolic solution (Chattaway, 1931). The synthesized substance was dissolved in 70% ethanolic solution and the radioactivity was adjusted to 9.5 µCi/10µl. G- ^{3}H 1-CTHH was synthesized as described (Susilo and Rommelspacher, 1985; spec. activity 1.38 Ci/mmol). Unlabelled compounds, namely NAT, harmalan and 1-carboxytetrahydroharman were synthesized in our laboratory and the identity was established by MS and NMR data. Other chemicals were obtained from commercial sources.

Chemical condensation reaction of acetaldehyde and pyruvic acid with tryptamine

6 nmoles ^{3}H tryptamine (spec. activity 31.2 Ci/nmol, NEN) was incubated with acetaldehyde (0.1-20 mM) or with pyruvic acid (0.5-50 mM) in phosphate buffer pH 7.4, at 37°C in a total volume of 100 ml. Other experiments were carried out in 0.1 M sodium acetate buffer pH 5.0. The reactions were stopped after 2 h by placing the tubes in ice water. The reactions were performed in duplicate. After addition of 100 µl methanol, an aliquot of the reaction mixture was subjected to TLC analysis (solvent system I and III).

In vivo experiments with ^{3}H N-acetyltryptamine

Intraventricular cannulae were implanted to male Wistar rats (3 animals, 190-200 g) 1 week before injection. Ten microlitres ^{3}H NAT (7 nmol) was injected intraventricularly and the animals were decapitated 1 h after administration.

Brain, liver, kidney and urine were analyzed by TLC.

Preparation of samples

The tissue was mechanically homogenized in ice-cold ethanol (1:10, w/vol) containing 100 μg of tetrahydroharman, harmalan, and harman. The homogenate was centrifuged (8000xg, 20 min, 4°C), the supernatant evaporated to dryness at 30°C under vacuum and the residue dissolved in methanol. Urine was diluted with ethanol (1:1) and directly spotted on TLC plates.

Chromatography

An aliquot of the methanol extract was spotted on silica gel F-254 TLC plates and developed in various solvent systems I, isopropanol-methylacetate-12.5%ammonia (45:35:20, vol/vol); II, n-butanol-glacial acetic acid-water (40:10:20, vol/vol); III, chloroform-acetone-diethylamine (60:30:10, vol/vol). The chromatography plates were scanned with a TLC-linear analyzer (Berthold, model LB 2821).

In vivo decarboxylation of ^{3}H 1-CTHH

0.7 nmol H 1-CTHH was injected intraventriculary into rats (male Wistar rats, 190-220 g), the animals were decapitated 1 h after injection. The brain, liver and urine were prepared as aforementioned (see "preparation of the samples") and analysis by TLC.

In vitro decarboxylation of ^{3}H 1-CTHH

^{3}H 1-CTHH (0.15 nmoles) and PALP (5 nmoles) were incubated with brain homogenate (15% w/vol) or liver homogenate (7.5% w/vol) for 1 h, 37°C in phospate buffer (0.1 M, pH 7.4) in a total volume of 500 μl.
Reaction blanks were prepared with boiled tissue homogenates. The reaction was terminated by adding 100 μl of 6% HCLO . The sample was then centrifuged. The supernatant was neutralized, evaporated under vacuum to dryness and the residue was taken up in methanol. After addition of ten μg of each reference compound (THH, harmalan, harman), the methanol extract was subjected to TLC analysis.

Results

Experiments concerning the route I

Only unchanged ^{3}H N-acetyltryptamine was detected in the brain of rats 1 h after intraventricular administration of the compound (fig. 2). Several metabolites of NAT were observed in the urine, most of them were extracted into ethylacetate phase at pH 10. The expected cyclisation products (harmalan, harman or tetrahydroharman) could not be detected (Fig 2). Tryptamine as a deacetylated product was not found as well. Due to the low amount of radioactivity, the metabolites of acetyltryptamine could not be analyzed in the kidney an liver. The chemical identity of the metabolites was not further investigated.

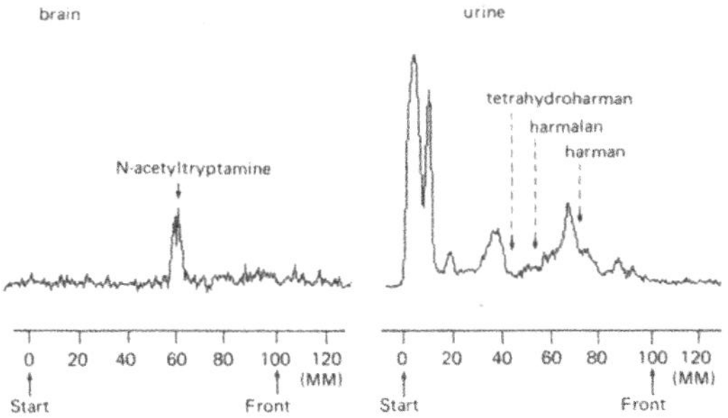

Fig.2 Radiochromatograms of a sample from rat brain (left panel) and urine (right panel). The animals had been treated with ^{3}H NAT intraventricularly one hour before decapitation (solvent system III).

Experiments concerning the route III

In vitro incubation of 0.3 µM ^{3}H 1-CTHH with brain or liver homogenate and 10 µM PALP in 0.1 M phosphate buffer at pH 7.4 resulted in the formation of harmalan. Incubation of 1-CTHH with boiled homogenate or only with PALP under the same conditions yielded no harmalan (Fig 3). The enzymatic decarboxylation of 1-CTHH depended on the PALP concentration. Without addition of PALP, only trace amounts of harmalan were detected.

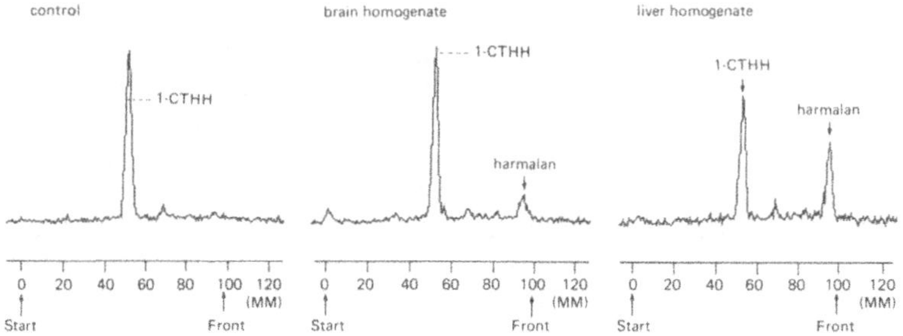

Fig.3 Radiochromatograms of samples after incubation of ^{3}H 1-CTHH with boiled brain homogenate (control, left panel), similar sample with brain homogenate (middle panel)as well as of liver homogenate (right panel) in solvent system I.

Intraventriculary administered ^{3}H 1-CTHH (0.7 nmole) was metabolized in the rat into THH, harmalan and harman. One hour after injection, nearly equal concentrations of THH and harmalan were measured in the brain, whereas harman was not detectable. All three oxidation states of the BC (tetrahydro-BC, dihydro-BC and the full aromatic BC) were found in the liver with harmalan as the major metabolite. In the urine traces of harmalan and harman were detected but not tetrahydroharman (Fig. 4).

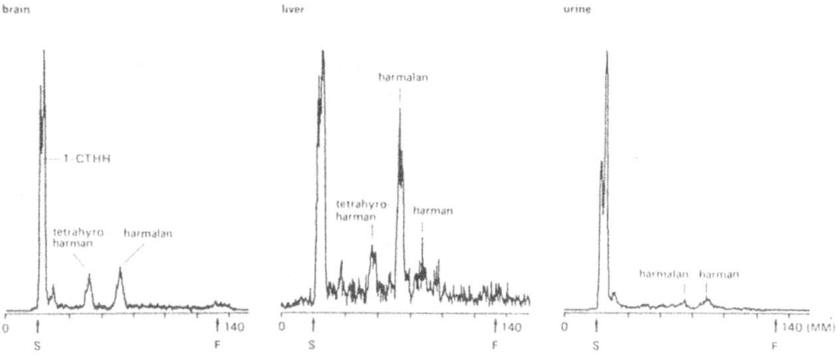

Fig.4 Radiochromatograms of samples from various organs of rats treated intraventriculary with ^{3}H 1-CTHH. The

animals were decapitated 1 h after injection (Solvent system III).

Reinvestigation of Hahn's experiments (route II and III) showed no chemical condensation reactions after 2 h incubation of ^3H tryptamine with various concentrations of acetaldehyde or with various concentration of pyruvic acid in buffer at pH 5.0 or 7.4 and 37°C (concentration range see 'Materials and Methods').

Tab. 1: Rf values of tryptamine derivatives, their metabolites and of ß-Carbolines in various solvent systems

Compound	Solvent system I	II	III
tryptamine	0.86	0.63	0.63
N-acetyltryptamine	0.92	0.86	0.62
N-acetylserotonin	0.90	0.84	0.29
indole acetic acid	0.52	0.90	0
1-CTHH	0.57	0.54	0
harmalan	0.95	0.50	0.49
tetrahydroharman	0.91	0.63	0.37
harman	0.98	0.58	0.63

I Isopropanol - methylacetate - 12.5% ammonia (45:35:20)
II n-Butanol - glacial acetic acid - water (40:10:20)
III Chloroform - acetone - diethylamine (60:30:10)

Discussion

Although the BC biosynthesis in plants via the cyclodehydration of N-acetyl-ß-indolylethylamine (route I) has been demonstrated (Slaytor and McFarlane,1968), we could not detect an analogeus reaction in the rat. Analysis of brain and urine of rat 1 h after intraventricular injection of ^3H NAT did not reveal the presence of the cyclisation product. High concentrations of unchanged NAT were found in the brain, whereas a trace of unchanged NAT was excreted into the urine. The major portion of the metabolites in the urine possessed polar properties indicating conjugate products, probably hydroxy-conjugates of sulfuric and glucuronic acids, similar the metabolism of N-acetyl-5-methoxy tryptamine (melatonin; Kveder and McIssac, 1961). Deacetylation of NAT does not occur apparently since neither tryptamine nor indole acetic acid was found. This observation is in agreement with the reported metabolism of melatonin (Kveder and McIssac, 1961).

Recently, we have investigated the in vivo condensation reaction of pyruvic acid and tryptamine to 1-CTHH in the rat brain (Susilo and Rommelspacher, 1985). The reaction is dependent on time and pyruvic acid concentration and inhibited by a high dose of pargyline suggesting an enzymatic process. The most relevant argument against the "pyruvic acid pathway" (route III) was the low reactivity of pyruvic acid and suitable mild conditions for the decarboxylation reaction of the first condensation product, in our example 1-CTHH (Whaley and Govindachari, 1951). However, it has been shown that 1-CTHH is an intact precursor for harman in the plant Passiflora edulis and for THH in Eleagnus angustifolia (Herbert and Mann, 1982). After intraventricular administration of ^3H 1-CTHH we observed that harmalan and tetrahydroharman was formed in the brain. Harmalan represents the first decarboxylation product of 1-CTHH, which is further hydrogenated to THH. This finding suggested that in the brain reductive reactions occur after the decarboxylation. We found in the liver that harman was formed in addition to harmalan and THH indicating that oxidation reactions of the reaction chain occur as well in vivo (route III). The amount of ^3H THH in the brain exceeded that in the liver by approximately 5 fold. Since THH, harmalan, harman and the precursor are able to cross the blood-brain barrier, it could not be assessed whether THH is formed mostly in the brain and harman in the periphery.

The chemical experiments of several authors under so-called physiological conditions (Hahn and Ludewig, 1936; Hahn et al., 1935; McIsaac, 1961) were carried out with very high substrate concentrations, which were unlikely to be reached in vivo (Hahn and Ludewig, 1934: 72 mM tryptamine, 136 mM acetaldehyde; Hahn et al., 1935: 255 mM tryptamine, 500 mM pyruvic acid; McIsaac, 1961: 2 mM 5-methoxytryptamine, 1.8 M acetaldehyde). We have reinvestigated the experiments with approximately endogenous levels of tryptamine (13 ng/ml; endogenous conc. 0.5-1 ng/g brain, Jones, 1982) and various concentrations of acetaldehyde (0.1 mM-20 mM; endogen. conc. 0-0.2 μM, Pikkarainen et al., 1979; Eriksson and Peachey, 1980) as well as with pyruvic acid (0.5-50 mM; endog. conc. 100 μM; Fujishima et al. , 1978). No condensation product could be detected under these conditions suggesting that the chemical condensation reactions occur to a substantial degree only in the presence of unphysiologically high substrate concentrations. Since the endogenous level of the substrates were too low for a chemical reaction, it seemed that a non-enzymatic formation of BC from tryptamine and acetaldehyde or pyruvic acid in vivo is rather unlikely.

The present study shows a possible biosynthesis pathway of BC via the "pyruvic acid route". Further investigations are necessary to substantiate the findings, and to get more detailed information about the mechanism of the reaction in mammals.

Acknowledgement

Supported by a grant of the Deutsche Forschungsgemeinschaft (Ro 422/4).

REFERENCES

Chattaway F.D. (1931) Acetylation in aqueous alkaline solutions. J. Chem. Soc. 2495-2496.

Collins M.A., Dahl K, Nijm W and Major L.F. (1982) Evidence for homologous families of dopamine and serotonin condensation products in CSF from monkeys. Soc. Neurosci. 8, 277

Eriksson C.J.P., Peachey J.E. (1980) Lack of difference in blood acetaldehyde of alcoholics and controls after ethanol ingestion. Pharmac. Biochem. Behav. 13, Suppl. 1, 101-105.

Fujishima M., Ogata J., Oniki H., Onoyama K., Morotomi Y. and Omae T. (1978) Lactate, pyruvate and ATP concentrations of the brain in two models of hypertension in rats. Brain Res. 152, 189-195.

Hahn G. and Ludewig H. (1934) Synthese von Tetrahydroharman-Derivaten unter physiologischen Bedingungen. Ber. 67, 2031-2035.

Hahn G., Bärwald L., Schales O. and Werner H. (1935) Synthese von Tetrahydroharman(4-Carbolin)-Derivaten unter physiologischen Bedingungen. Justus Liebigs Annal. Chem. 520, 107-123.

Herbert R.B. and Mann J. (1982) The biosynthesis of the ß-carboline alcaloids, harman and eleagnine. J. Chem. Soc. Perkin Trans. I, 1523-1525.

Honecker H. and Rommelspacher H. (1978) Tetrahydronorharman (tetrahydro-ß-carboline), a physiologically occurring compound of indole metabolism. Naunyn-Schmiedeberg's Arch. Pharmacol. 305, 135-141.

Johnson J.V., Yost R.A. and Faull K.F. (1984) Tandem mass spectrometry for the trace determination of tryptolines in crude brain extracts. Anal. Chem. 56, 1655-1661.

Jones R.S.G. (1983) Tryptamine: a neuromodulator or neurotransmitter in mammalian brain. Progr. in Neurobiology, 19, 117-139.

Kveder S. and McIsaac W.M. (1961) The metabolism of melatonin (N-acetyl-5-methoxytryptamine) and 5-methoxytryptamine. J. Biol. Chem. 236, 3214-3220.

McIsaac W.M. (1961) Formation of 1-methyl-6-methoxy-1,2,3,4-tetrahydro-2-carboline under physiological conditions. Biochim. Biophys. Acta 52, 607-609.

Peura P., Kari I. and Airaksinen M.M. (1980) Identification by selective ion monitoring of 1-methyl-1,2,3,4-tetrahydro-ß-carboline in human platelets and plasma after ethanol intake. Biomed. Mass Spectrom. 7, 553-555.

Pikkarainen P.H., Salaspuro M.P. and Lieber C.S. (1979) A method for the determination of free acetaldehyde in the plasma. Alcoholism 3, 259-261

Rommelspacher H., Strauss S. and Lindemann J. (1980) Excretion of tetrahydroharman and harmane into the urine of man and rat after load with ethanol. FEBS Lett. 109, 207-212

Rommelspacher H., Damm H., Strauss S. and Schmidt G. (1984) Ethanol induces an increase of harman in the brain and urine of rats. Naunyn-Schmiedeberg's Arch. Pharmacol. 327, 107-113.

Schöpf C. and Salzer W. (1940) Die Synthese des 1-(3,4-Methylendioxybenzyl)6,7-dioxy-1,2,3,4-tetrahydroisochinolins unter zellmöglichen Bedingungen. Annal. Chem. 544, 1-30.

Slaytor M. and McFarlane I.J (1968) The biosynthesis and metabolism of harman in Passiflora edulis. Phytochemistry, 7, 605-611.

Susilo R. and Rommelspacher H. (1985) In vivo formation of a ß-carboline (1,2,3,4-Tetrahydro-1-methyl-ß-carboline-1-carboxylic acid) from tryptamine and pyruvic acid in rat

brain, submitted.
Whaley W.M. and Govindachari T.R. (1951a) The preparation of 3,4-dihydroisoquinolines and related Compounds by the Bischler-Napieralski reaction. "Organic Reactions", Vol VI, John Wiley and Sons, Inc., New York, pp 74-123.
Whaley W.M. and Govindachari T.R. (1951b) Pictet-Spengler synthesis of tetrahydroisoquinolines and related compounds. Ibid., pp 151-190.

Brain octopamine and strain Differences in Avoidance Behavior. Prenatal ontogenesis of phenolamines.

J.C. DAVID, J. DELACOUR, J.F. COULON and

C. GUENAIRE

Laboratoire de Biochimie du Développement, L.A. 256 C.N.R.S., Université Rennes I, 35042 RENNES CEDEX, France, and Laboratoire de Psychophysiologie, Université Paris 7, Paris, France.

During the last two decades phenolamines have received increasing attention as neurotransmitters, neuroregulators or neuromodulators. Since 1976 (Robertson and Juorio), two reviews mainly devoted to octopamine have appeared (Robertson, 1981 ; David and Coulon, 1985). However, even in this last article it appears that the extent of work and results obtained with Invertebrates exceeds by far the studies in Vertebrates. For about five years our Group has became interested in the possible involvement of octopamines and related amines in neurovegetative processes (David and Delacour, 1980). Some involvements have already been found in depression (Sandler et al., 1979) and hypertension (David, 1979). We studied avoidance conditioning through a genetic approach based upon Roman strains of rats. Originally developped by Broadhurst and Bignami (1965) these strains are divided into two groups : Roman High Avoidance (RHA) selected for high number of avoidance responses and Roman Low Avoidance (RLA) for low ones in a shuttle box using light as a conditioned stimulus. The Roman Control Avoidance (RCA) rats were a non selected group originating from the same Wistar stock.

RESULTS AND DISCUSSION

Behavioural experiments :

Locomotor activity and avoidance response were recorded. Locomotor activity was measured by the number of spontaneous crossings between the two compartments. An avoidance conditioning session consisted of 40 trials separated by an interval of 70 S. Five seconds after onset of the light signal, an electric shock of 1 mA AC was delivered to the floor. A complete crossing of the barrier during the signal shock interval terminated the signal and no shock was delivered (avoidance response). Crossings made more than 5 S after the signal ended the signal and the shock simultaneously (escape response).

The animals were divided into two groups : one received 250 µg of para-octopamine (Sigma) and the other 50 µg of the same compound. Controls received the vehicle (isotonic saline) intracerebroventricularly through an Hamilton syringe adaptated to a catheter in 10 µl.

The effects of administrations of respectively 50 µg and 250 µg of p-octopamine on locomotor activity of unshocked rats are reported in Table I.

TABLE I

Effects of 50 and 250 µg of p-octopamine on locomotor activity of unshocked rats (n = 4)

Group	Strain RHA	RCA	RLA
P-octopamine 50 µg			
Experimental	69.8 ± 19	62.4 ± 20	63 ± 16
Control	66.4 ± 13	60 ± 14	57 ± 15
P-octopamine 250 µg			
Experimental	24.5 ± 6	27 ± 7	20 ± 5
Control	78 ± 12	60 ± 17	64 ± 15

Consecutively to an administration of 50 µg of p-octopamine no effect was observed on spontaneous locomotor activity, avoidance conditioning and intertrial crossings. for 250 µg, octopamine depressed significantly locomotor activity. Avoidance responding gives a different pattern of results since strain and drug effects were very significant but the differences disappear when para-octopamine was administered (shown at the oral presentation).

Neurochemical experiments :

At the end ot fhe conditioning session, the animals were killed and octopamines (meta and para) and catecholamines (dopamine and noradrenaline) determined radioenzymatically in hypothalamus and brain stem. No difference was found in the dopamine and noradrenaline concentrations after p-octopamine administration in the three different strains of rats. However though both in brain stem and hypothalamus the p-octopamine were increased by its administration, the strain difference consecutively disappeared.

Several aspects must however be pointed out. First of them is that the p-octopamine dose presently used is high and its neurochemical effects are moderate and rather specific. This dose produced a moderate increase of p-octopamine in the brain but m-octopamine and catecholamines were unaffected. The behavioural consequences of this administration were remarkably parallel on avoidance conditioning and intertrial crossings. A dose of 50 µg p-octopamine had no effect. Eventually several arguments seem to exclude that the behavioral effects are due to catecholamines. In the absence of shock, octopamine produced a decrease of locomotor activity. On the contrary, intraventricular infusion of dopamine (20-240 µg/h), noradrenaline (20-120µg/h) and amphetamine (50-500 µg) (Geyer et al., 1972 ; Geyer et al., 1973 ; Segal et al., 1974) increased locomotor activity of unshocked rats. From this point of view, octopamine appears to be a critical feature of the Roman strains. RLA have been shown to have lower tyrosine hydroxylase (TH) and dopamine β hydroxylase (DβH) activity than the RHA at the adrenal level but a higher DβH activity at the brain level (Coyle et al., 1973). Our results are the first direct evidence of a neurochemical (octopamine) basis of the main behavioral features of these strains. One of the major questions concerns the neurochemical ontogeny of the Roman nervous system in comparison to normal Wistar rats. The basic hypothesis of this work is a possible developmental difference between brain phenolamines and catecholamines in the respective Wistar and Roman strains.

For this purpose, brain embryonic development for phenolamines and catecholamines has been studied in relation to their metabolizing enzymes (TH, DβH and MAO). The same approach is under investigations in the Roman strains. Embryonic brains have been dissected from 13 days to birth into diencephalon and stem and more recently into hypothalamus and stem (David, 1984). In the diencephalon an important peak of p-octopamine is observed at 13 days of fetal life (57 ng/g) and this peak is mainly localized in hypothalamus where it reaches 420 ng/g. In both cases for the Wistar rats this amine concentration abruptly decreases at 18 days of gestation to levels comparable to 13 days. Meanwhile m-octopamine and catecholamines (dopamine and noradrenaline) concentrations were studied in the developing brain (diencephalon and stem) and heart.

As shown in Table 2, m-octopamine was not detected in the brain before 16 days of fetal life and catecholamines not before 15 days using the radioenzymatic assay. The important feature, however, concerning these three amines is that after their appearance their levels display regular increase without any apparent decrease. A similar observation can be made in the heart. Many hypothesis may of course arise about this difference between the ontogeny of p-octopamine and catecholamines + m-octopamine in the brain. The most evident is the selective inactivation of p-octopamine by enzymes expressed lately in embryonic life. Investigations were made to follow the appearance of TH, Dopa decarboxylase (DD), DβH and MAO in the embryonic brain using the enzymatic activity assay. TH and DD activities are expressed in pmol $^{14}CO_2$/min/mg protein. DβH activity is expressed in pmol/min/mg protein as octopamine formed. Each determination was performed on a pool of six tissues.

As shown in Table 3, developmental characteristics of TH, DD and DβH are firstly that DβH activity is detected as early as day 13 of embryonic life while DD is latter detected in the diencephalon. TH activity is observed only after 14 days and afterwards. Secundly, the increase of activity is regular for the three enzymes until birth. MAO activity was detected only after 18 days in these parts of the brain studied and this can account for the decrease of p-octopamine since this amine has a high turnover rate (Kakimoto and Armstrong, 1962 ; Brandau and Axelrod, 1972 ; Molinoff and Axelrod, 1972). It seems likely that m-octopamine turnover is more closely associated to catecholamines than to p-octopamine. It must be noticed, however, that the decrease of p-octopamine in Wistar rat brain is under the control of some intrinsic or environmental factors since it

TABLE 2

Developmental appearance and changes of m-octopamine and catecholamines in the developing Wistar rat brain and heart (mean of 5 determinations ± SEM)

	Age of embryos (in days)						
	14	15	16	17	18	20	21
m-octopamine (ng/g)							
diencephalon	n.d.	n.d.	n.d.	0.12 ± 0.02	0.20 ± 0.03	0.30 ± 0.05	0.35 ± 0.04
brain stem	n.d.	n.d.	n.d.	n.d.	0.15 ± 0.02	0.20 ± 0.03	0.30 ± 0.04
dopamine (µg/g)							
diencephalon	n.d.	n.d.	-	0.03±0.005	0.04 ± 0.01	0.05 ± 0.01	0.10 ± 0.01
brain stem	n.d.	0.01 ± 0.003	-	0.05±0.006	0.1 ± 0.01	0.4 ± 0.05	0.5 ± 0.1
heart	n.d.	0.02 ± 0.002	-	0.07±0.01	0.35 ± 0.04	0.8 ± 0.01	1.2 ± 0.15
noradrenaline (µgLg)							
diencephalon	n.d.	0.02 ± 0.005	-	0.05±0.01	0.08 ± 0.02	0.25 ± 0.05	0.6 ± 0.1
brain stem	n.d.	n.d.	-	0.02±0.005	0.04 ± 0.01	0.10 ± 0.02	0.25 ± 0.05
heart	n.d.	0.03 ± 0.005	-	0.07±0.01	0.11 ± 0.02	0.15 ± 0.02	0.30 ± 0.1

n.d. not detected

TABLE 3

Developmental appearance and changes of TH, DD and DβH in the embryonic rat brain (mean of five determinations ± SEM)

Enzymes	Age of embryos (in days)			
	13	14	15	17
TH				
Diencephalon	n.d.	n.d.	1.3 ± 0.1	5 ± 0.5
Stem	n.d.	0.5 ± 0.07	2 ± 0.2	5 ± 0.6
DD				
Diencephalon	n.d.	7 ± 0.8	10 ± 1	27 ± 4
Stem	12 ± 2	18 ± 2	26 ± 2	37 ± 5
DβH				
Diencephalon	3.5 ± 0.2	5 ± 0.5	8.5 ± 0.5	13 ± 2
Stem	4.2 ± 0.5	8 ± 0.9	12 ± 1	25 ± 5

is not observed in cultured brains (David, 1984).

In itself this problem desserves futher studies to better underline the relationship between octopamine(s) and catecholamines embryonic metabolism. One of the major route of investigation might be at the enzymatic level with a better understanding of the differential triggering of TH and DD or DβH activity. This work is in fact the beginning of present investigations concerning octopamine metabolism in Roman strains embryonic brain.

References

- BRANDAU, K. and AXELROD, J. 1972. The biosynthesis of octopamine. *Naunyn Schmiedebergs Arch. Pharmacol.* 273, 123-133.

- BROADHURST, P.L. and BIGNAMI, C. 1965. Correlative effects of psychogenetic selection. A study of the Roman High and Low Avoidance strains of rats. *Behav. Res. Therap.* 2, 273-280.

- COYLE, J.T., WENDER, P. and LIDIKY, A. 1973. Avoidance conditioning in different strains of rats : Neurochemical correlates. *Psychopharmacologia* 31, 25-34.

- DAVID, J.C. 1979. Augmentation des taux de phényléthanolamine, m-octopamine et p-octopamine au niveau de l'hypothalamus et du tronc cérébral de rats génétiquement hypertendus (SHR Kyoto). *C. R. Acad. Sci. (Paris), sér. D,* 287, 1293-1295.

- DAVID, J.C. and DELACOUR, J. 1980. Brain contents of phenylethanolamine, m-octopamine and p-octopamine in the Roman strains of rats. *Brain Research* 195, 231-235.

- DAVID, J.C. 1984. Relationship between phenolamines and catecholamines during rat embryonic development *in vivo* and *in vitro*. *J. Neurochem.* 43, 668-674.

- DAVID, J.C. and COULON J.F. 1985. Octopamine in Invertebrates and Vertebrates. *Progress in Neurobiology, in press.*

- GEYER, M.A., SEGAL, D.S. and MANDELL, A.J. 1972. Effect of intraventricular infusion of dopamine and norepinephrine on motor activity. *Physiol. Behav.* 8, 653-658.

- GEYER, M.A. and SEGAL, D.S. 1973. Differential effects of reserpine and α methylparatyrosine on norepinephrine and dopamine induced behavioral activity. *Psychopharmacologia*

- KAKIMOTO, Y. and ARMSTRONG, M.D. 1962. On the identification of octopamine in mammals. *J. Biol. Chem.* 237, 422-427.

- MOLINOFF, P.B. and AXELROD, J. 1972. Distribution and turnover of octopamine in tissues. *J. Neurochem.* 19, 157-163.

- ROBERTSON, H.A. and JUORIO, A.V. 1976. Octopamine and some related non catecholic amines in Invertebrate nervous systems. *Int. Rev. Neurobiol.* 19, 173-224.

- ROBERTSON, H.A. 1981. Octopamine - After a decade as a putative neuroregulator. *Essays Neurochem. Neuropharmacol.* 5, 47-73.

- SANDLER, M., RUTHVEN, C.R.J., GOODWIN, G.L., REYNOLDS, G. P., RAO, V.A.R. and COPPEN A. 1979. Deficient production of tyramine and octopamine in case of depression. *Nature (London)* 278, 357-358.

- SEGAL, D.S., MC ALLISTER, C. and GEYER, M.A. 1974. Ventricular infusion of norepinephrine and amphetamine, direct versus indirect action. *Pharmacol. Biochem. Behav.* 2, 79-86.

RELATIONS BETWEEN MPTP (1-METHYL-4-PHENYL-1,2,3,6-TETRAHYDROPYRIDINE), THE NEUROTOXIN MPP+ (1-METHYL-4-PHENYLPYRIDINIUM ION), AND MAO IN RAT BRAIN

Hiroyasu Kinemuchi, Yuichiro Arai, Yoshie Toyoshima, Takeshi Tadano* and Kensuke Kisara*
Department of Pharmacology, School of Medicine, Showa University, Tokyo 142 and *Department of Pharmacology, Tohoku College of Pharmacy, Sendai 983, Japan.

A fundamental biological characteristic of Parkinson's disease is a damaged dopamine (DA) system between the substantia nigra and the corpus striata. This disease is a slowly progressive-neurodegenerative disorder pathologically characterized by the loss of neurons in the zona compacta of the substantia nigra. The etiology of this disease has been extensively investigated, but the precise reason for the loss of DA producing cells in this region remains obscure. Previous animal models of this disease were made by nigrostriatal lesion, depletion of DA with reserpine or α-methyl-p-tyrosine, blockage of the DA receptor with neuroleptics such as haloperidol, destruction of the substantia nigra with 6-hydroxy-DA (see review, Zigmond and Striker, 1984) or manganese intoxication (Donaldson *et al.*, 1982).

The compound, 1-methyl-4-phenyl-1,2,3,6-tetrahydropyridine (MPTP) induces a persistent neurological syndrome similar to idiopathic Parkinson's disease in human and non-human-primates (see review, Langston, 1985). Other animals such as mice and rats are relatively less sensitive to this neurotoxic effect and large and/or a long-term administration of MPTP are required for the neurotoxicity (Hallman *et al.*, 1984; Heikkila *et al.*, 1984). This neurotoxicity is accompanied by selective destruction of the nigrostriatal

DA system, as has been observed in human Parkinson's disease. This action of MPTP on animals, thus, provide a useful model for insight into the development of Parkinson's disease and this model may offer an opportunity to study screening of antiparkinsonian drugs and, experimentally, the side-effects of long-term DA precursor and agonist therapy.

The neurotoxicity seems to be associated with an oxidized neurotoxic metabolite of MPTP, 1-methyl-4-phenylpyridinium ion (MPP+), which is probably biotransformed in the brain by one of the two forms of monoamine oxidase (MAO, EC 1.4.3.4), MAO-B, but not by MAO-A. This ionized compound will then be trapped and accumulated in brain cells for a long time. This conclusion on biotranformation comes from findings that *in vivo* in mouse (Markey *et al.*, 1984) and monkey brain (Langston *et al.*, 1984) and *in vitro* in rat brain mitochondria (Chiba *et al.*, 1984), pargyline and deprenyl, both selective MAO-B inhibitors, can block metabolism *in vitro* and thus by preventing production of MPP+, might block the *in vivo* neurotoxicity of systemically administered MPTP. Clorgyline, a selective MAO-A inhibitor, has no such effect. In primates, the blocking of conversion

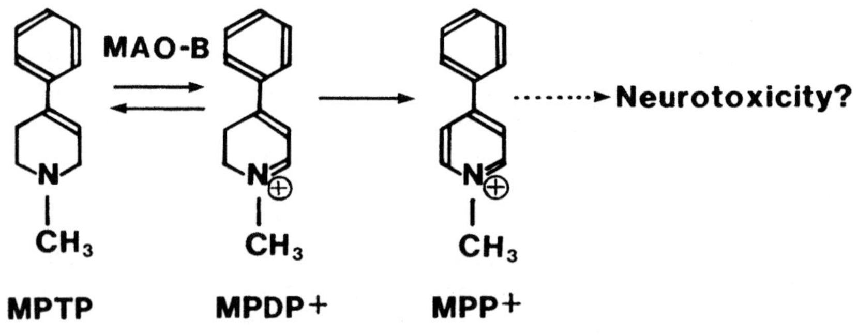

Fig.1. Proposed pathway of MPTP to MPP+

to MPP+ causes much higher concentrations of MPTP in the brain, so these animals remain normal. This suggests that MPTP itself is not toxic, but the enzymic product, MPP+, is. These results confirm that MAO-B is primarily responsible for the conversion of MPTP to MPP+, probably via formation of an intermediate ion, 1-methyl-4-phenyl-2,3-dihydropyridinium (MPDP+), from MPTP(Fig. 1). In this step, MAO-B acts only as a catalyst to produce MPDP+ which in turn spontaneously yields MPTP and MPP+ (Castagnoli et al., 1985). Heikkila et al. (1985a) have shown that MPTP acts as a MAO-B substrate and its oxidation rate is similar to that of another selective MAO-B substrate, benzylamine (BZ).

Recently, Del Zompo et al.(1984) and Wieczorek et al. (1984) reported the existence of specific, unwashable [^{3}H]MPTP binding sites in monkey and rat brain. These were also found in human brain, but with much higher binding densities in the caudate, substantia nigra and locus coeruleus than those in rat brain (Javitch et al., 1984b). In rat the binding sites correspond in anatomical localization of MAO in the brain and peripheral tissue (Parsons and Rainbow, 1984). This binding was specifically suppressed by deprenyl in monkey brain (Del Zompo et al., 1984), but in contrast, it was suppressed by clorgyline in rat brain (Parsons and Rainbow, 1984). MPTP has recently been reported to be an MAO inhibitor (Parsons and Rainbow, 1984; Salach et al., 1984; Markstein and Lahaye, 1985; Fuller and Steranka, 1985). It thus seems to be important to determine if MPTP is a substrate or an inhibitor for MAO, especially for MAO-B; and to study interactions of the neurotoxic product, MPP+ with both MAO-A and MAO-B, in rat brain.

We first estimated metabolism of MPTP by MAO in rat (Sprague-Dawley strain, weighing about 300g) forebrain homogenates using a fluorimetric method to determine the H_2O_2 production by coupling this to peroxidase-catalysed oxidation of homovanillic acid (Tipton, 1969). Results indicate that, at both concentrations tested (1 and 0.1 mM),MPTP was oxidized by both MAO-A (10^{-7}M l-deprenyl-treated preparations) and MAO-B(10^{-7}M clorgyline-treated preparation). The rate of oxidation by MAO-B was more rapid than that by MAO-A, and the rate by MAO-B was similar to that of BZ, a selective MAO-B substrate.

In preliminary experiments, deamination of 5-HT (0.1 mM, final concentration after dilution with unlabelled 5-HT) and BZ (0.1 mM) was almost completely inhibited by 10^{-7}M clorgyline and 10^{-7}M 1-deprenyl, respectively, in rat striatum and forebrain homogenates. This indicated that the bulk of either substrate, at the concentration used, was essentially metabolized by either MAO-A (5-HT) or MAO-B (BZ) alone. These results agree with the literature (see, e.g. Kinemuchi et al., 1984). Next, inhibitory effects on MAO-A (0.1 mM 5-HT) and MAO-B (0.1 mM BZ) activity were studied without preincubation with these compounds. In spite of in vitro metabolism of MPTP by both forms of MAO, results revealed that MPTP dose-dependently inhibited both MAO-A and MAO-B activity with a higher potency towards MAO-A. IC_{50} values, estimated from the simple inhibition curves, are listed in Table 1, together with values obtained with 0.1 mM DA as a substrate for the striatum homogenates only. This MAO-A selectivity was also evident in the inhibition of MAO by MPP+, determined with these three substrates (Table 1). These results strongly suggest that inhibition of MAO, as evidenced by MPTP and MPP+, may be interpreted as being due to an essentially qualitative similarity between the respective forms of MAO in the two brain regions. Thus, in further experiments, we mainly used rat forebrain homogenates as enzyme sources. The nature of the initial inhibition of MAO-A and MAO-B by MPTP and MPP+ was further studied using different concentrations of 5-HT or BZ as the respective substrate. In contrast to previous reports that MPTP was a non-competitive inhibitor of MAO-B, but competitive for MAO-A (Fuller and Steranka, 1985), in the present study, both MPTP and MPP+ competitively inhibited both MAO-A and MAO-B activities. In support of the present results, linear slope replots against various concentrations of these two compounds showed higher MAO-A selectivity for MPP+ than for MPTP (Table 1). A similar Ki value (9 μM) of MAO-A for MPTP was previously reported (Fuller and Steranka, 1985). The Ki values obtained in this study indicate that MPP+ is a more potent MAO-A inhibitor than MPTP, while for inhibition of MAO-B, MPTP is more potent than MPP+ (116 μM vs 180 μM). As can be determined from the values listed in Table 1, the ratios, (Ki for MAO-A)/(Ki for MAO-B) were 1:25.8 for MPTP and 1:90 for MPP+, confirming a much greater MAO-A selectivity for MPP+.

<u>Table 1</u>. IC_{50} and K_i values for inhibition of MAO-A and MAO-B by MPTP and MPP+.

	IC_{50} value (μM)*		K_i values (μM)**		
	MAO-A	MAO-B	MAO-A	MAO-B	K_i^A/K_i^B
MPTP					
(forebrain)	8.6	260	4.5	116	1:25.8
(striatum)	12.5	200	n.d.	n.d.	--
(striatum, 0.1 mM DA)	20.0				
MPP+					
(forebrain)	4.3	405	2.0	180	1:90
(striatum)	4.2	580	n.d.	n.d.	--
(striatum, 0.1 mM DA)	5.1				

*IC_{50} values calculated, without preincubation, from plots of percentage inhibition of MAO-A (with 0.1 mM 5-HT as a substrate) and MAO-B (with 0.1 mM BZ).

**K_i values were determined from slope replots against various concentrations of MPTP and MPP+, without preincubation. For all estimates of MAO activity, a short incubation time (5 min) was used to ensure no significant irreversible inhibition (see below) during assay period.

n.d.; not determined.

The following results may preclude the possibility that MPTP competitively inhibits both forms of MAO by acting as their substrates. Preincubation of MAO and MPTP was necessary in order to achieve full inhibition. The time-dependent inhibition of MAO activities was studied by preincubating the enzyme with MPTP for various periods at 37°C before assaying the remaining activity towards 5-HT or BZ. The degrees of inhibition of MAO-A and MAO-B by this compound at a concentration range 5 x 10^{-5} to 10^{-6}M for MAO-A, and 5 x 10^{-4} to 10^{-5}M for MAO-B increased to nearly reach maximum inhibition in 20-30 min. However, after that preincubation period, slight increase in inhibition was still observed with time up to 80 min of preincubation, especially with lower concentrations of MPTP. In contrast, inhibition of either form of MAO by MPP+ was not time- dependent, indicating MPP+ to be a reversible inhibitor of both MAO-A and MAO-B. Dialysis (24 hrs at 4°C) of samples pretreated 1 hr at 37 ℃ with either 10 μM MPTP for MAO-A (which caused about 50% inhibition of MAO-A) or 100 μM MPTP for MAO-B (causing about 65% inhibition of MAO-B) completely reversed the inhibition of MAO-A, but only partially reversed that of MAO-B. This verifies that MAO-B inhibition by MPTP was essentially irreversible, but MAO-A inhibition by MPTP was reversible. This latter finding is compatible with the result of Fuller and Steranaka (1985), who reported MPTP to be a reversible MAO-A inhibitor. Our conclusion disagrees in part, however, with the recent findings of Salach _et al_. (1984), who reported MPTP to be an irreversible inhibitor of both MAO-A and MAO-B. Despite of time-dependence of MAO-A inhibition by MPTP in untreated forebrain homogenates, there was less marked time-dependent MAO-A inhibition by MPTP in preparations where MAO-B had been selectively inhibited by 10^{-7}M 1-deprenyl-pretreatment. A time-dependent inhibition of MAO-B, in contrast, was still observed in the samples where MAO-A had been selectively inhibited by 10^{-7}M clorgyline-pretreatment. These results suggest that the time-dependent MAO-A inhibition by MPTP in the untreated preparations is mainly due to the rate of MPP+ formation produced by MAO-B. Data obtained before and after preincubation of rat forebrain homogenates with MPTP are summarized below. Briefly, after preincubation, IC_{50} values of oxidation of BZ (MAO-B) were greatly lowered (260 to 20 μM), but decrease of 5-HT (MAO-A) was less marked (8.6to 3.5 μM). Of course, since it is a reversible inhibitor of MAO, MPP+ did not change the values or the modes of inhibition of either form of MAO, regardless of preincubation. After preincubation, competitive inhibition of MAO-B by MPTP changed to a mixed type, but that of MAO-A remained competitive. This mixed type of MAO-B inhibition is consistent with the appearance of an irreversible component during the preincubation period.

All results presented here suggest that MPTP may act as a simple substrate for MAO-A, while it might act as both a substrate for and an enzyme-activated ("suicide") irreversible inhibitor of MAO-B, according to the following scheme (Waley, 1980):

$$E + I \underset{k_{-1}}{\overset{k_1}{\rightleftharpoons}} E{\cdot}I \xrightarrow{k_2} (EI) \xrightarrow{k_4} E\text{-}I^*$$
$$(EI) \xrightarrow{k_3} E + \text{product}$$

where E and I represent the enzyme and MPTP, respectively; E•I and E-I represent the reversible and irreversible enzyme-inhibitor complexes, respectively; and (EI) is a complex involving an activated form of MPTP that can break down into a final product, MPP+, in this case, or react to irreversibly bind to the enzyme (E-I*). The most plausible candidate for (EI) is a MAO-B complex formed with the intermediate between MPTP to MPP+, MPDP+ (Chiba et al.,1984; Castagnoli et al, 1985). Even though MPTP and MPP+ both inhibit MAO-A and MAO-B, MPTP inhibits MAO-B activity very slowly, compared with the rapid oxidation of MPTP by MAO-B, indicating that the partition coefficient (k3/k4) in the scheme will be large. The inhibition of MAO by MPP+ is not consistent with simple "product inhibition", since this type of inhibition would not give a rapid onset of inhibition at high substrate concentrations and thus it is usually time-dependent (Orsi and Tipton, 1979; Kinemuchi et al., 1982). The time-dependent, but reversible inhibition of MAO-A by MPTP thus may exclude the possibility that MPTP is an irreversible "suicide" inhibitor of both MAO-A and MAO-B (Salach et al., 1984). The inhibition of MAO-A by MPTP might be simply due to its action as a substrate and the inhibition by MPP+ will be an inhibitor of MAO-B as well. Our present data suggest that MAO-B is responsible for most of the MPTP oxidation, and this enzyme is only sensitive to time-dependent irreversible inhibition. The predominant, but not exclusive extra-neuronal localization of MAO-B in brain (see e.g., Oreland et al., 1983; Fowler et al., 1984) suggests that MPP+ is predominantly produced by MAO-B extraneuronally and then more highly taken up into the neurons than MPTP by, in the case of the nigrostriatal DA-containing neurons, a DA uptake system (Javitch and Snyder, 1984a; Heikkila et al., 1985b). In the neurons, MPP+ then reaches a concentration sufficient to inhibit intra-neuronal MAO. One possible explanation for the difference between the MPTP sensitive species, primates, and the relatively less sensitive rodents may be, in part, the difference in the MAO-A/MAO-B activity ratios in these brains. Support for this hypothesis comes from the findings of a low A/B ratio (1:3-4) in the brain of human (see, e.g. Oreland et al., 1983) and monkey (Murphy et al., 1979) and a high ratio (4:1) in the brain of rodents, such as rat (see, e.g. Oreland et al., 1983).

Intracerebroventricular (i.c.v.) injection of MPP+ into mice (2.5, 5, 10, 20 and 40 μg/mouse) dose-dependently elicited, initially, decreased locomotor activity (LMA), sedation, and hypothermia; and later in contrast, increased plus various behavioural effects, such as circling, licking, self-biting, scratching and head-twitch. The decrease of LMA induced by MPP+ was suppressed by pretreatment with either nomifensine, amphetamine or pargyline, suggesting possible change in presynaptic DA neurons, as proposed by Heikkila et al. (1984). The mechanism of injected MPP+ activity in our study was not quite the same as in this proposal, since the decreased LMA induced by MPP+ was not completely prevented even after pretreatment with desmethylimipramine (5 mg/kg, i.p.) plus 6-hydroxy-DA (50 or 100 μg/mouse, i.c.v.). Thus, the effect of MPP+ on LMA may depend on other factors, such as its direct action on post synaptic DA receptors, the 5-HT system and/or the noradrenergic system. In the later stage, increased LMA may be due to various monoamines released in the cleft. The head-twitch response induced by MPP+ is most likely due to activation of the brain 5-HT system (see Kisara et al., this symposium), since the response was completely inhibited by p-chlorophenylalanine, an inhibitor of 5-HT biosynthesis.

REFERENCES

Castagnoli Jr. N., Chiba K., and Trevor A.J. (1985) Potential bioactivation pathways for the neurotoxin 1-methyl-4-phenyl-1,2,3,6-tetrahydropyridine (MPTP). Life Sci. 36 225-230.

Chiba K., Trevor A., and Castagnoli Jr.N. (1984) Metabolism of the neurotoxic tertiary amine, MPTP, by brain monoamine oxidase. Biochem. Biophys. Res. Commun. 120 574-578.

Del Zompo M., Pintus S., Zuddas A., and Corsini G.V. (1984) Deprenyl selectively inhibits [^{3}H]MPTP binding sites in monkey brain. Eur. J. Pharmacol. 107 285-286.

Donaldson J., McGregor D., and LaBella F. (1982) Manganese neurotoxicity: A model for free radical mediated neurodegeneration ? Can. J. Pharmacol. 107 1398-1405.

Fowler C.J., Magnusson O., and Ross S.B. (1984) Intra- and extra-neuronal monoamine oxidase. Blood vessel 21 126-131.

Fuller R.W. and Steranka L.R. (1985) Central and peripheral catecholamine depletion by 1-methyl-4-phenyl-tetrahydropyridine (MPTP) in rodents. Life Sci. 36 243-247.

Hallman H., Olson L., and Jonsson G. (1984) Neurotoxicity of the meperidine analogue N-methyl-4-phenyl-1,2,3,6-tetrahydropyridine on brain catecholamine neurons in the mouse. Eur. J. Pharmacol. 97 133-136.

Heikkila R.E., Hess A., and Duvoisin R.C.(1984) Dopaminergic neurotoxicity of 1-methyl-4-phenyl-1,2,3,6-tetrahydropyridine in mice. Science 244 1451-1453.

Heikkila R.E., Hess A., and Duvoisin R.C.(1985a)

Dopaminergic neurotoxicity of 1-methyl-4-phenyl-1,2,3,6-tetrahydropyridine (MPTP) in mouse: Relationship between monoamine oxidase, MPTP metabolism and neurotoxicity. Life Sci. 36 231-236.

Heikkila R.E., Youngster S.K., Manzino L., Cabbat F.S., and Duvoisin R.C. (1985b) Effect of 1-methyl-4-phenyl-1,2,3,6-tetrahydropyridine and related compounds on the uptake of [^{3}H]3,4-dihydroxyphenylethylamine and [^{3}H]5-hydroxytryptamine in neostriatal synaptosomal preparations. J. Neurochem. 44 310-313

JavitchJ.A. and Snyder S.H. (1984a) Uptake of MPP+ by domapine neurons explains selectivity of parkinsonism-inducing neurotoxin, MPTP. Eur. J. Pharmacol. 106 455-456

Javitch J.A., Uhl G.R., and Snyder S.H. (1984b) Parkinsonism-inducing neurotoxin, N-methyl-4-phenyl-1,2,3,6-tetrahydropyridine: Characterization and localization of receptor binding sites in rat and human brain. Proc. Natl. Acad. Sci. USA 81 4591-4595.

Kinemuchi H.,Arai Y.,Oreland L., Tipton K.F., and Fowler C.J. (1982) Time-dependent inhibition of monoamine oxidase by β-phenylethylamine. Biochem. Pharmacol. 31 959-964.

Kinemuchi H., Fowler C.J., and Tipton K.F. (1984) Substrate specificities of two forms of monoamine oxidase. In Monoamine oxidase and disease (Tipton K.F., Dostert P., and Strolin-Benedetti M. eds), pp.53-62.

Langston J.W., Irwin I., and Langston E.B. (1984) Pargyline prevents MPTP-induced Parkinsonism in primates. Science 225 1480-1482.

Langston J.W. (1985) MPTP and Parkinson's disease. Trend in NeuroScience 8 79-83.

Markey S.P., Johannessen J.N., Chiueh C.C., Burns R.S., and HerkenhamM.A. (1984) Intraneuronal generation of a pyridinium metabolite may cause drug-induced Parkinsonism. Nature (London) 311 464-467.

Markstein R., and Lahaye D. (1985) Neurochemical investigations in vitro with 1-methyl-4-phenyl-1,2,3,6-tetrahydropyridine (MPTP) in preparations of rat brain. Eur. J. Pharmacol. 106 301-311.

Murphy D.L., Redmond D.E., Garrick N., and Baulu J. (1979)

Monoamine oxidase type A and B activities in the vervet monkey. Neurochem. Res. 4 53-62

Oreland L., Arai Y., Stenstrom A., and Fowler C.J. (1983) Monoamine oxidase and localisation in the brain and the activity in relation to psychiatric disorder. In Modern problems of pharmacopsychiatry (Beckman H., and Riederer P., eds) pp. 246-254, Karger, Basel.

Orsi B.A. and Tipton K.F. (1979) Kinetic analysis of progress curves. In Methods in enzymology, 63A pp.159-183.

Parsons B.,and Rainbow T.C. (1984) High-affinity binding sites for [^{3}H]MPTP may correspond to monoamine oxidase. Eur. J. Pharmacol. 102 375-377.

Salach J.I., Singer T.P., Castagnol Jr.N., and Trevor A. (1984) Oxidation of the neurotoxic amine 1-methyl-4-phenyl-1,2,3,6-tetrahydropyridine (MPTP) by monoamine oxidase A and B and suicide inactivation of enzymes by MPTP. Biochem. Biophys. Res. Commun. 125 831-835.

Tipton K.F. (1969) A sensitive fluorometric assay for monoamine oxidase. Analyt. Biochem. 28 318-325.

Waley S,G. (1980) Kinetics of suicide substrate. Biochem. J. 185 771-773.

Wieczorek C.M., Parsons B., and Rainbow T.C. (1984) Quantitative autoradiography of [^{3}H]MPTP binding sites in rat brain. Eur. J. Pharmacol. 98 453-454.

Zigmond M.J. and Stricker E.M. (1984) Parkinson's disease: Studies with an animal model. Life Sci. 35 5-18.

Free Communications

PRODRUGS OF TRACE AMINES

Ronald T. Coutts, Glen B. Baker, Adil J. Nazarali, Tadimeti S. Rao, Ronald G. Micetich and T. W. Eric Hall.

PMHAC Research Unit and Neurochemical Research Unit, Faculty of Pharmacy and Pharmaceutical Sciences and Department of Psychiatry, University of Alberta, Edmonton, Alberta, Canada.

INTRODUCTION

The "trace" amines 2-phenylethylamine (Ia; PEA) and tryptamine (IIa; T) are endogenous CNS stimulants with rapid turnover rates. It has been suggested (Dewhurst, 1968; Sabelli and Mosnaim, 1974) that in depression there is a functional deficiency of one or both of these amines in the brain. Brain levels of PEA and T can be greatly increased in laboratory animals by the administration of monoamine oxidase (MAO)-inhibiting antidepressants (e.g. phenelzine, tranylcypromine) (Philips and Boulton, 1979) but this treatment is not selective and increases brain levels of other endogenous amines. To study the role of PEA and T in affective disorders, it is necessary to increase brain levels of individual endogenous amines and observe the biochemical and behavioural changes that result. Since MAO inhibitors which selectively inhibit the metabolism of individual brain amines do not exist, alternative methods of increasing brain levels of amines must be utilized. In our laboratory, we have investigated the usefulness of prodrugs with the hope that they may alter advantageously the pharmacokinetics of the amine released by the prodrug, such that brain levels of amine are increased, prolonged or better controlled. All the prodrugs synthesized and investigated by us (Figure 1) contain N-substituents (e.g. ethyl, *n*-propyl, 3-chloropropyl, 2-cyanoethyl and ethoxycarbonyl) which are metabolically removed in the body at different rates to yield the desired drug (Nazarali et al., 1983; Baker et al., 1984). The prodrugs of interest in the present study are the cyanoethyl derivatives of PEA and T (Ib; CE-PEA, and IIb; CE-T, respectively) and the ethoxycarbonyl derivative of PEA (Ic, CAR-PEA). All prodrugs were synthesized and characterized in our laboratories.

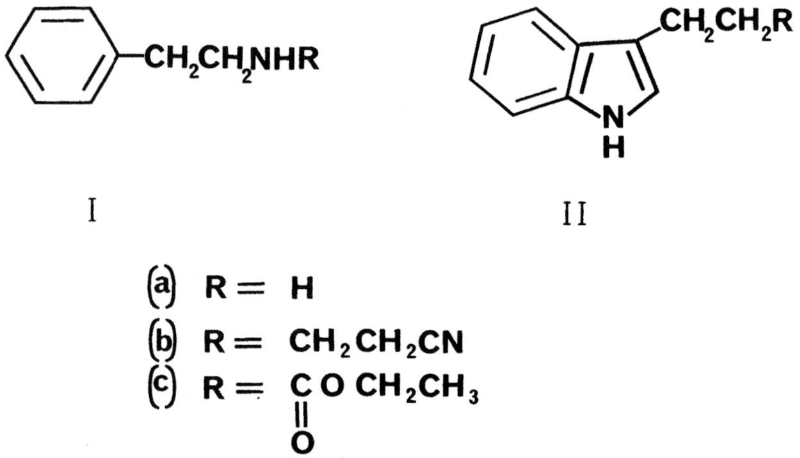

Fig. 1. Structures of compounds of interest to this study.

In this study, we have determined the brain levels of PEA and T in rats after administration of each amine and the appropriate prodrugs.

METHODS

a) Groups of male Sprague-Dawley rats (180-200 g) were pretreated with the MAO-inhibitor pargyline (60 mg/kg i.p.) then injected i.p. with 0.1 mmol/kg of drug (PEA or T) or prodrug (CE-PEA or CE-T), or with 0.5 mmol/kg of CAR-PEA, and 5 or 6 rats were sacrificed at appropriate time intervals (5-180 min) after administration of drug or prodrug. Brains were rapidly removed, frozen solid in isopentane on dry ice, and stored at -60°C until analyzed.

b) method a) was repeated except that pargyline pretreatment was omitted.

c) analysis. PEA and CE-PEA were converted to their pentafluorobenzoyl (PFB) derivatives using a modification of the procedure of Cristofoli _et al._ (1982). T and CE-T were reacted with pentafluoropropionic anhydride (PFPA) after basic extraction with ethyl acetate or were reacted with pentafluorobenzenesulfonyl chloride following basification. CAR-PEA was reacted with PFPA after basification and extraction. These derivatives were then separated and quantitated using a Hewlett-Packard model 5880A gas chromatograph equipped with either a 2 m packed column (3% OV-17) or a 25 m fused silica capillary column (5% phenylmethylsilicone) and an electron-capture detector and recorder/integrator. Baseline separation of all derivatives was achieved. The structures of all the perfluoroacylated derivatives were confirmed by combined gas chromatography-mass spectrometry.

RESULTS

Because levels of T in untreated rats are less than 3.5 pmol/g and difficult to measure accurately by GC-ECD, and as PEA and T are rapidly metabolized by MAO, rats were initially pretreated with the MAO inhibitor pargyline prior to administration of CE-PEA or CE-T to ensure that measurable quantities of PEA and T would be encountered. Inspection of the data in Table 1 indicates that rat brain levels of PEA and T one hour after injection of CE-PEA and CE-T were significantly higher than in rats treated with only pargyline, or in untreated rats. The brain levels of T, for example, after CE-T administration were 215 times as high as control values (pargyline only) and about 1350 times the values observed in untreated rats.

In subsequent experiments, pretreatment with pargyline was omitted.

CAR-PEA was evaluated as a prodrug by administering 0.5 mmol/kg i.p. to rats and measuring brain levels 1 h after injection. At that time brain levels of CAR-PEA and PEA were 64.2 $\pm$ 8.2 nmol/kg and 190.0 $\pm$ 36.1 pmol/g, respectively (Baker *et al.*, 1984). Control brain levels of PEA are 10-15 pmol/g. The administration of the carbamate prodrug (CAR-PEA) may represent a means of selectively increasing brain concentrations of PEA without affecting levels of other biogenic amines.

Concentration/time profiles of brain levels of PEA and T were constructed after the administration to the rats of 0.1 mmol of PEA, T, CE-PEA or CE-T per kg. Results obtained with PEA and CE-PEA are shown in Figure 1. After administration of PEA itself, very high brain levels are attained, but by 5 min, values are down almost to control levels again. However, after injection of CE-PEA, levels of PEA

Table I. Concentrations (means ± S.E.M.) of prodrugs (CE-PEA, CE-T) and amines (PEA, T) in rat brain 1 h after i.p. injection of prodrug[a]. (Rats were pretreated with pargyline, 60 mg/kg i.p.).

Dose of Prodrug (mmol/kg)	Brain Concentration of Prodrug (nmol/g)	Brain Concentration of Amine (nmol/g)[c]
CE-PEA, 0.1	CE-PEA, 7.93 ± 1.32[b]	PEA, 26.28 ± 0.99
CE-T, 0.1	CE-T, 69.09 ± 5.39	T, 4.73 ± 0.18

[a]Abbreviations are identified in text. [b]Six rats were used in each experiment. [c]Brain levels of PEA and T in rats treated with pargyline only (no prodrug) were 1.09 ± 0.13 nmol/g and 21.8 ± 2.2 pmol/g, respectively; in untreated rats, brain levels of PEA and T were 12.0 ± 0.18 and <3.5 pmol/g, respectively.

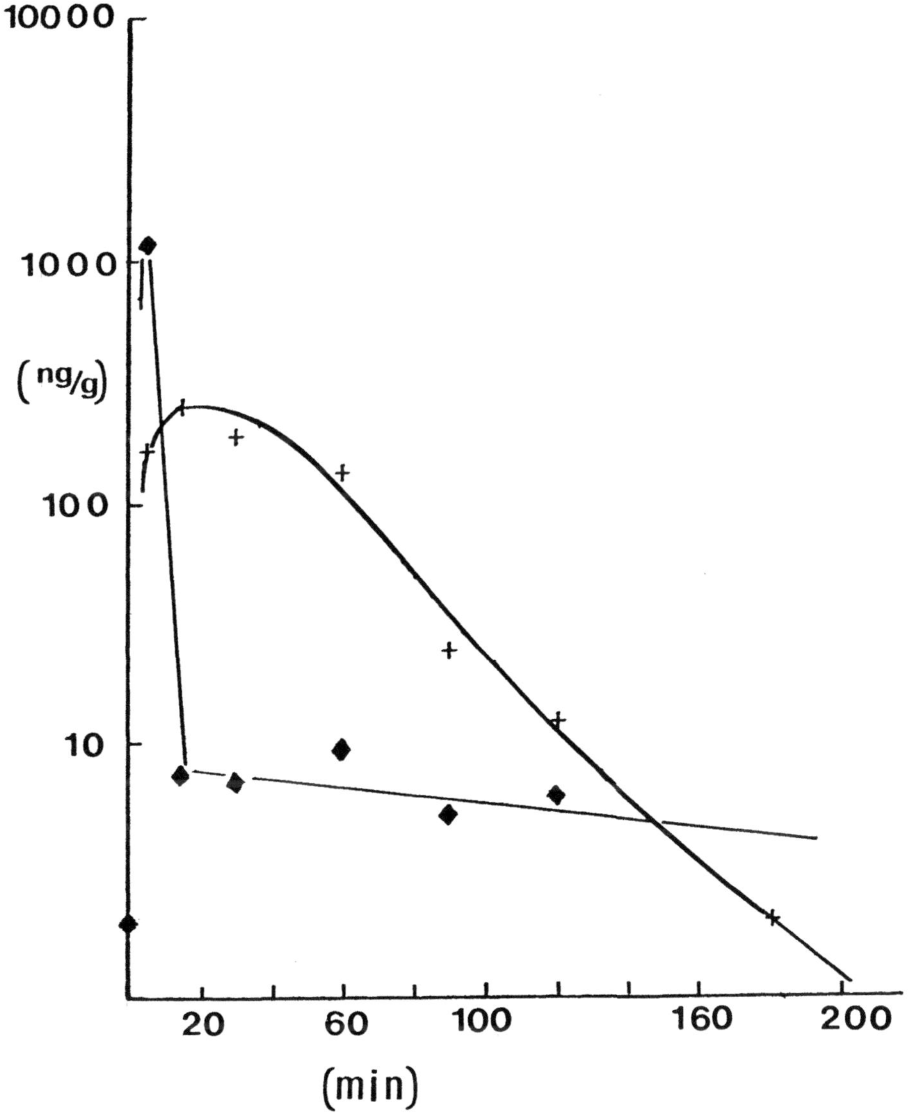

Fig. 2. Concentration-time profiles of brain levels of PEA after administration of 0.1 mmol/kg doses of PEA (◆), and CE-PEA (+).

attain a peak concentration of almost 300 ng/g and remain above control levels for 2h. Preliminary studies with T and CE-T have shown a similar situation, although the levels of T attained after prodrug administration are not as dramatic as with PEA formed from CE-PEA.

These findings indicate that prodrugs of trace amines offer a promising means of selectively elevating these amines in the central nervous system. Further studies are in progress to evaluate whether chronic administration of low doses of CE-PEA or CE-T to rats will result in steady-state elevated levels of PEA or T.

ACKNOWLEDGEMENTS

Funds were provided by the Provincial Mental Health Advisory Council (PMHAC), the Medical Research Council of Canada and the Alberta Heritage Foundation for Medical Research. The authors are grateful to Dr. T. J. Danielson for helpful discussions.

REFERENCES

Baker G. B., Coutts R. T., Nazarali A. J., Danielson T. J. and Rubens M. (1984) Carbamate prodrugs of phenylethylamine: a neurochemical investigation. Proc. West Pharmacol. Soc., **27**, 523-525.

Cristofoli W. A., Baker G. B., Coutts R. T., and Benderly A. (1982) Analysis of a monofluorinated analogue of amphetamine in brain tissue using gas chromatography with electron-capture detection. Progr. Neuro-Psychopharmacol. & Biol. Psychiat., **6**, 373-376.

Dewhurst W. G. (1968) New theory of cerebral amine function and its clinical application. Nature (London), **218**, 1130-1133.

Nazarali A. J., Baker G. B., Coutts R. T., and Pasutto F. M. (1983) Amphetamine in rat brain after intraperitoneal injecton of N-alkylated analogues. Progr. Neuro-Psychopharmacol. & Biol. Psychiat., **7**, 813-816.

Sabelli H. C., and Mosnaim, A. D. (1974) Phenylethylamine hypothesis of affective behaviour. Am. J. Psychiatry, **131**, 695-699.

THE EFFECT OF DEUTERIUM LABELLING ON IONIZATION CONSTANTS AND LIPOPHILICITY OF TRACE AMINES

D. A. Durden

Psychiatric Research Division
Cancer and Medical Research Building
University of Saskatchewan
Saskatoon, Saskatchewan
Canada S7N 0W0

INTRODUCTION

In recent work in this laboratory it was observed that when equimolar mixtures of phenylethylamine (PE) and $\alpha,\alpha,\beta,\beta$-tetradeutero phenylethylamine (PE-d_4) were injected intraperitoneally into the rat, the amount of PE-d_4 in the rat brain greatly exceeded that of PE (Dyck et al.). This could be explained by the resistance of PE-d_4 to oxidation by MAO or perhaps a higher transport rate of PE-d_4 into the brain. The transport of an amine across membrane barriers may be approximately related to its distribution coefficient (D) into a lipophilic solvent such as an 1-octanol system (Leo et al., 1971). The distribution coefficient is in turn related to the partition coefficient (P), the dissociation constant (pK_a) and the pH of the aqueous solution (Mack and Bonisch, 1979). Labelling of the alkyl chain of alkyl amines and acids tends to increase the pK_a value (Halevi et al., 1963; Vree et al., 1971). This in turn, assuming constant partition coefficient, would reduce the distribution coefficient and hence the relative transport of a deuterium labelled trace amine into the lipophilic solvent. The aim of the present work was to measure the ionization constants of a series of trace amines, PE, tryptamine (TRYPT) and meta- and para-tyramine (*m*-TA, *p*-TA) and of their deuterium isotopomers and the relative distribution coefficients and partition coefficient in the octanol-water system, and test the above hypothesis.

METHOD

pK_a Values

The pK_a values of deuterium labelled phenylethylamine (PE-α-d_1, PE-α,α-d_2, PE-β,β-d_2, PE-α,α,β,β-d_4), tryptamine (trypt-α,α,β,β-d_4), meta-tyramine (m-TA-α,α,β,β-d_4), and para-tyramine (p-TA-α,α,β,β-d_4) and their proteo analogues were measured by the procedure of Albert and Serjeant (1984). Profiles of pH versus titration volume were measured for a deuterated amine and its proteo analog consecutively with pH buffer controls before and after. The multiple pK_a values (m-TA and p-TA) were determined by an iterative computer program (Albert and Serjeant, 1984). The proteo and deutero compound from each set of measurements were treated as pairs to maximise precision and reduce the influence of long term drift of the pH meter.

Distribution Coefficients

To obtain relative distribution coefficients, approximately equimolar quantities of the amine and its d_4 isotopomer were extracted from pH 7.41 buffer (saturated with 1-octanol) into 1-octanol (saturated with buffer) by inversion mixing for 5 min (Mack and Bonisch, 1979). The amines were converted into their dansyl derivatives and extracted into toluene. The relative ratio of the intensities of their molecular ions were determined by high resolution mass spectrometry (HRMS) (Durden et al., 1974).

To obtain absolute distribution coefficients (not reported here), the proteo compound alone was extracted into octanol and after separation of the phases, the d_4 amine was added to each phase as internal standard. The amounts of proteo amine were determined by HRMS of the dansyl derivatives. In the case of PE and PE-d_4, PE-d_9 was added as an internal standard to obtain values for both compounds simultaneously.

RESULTS

Deuterium labelling with deuterium atoms on the α-carbon caused the pK_a to increase 0.04 to .05 units, whereas labelling the β position caused essentially no increase (Table 1). For the tyramines the major increase in pK_a occurred at the higher value, i.e. that due to ionization of the phenolic group. The relative distribution coefficients of the amines are given in Table 2. As predicted, less deuterium labelled amine than proteo amine was extracted into octanol. Also shown in Table 2 is the calculated ratio which is based on the assumption that the partition coefficient is constant and that the effect is only due to increases in pK_a. The calculated partition coefficient of PE-d_4 is 5% smaller than that of PE (Table 3).

Table 1

Increase of pK_a Values of Trace Amines by Alkyl Chain Deuterium Labelling

Compound		$pK_{ad} - pK_{ah}$	
PE-α-d_1		0.032, 0.046 (2)	
PE-α,α-d_2		0.040±.005 (4)	**
PE-α,α,β,β-d_4		0.051±.010 (4)	**
PE-β,β-d_2		0.004±.003 (3)	
Trypt-α,α,β,β-d_4		0.056±.011 (3)	**
_m_-TA-α,α,β,β-d_4	pK_{a1}	0.036±.014 (4)	*
	pK_{a2}	0.070±.002 (4)	**
_p_-TA-α,α,β,β-d_4	pK_{a1}	0.022±.011 (4)	
	pK_{a2}	0.065±.003 (4)	**

Values are means ±S.E.M. (number of measurements).
* Significantly different from zero at $p<.05$.
** Significantly different from zero at $p<.005$.

Table 2

Effect of Deuterium Labelling on Extraction of Trace Amines into 1-Octanol at pH 7.41

Compound	Ratio of Deuterium: Protium Compound			
	Original Solution	After Extraction	In Octanol	Predicted from D
PE	1.0	1.050±0.11 (5)	0.945±.009 (5)	0.92[1] 0.90[2]
TRYPT	1.0	1.038±.020 (5)	0.951±.007 (5)	0.91[1] 0.89[3]
m-TA	1.0	1.014±.002 (5)	0.905±.007 (5)	0.93[1]
p-TA	1.0	1.008±.005 (3)	0.893±.021 (3)	0.95[1] 0.94[3]

Values are means ± S.E.M. (number of measurements). [1]Based on measured distribution coefficients. [2]Based on values of Mack and Bonish (1979) and measured differences. [3]Based on values of Leo et al. (1971) and measured differences.

Table 3

Lipophilicity of PE and PE-d_4

Compound	Measured D	Log D	Log F	Log P	Ratio D_d/D_h
PE	0.440	-0.36	2.51	2.15	-
PE-d_4	0.370	-0.43	2.56	2.13	0.95

Distribution coefficient D was measured for PE and PE-d_4 using PE-d_9 internal standard.

DISCUSSION

The pK_a values increased in accordance with previous observations and the increase of about 0.05 units due to deuterium atoms in the α-position of PE and TRYPT is consistent with measurements for benzylamine (Halevi et al., 1963) and for alkyl substituted amphetamines (Vree et al., 1971). The small increase due to β-labelling also agrees with a previous observation for propionic acid (Halevi et al., 1963). The larger effect on pK_a for the tyramines indicates that at intermediate pH (i.e. 8-9), these compounds exist primarily as zwitterions in aqueous media (Armstrong and Barlow, 1976).

The following equations for partition coefficient (P), distribution coefficient (D) and fraction of uncharged species (F) were used to predict the relative transport of amines and their deutero-isotopomers (Mack and Bonish, 1979). Log P = Log D - Log F where the fraction of unionized species F is given by:
$F = 1/[1 + \text{Antilog}\ (pK_a - pH)]$ for simple amines or
$F = 1/[1 + \text{Antilog}\ (pK_{a1} - pH) + \text{Antilog}\ (pH - pK_{a2}]$
for phenolic amines. The distribution coefficient was calculated from the measured amounts of the amine in the 1-octanol (A) and buffer (B) and r the ratio of octanol to buffer volumes from $D = A/[(B)\cdot r]$. Thus we can predict that the effect of a small change in pK_a on D is maximized if r is small (1.0) and pH is less than pK_a. This is the case for amines as the pK_a is usually 2 to 3 units above physiological pH. Secondly, an increase in pK_a reduces D and thus the amount (A) extracted into octanol. Finally for phenolic amines a change in pK_{a1} has a much greater effect on D than does an equivalent change in pK_{a2}. The extractions were done at physiological pH (7.41) using a 1:1 ratio of octanol to water to maximize the discrimination.

As was predicted from the theoretical calculations, less deuterium isotopomer is extracted into the lipophilic (octanol) layer. The amount is somewhat less than predicted if we assume P constant for both compounds and that changes in pK_a are the only variable (Table 1). From Table 3 we can see that there is, however, a decrease of 5% in the partition coefficient of PE-d_4 when compared to PE. This decrease tends to counteract the effect of the increase of pK^a which reduces the distribution coefficient.

CONCLUSION

An increase in pK_a due to deuterium labelling of the trace amines at the carbon α to the amino group causes a reduction of transport into a lipophilic phase and for the case of the relative levels of PE and PE-d_4 observed in rat brain the discriminating effect is probably due to inhibition of enzyme activity towards the deuterium labelled compound.

ACKNOWLEDGEMENTS

I would like to thank Dr. B.A. Davis for synthesis of the deuterium labelled compounds, M. Mizuno for performing the MS analysis and Saskatchewan Health for continuing financial support.

REFERENCES

Albert A. and Serjeant E.P. (1984) The Determination of Ionization Constants. 3rd Ed. Chapman and Hall, London.

Armstrong J. and Barlow R.B. (1976) The ionization of phenolic amines, including apomorphine, dopamine and catecholamines and an assessment of zwitterion constants. Br. J. Pharmac. 57, 501-516.

Durden D.A., Davis B.A. and Boulton A.A. (1974) Qualitative and quantitative mass spectrometry of some non-catecholic biogenic amines and related compounds as their dansyl derivatives. Biomed. Mass Spectrom. 1, 83-95.

Dyck L.E., Durden D.A. and Boulton A.A. (1985) The effects of deuterium substitution on the catabolism of β-phenylethylamine: an in vivo study. J. Neurochem. (submitted).

Halevi E.A., Nussim M. and Ron A. (1963) Secondary hydrogen isotope effects. Part I. Strengths of α-deuterated carboxylic acids and amines. J. Chem. Soc. 866-875.

Leo Albert, Hansch Corwin and Elkins David. (1971) Partition coefficients and their uses. Chem. Rev. 71 (6), 525-616.

Mack F. and Bonisch H. (1979) Dissociation constants and lipophilicity of catecholamines and related compounds. Nauyn-Schmiedeberg's Arch. Pharmacol. 310, 1-9.

Vree T.B., Gorgels J.P.M.C., Muskens A.Th.J.M. and Van Rossum J.M. (1971) Deuterium isotope effects in the metabolism of N-alkylsubstituted amphetamines in man. Clin. Chim. Acta 34, 333-344.

β-PHENYLETHYLAMINE AND AMPHETAMINE COMPARED IN TESTS OF ANOREXIA AND PLACE-PREFERENCE CONDITIONING

D. B. Gilbert and S. J. Cooper

Department of Psychology

University of Birmingham
Birmingham B15 2TT, U.K.

ABSTRACT

Adult, male non-deprived rats were accustomed to eating a highly-palatable diet in a 30 min period. Administration of β-phenylethylamine (PE) immediately before the feeding test produced a dose-related (3-30 mg/kg, i.p.) reduction in intake over the first 15 min. A significant anorectic effect was found following 10 and 30 mg/kg of PE. The effect of PE was short-lived, since over the complete 30 min, the reduction in the consumption of the diet was then considerably attenuated. Both d- and l-amphetamine (AMPH) produced dose-related anorectic effects in the same test, and 30 min intake was significantly reduced after 0.3 mg/kg d-AMPH and 1.0 mg/kg l-AMPH. Comparisons with other data show that the anorectic action of PE can be distinguished from its hypodipsic effect, its reinforcing effect as determined in the conditioned place-preference test, and from its ability to induce behavioural stereotypy. The data indicate, therefore, a relatively selective behavioural effect of PE to reduce the consumption of a palatable diet.

INTRODUCTION

PE is an endogenous trace amine found in mammalian neural tissue, including human brain (Boulton and Juorio,

1982). In the rat, highest concentrations are found in the hypothalamus and caudate nucleus. PE is structurally similar to AMPH, lacking only the α-methyl group in the alkyl side chain. The in vivo biochemical effects of d-AMPH and PE have been determined on the rate of dopamine (DA) synthesis in brain regions rich in DA terminals, and on the efflux of the DA metabolite, DOPAC, in cerebroventricular perfusates of freely-moving rats (Nielsen et al., 1983). In small doses, d-AMPH (1 mg/kg, s.c.) and PE (30 mg/kg, s.c.) increased the efflux of DOPAC and the rate of DA synthesis in the striatum. Higher doses had opposite effects. The authors suggested that small doses of d-AMPH and PE facilitate the neuronal release of DA.

In behavioural studies, acute administration of large doses of either d-AMPH or PE elicits a characteristic syndrome of behaviour stereotypy in rats. Components of the syndrome include sniffing, head-weaving and- bobbing, splayed hindlimbs, forepaw padding and backward walking (Dourish, 1984), and can be elicited at 25 or 50 mg/kg of PE (Cooper and Dourish, 1984). At doses of PE below 25 mg/kg, signs of the behavioural stereotypy syndrome were absent. The stereotyped behaviour which is induced by AMPH treatments is preceded by a phase of heightened locomotor activity (Robbins and Sahakian, 1981). Increased locomotion and rearing in rats has also been detected after injection of 12.5 or 25 mg/kg of PE (Dourish and Cooper, 1984b). PE in combination with naloxone, an opiate receptor antagonist, produced a marked stimulation of locomotor activity (Dourish and Cooper, 1984a). In general, the psychomotor stimulant effects of PE and AMPH are similar.

PE AND AMPH AS REINFORCERS

Drugs with abuse potential in humans act as reinforcers in behavioural tests using laboratory animals. One method of demonstrating the reinforcement due to drug delivery is the self-administration procedure. Intravenous AMPH self-administration has been described in the rat (Yokel and Pickens, 1973), and intravenous PE self-administration in the dog (Risner and Jones, 1977; Shannon and DeGregorio, 1982). Place-preference conditioning offers an alternative method to assess the reinforcing effect of drug treatments. Using this paradigm, a conditioning of place-preference with d-AMPH has been demonstrated in the rat (Spyraki et al., 1982). We carried out a similar study to compare the potencies of the two isomers of AMPH, and of PE (Gilbert and Cooper, 1983). Our data showed significant place-preference conditioning with all three drugs, with a potency relationship as follows: d-AMPH>l-AMPH>PE. To detect an effect with PE, the least potent of the three, a dose of 75 mg/kg was required. At that dose, the behavioural stereotypy syndrome was clearly in evidence. Hence, a reinforcing effect of PE in the rat could be demonstrated, but the dose required to achieve it was very high.

ANOREXIA AND HYPODIPSIA

In addition to their psychomotor stimulant effects and their reinforcing property, the two isomers of AMPH reduce the consumption of food (Baez, 1971) and reduce the ingestion of water (Stolerman and D'Mello, 1978). PE also reduces the intake of water in rehydrating rats, after administration of either 25 or 50 mg/kg, and within the first 15 min after its injection (Cooper and Dourish, 1984). The decrease in water intake coincided with the induction of the behavioural stereotypy syndrome. The experiment we now report deals with an immediate and transient anorectic effect of PE, and compares it with the effects of the two isomers of AMPH. A previous study reported a decrease in 24 hr food intake of rats after administration of 50 mg/kg of PE (Dourish and Boulton, 1981).

Methods

The subjects were 128 adult, male, blackhooded rats (General strain, bred in our laboratory; 250-400g). They were housed individually in stainless-steel cages with ad libitum access to standard rat diet and water. D-AMPH sulphate and l-AMPH sulphate (kindly provided by Smith Kline and French Ltd.) and PE hydrochloride (purchased from Sigma, Poole, Dorset) were dissolved in isotonic saline and injected i.p. in a volume of 1 ml/kg. AMPH isomers were injected 15-20 min before and PE less than 5 min before the food intake test. Eight rats were assigned at random to each dose condition for each drug. The feeding test has been described in detail elsewhere (Gilbert and Cooper, 1985); non-deprived animals were first thoroughly familiarized with eating a highly palatable sweetened diet in a 30 min test, in a test cage identical to the home cage except that the standard food pellets and water were not available. A weighed quantity of the palatable diet was introduced and the subsequent intake was measured to the nearest 0.1g (taking account of any spillage) over a 15 min or 30 min period. Two separate experiments were conducted with PE; in the first, food intake was measured over the first 15 min and in the second, it was measured over the full 30 min. For the two isomers of AMPH, 30 min intake data were collected. The results were analysed using a one-way analysis of variance and Dunnett's _t_ test.

Results

As Table 1 shows, in the 15 min test, food intake was significantly reduced after injection of PE ($F_{3,28}=16.57$, $P<0.001$). Following 3 mg/kg of PE, there was a 25.8% reduction in food intake, and after 10 mg/kg, there was a highly significant 50% reduction. The anorectic effect of PE was evidently very short-lived, because it was markedly attenuated when the 30 min test was conducted. Both d-AMPH and l-AMPH had dose-dependent effects on food intake, and a significant decrease was detected after 0.3 mg/kg d-AMPH (58.5% reduction) and after 1.0 mg/kg l-AMPH (47% reduction). Potency relationships were: d-AMPH>l-AMPH>PE.

Discussion and Conclusions

These data show, for the first time, an immediate but transient anorectic effect of PE in the non-deprived rat. Furthermore PE was less potent than the two isomers of AMPH. We suggest that the anorectic effect of PE represents a relatively specific effect on the control of food intake. Previous data indicate that the hypodipsic effect, the reinforcing effect, and the behavioural stereotypy syndrome induced by PE, are all obtained at doses which are several times greater than the doses which reduced food intake. In addition, negative findings in a conditioned taste aversion test (Greenshaw and Dourish, 1984) rule out the possibility that the anorexia was a consequence of an aversive effect of

TABLE 1

The consumption of a palatable diet by non-deprived male rats following injection of β-phenylethylamine (3-30 mg/kg, i.p.), in a 15 min and a 30 min intake test. Comparison with the anorectic effects of d-amphetamine (0.1-1 mg/kg, i.p.) and l-amphetamine (0.3-3 mg/kg, i.p.) over a 30 min test. Results are shown as mean ± S.E.M. food intake (g). N=8 per group. *P <0.01 in comparison with control level of intake (Dunnett's t test).

	β-phenylethylamine (mg/kg)			
	0	3	10	30
15 min test	12.0 ± 1.1	8.9 ±1.5	6.0* ±1.2	0.6* ±0.2
30 min test	15.5 ± 1.3	14.6 ± 1.0	15.8 ± 1.1	9.4* ±1.3

	d-amphetamine (mg/kg)			
	0	0.1	0.3	1.0
30 min test	17.6 ± 1.0	13.1 ± 1.5	7.3* ±1.3	5.1* ±1.5

	l-amphetamine (mg/kg)			
	0	0.3	1.0	3.0
30 min test	18.1 ± 0.8	14.5 ± 1.8	9.6* ±1.2	6.3* ±1.7

PE administration. Additional work is required to determine the relationship of the anorectic effect of PE with its effects on locomotion and rearing (Dourish and Cooper, 1984b). Nevertheless, our data encourage the view that there is a direct effect of PE on feeding mechanisms. The anorectic effect of a small dose of d-AMPH appears to be due to an action of DA at D-1 receptors (Gilbert and Cooper, 1985). Future work should consider the possibility that PE-induced anorexia also follows from increased DA activity at D-1 receptors.

REFERENCES

Baez, L.A. (1974) Role of catecholamines in the anorectic effects of amphetamine in rats. Psychopharmacologia(Berl.) 35: 91-98.

Boulton, A.A. and Juorio, A.V. (1982) Brain trace amines. In: Handbook of Neurochemistry, Vol.1 (Lajtha, A., ed), pp. 189-222. Plenum, New York.

Cooper, S.J. and Dourish, C.T. (1984) Hypodipsia, stereotypy and hyperactivity induced by β-phenylethylamine in the water-deprived rat. Pharmacol. Biochem. Behav. 20: 1-7.

Dourish, C.T. (1984) Studies on the mechanism of action of β-phenylethylamine stereotypy in rodents: Implications for a β-phenylethylamine animal model of schizophrenia. In: Neurobiology of the Trace Amines (Boulton, A.A., Baker, G.B., Dewhurst, W.G. and Sandler, M., eds), pp. 389-411. Humana Press, Clifton NJ.

Dourish, C.T. and Boulton, A.A. (1981) The effects of acute and chronic administration of β-phenylethylamine on food intake and body weight in rats. Prog. Neuropsychopharmacol. 5: 411-414.

Dourish, C.T. and Cooper, S.J. (1984a) Potentiation of total horizontal activity and ambulation in rats treated with combinations of β-phenylethylamine and naloxone. Neuropharmacology 23: 1059-1064.

Dourish, C.T. and Cooper, S.J. (1984b) Environmental experience produces qualitative changes in the stimulant effects of β-phenylethylamine in rats. Psychopharmacology 84: 132-135.

Gilbert, D.B. and Cooper, S.J. (1983) β-phenylethylamine-, d-amphetamine- and l-amphetamine-induced place preference conditioning in rats. Europ. J. Pharmacol. 95: 311-314.

Gilbert, D.B. and Cooper, S.J. (1985) Analysis of dopamine D1 and D2 receptor involvement in d- and l-amphetamine-induced anorexia in rats. Brain Res. Bull. (in press).

Greenshaw, A.J. and Dourish, C.T. (1984) Differential aversive stimulus properties of β-phenylethylamine and of d-amphetamine. Psychopharmacology 82: 189-193.

Nielsen, J.A., Chapin, D.S. and Moore, K.E. (1983) Differential effects of d-amphetamine, β-phenylethylamine, cocaine and methylphenidate on the rate of dopamine synthesis in terminals of nigrostriatal and mesolimbic neurons and on the efflux of dopamine metabolites into cerebroventricular perfusates of rats. Life Sci. 33: 1899-1907.

Risner, M.E. and Jones, B.E. (1977) Characteristics of β-phenylethylamine self-administration by dog. Pharmacol. Biochem. Behav. 6: 689-696.

Robbins, T.W. and Sahakian, B.J. (1981) Behavioural and neurochemical determinants of drug-induced stereotypy. In: Metabolic Disorders of the Nervous System (Rose, F.C., ed), pp. 244-291. Pitman, London.

Shannon, H.E. and DeGregorio, C.M. (1982) Self-administration of the endogenous trace amines β-phenylethylamine, N-methyl phenylethylamine and phenylethanolamine in dogs. J. Pharmacol. Exp. Ther. 222: 52-60.

Spyraki, C., Fibiger, H. and Phillips, A. (1982) Dopaminergic substrates of amphetamine-induced place preference conditioning. Brain Res. 253: 185-193.

Stolerman, I.P. and D'Mello, G.D. (1978) Amphetamine-induced hypodipsia and its implications for conditioned taste aversion in rats. Pharmacol. Biochem. Behav. 8: 333-338.

Yokel, R.A. and Pickens, R. (1973) Self-administration of optical isomers of amphetamine and methylamphetamine by rats. J. Pharmacol. Exp. Ther. 187: 27-33.

INTERACTION OF TRACE AMINES WITH SEROTONIN RECOGNITION SITES IN RAT BRAIN

R. A. Locock, G. B. Baker, R. T. Coutts, W. G. Dewhurst and T. T. S. Liu.

Neurochemical Research Unit, Faculty of Pharmacy and Pharmaceutical Sciences and Department of Psychiatry, University of Alberta, Edmonton, Alberta, Canada.

INTRODUCTION

The "trace" amines such as 2-phenylethylamine, *m*- and *p*-tyramine and tryptamine accumulate in brain after administration of monoamine oxidase inhibitor antidepressant drugs (Boulton and Juorio, 1982). Accumulated quantities of these substances, in addition to having effects by themselves, may have important interactions with neurotransmitter receptor systems. In the study reported here we have examined the *in vitro* interactions of these trace amines with serotonin (5-hydroxytryptamine; 5-HT) recognition sites ($5\text{-}HT_1$ and $5\text{-}HT_2$) of rat cerebral cortex. Modulation of serotonin receptors by the central neurotransmitter, norepinephrine, has been recognized recently (Scott and Crews, 1985); the trace amines also may have important interactions with the serotonin receptors.

METHODS

Drugs and Reagents

[^{3}H] - 5-Hydroxytryptamine creatinine sulfate (23.4 Ci/mmol) and [^{3}H]-spiperone (26.3 Ci/mmol) were purchased from NEN Products, Dupont Canada, Lachine, Quebec. Cold 5-hydroxytryptamine creatinine sulfate, 2-phenylethylamine hydrochloride, tryptamine hydrochloride, *p*-tyramine hydrochloride, pargyline hydrochloride and butyl-PBD were obtained from Sigma Chemical, St. Louis, Missouri. *Meta*-

tyramine was obtained from Vega Biotechnologies, Tuson, Arizona. The isomers (+)- and (-)-α-methyltryptamine hydrochloride were prepared by fractional crystallization of the resolved (-)-tartaric acid salts of D,L-α-methyltryptamine purchased from the Aldrich Chemical Co., Milwaukee, Wisconsin, isolation of the free bases from the tartrates by treatment with alkali, and formation of the hydrochloride salts. Mianserin hydrochloride was generously given by Organon Canada Ltd., Westhill, Ontario. All other chemicals were reagent grade and were obtained from Fisher Scientific Co., Fairlawn, New Jersey. The scintillation fluor was a 4% solution of butyl-PBD in 677 ml toluene and 333 ml of Triton X-100.

Tissue Preparation

Adult male Sprague Dawley rats (150-200 g) were killed by cervical dislocation, decapitated and the frontal cortex removed following the procedure of Glowinski and Iversen, (1966). The cortex was homogenized in 10 volumes of ice-cold 0.32M sucrose in a Potter-Elvejhem apparatus with a teflon pestle. The homogenate was centrifuged at 700 x g for 10 min at 4°C. The supernatant was decanted and centrifuged at 50,000 x g for 10 min at 4°C. The resulting pellet was resuspended in 10 volumes of 50 mM Tris-HCl buffer, (pH 7.5), incubated at 37° for 10 min and recentrifuged at 50,000 x g for 10 min. The pellet obtained was resuspended in 50 mM Tri-HCl buffer (pH 7.7) containing 4 mM calcium chloride, 10 μM pargyline hydrocloride and 0.1% ascorbic acid. The final tissue suspension was incubated at 37° for 15 min and then stored on ice until used. The final tissue concentration of rat frontal cortex was 10 mg per ml.

Binding Assays

Incubation tubes in triplicate contained 0.8 ml of the tissue suspension, 100 μl of the [^{3}H] ligands and 100 μl of the various trace amines at 7 to 10 different concentrations in the 50 mM Tris-HCl buffer, (pH 7.7). The [^{3}H] ligand concentrations were 2 nM [^{3}H]-5-HT (5-HT_1 sites) and 0.24 nM [^{3}H]-spiperone (5-HT_2 sites). The tubes were incubated at 37° for 10 min for the 5-HT_1 determinations and 15 min for the 5-HT_2 assays. Rapid filtration under vacuum through 2.4 cm diameter Whatman GF/B glass microfiber filters (Whatman Inc., Clifton, New Jersey) terminated the incubation and was followed by three 5 ml rinses of ice-cold 50 mM Tris-HCl buffer (pH 7.7). The filters were placed in

Mini-vials (Fisher Scientific Co.) 5 ml of fluor added, and the samples were counted using a Beckman LS Liquid Scintillation System (Beckman Instruments Inc., Irvine California) after extraction for 18 h at 4°. Non-specific binding was determined by incubation in the presence of 3 μM 5-HT for the 5-HT_1 sites and 3 μM mianserin for the 5-HT_2 sites.

Data Treatment

The results from binding assays were plotted in the form of Hill plots, log [% bound/(100-% bound)] vs. log concentration of the trace amine inhibitor. A computer program ("paired sample analysis") for the Hewlett Packard 86 microcomputer (Hewlett Packard, Corvallis, Oregon) was used to fit the data to a straight line by the method of least squares. The concentration for 50% inhibition of specific binding, the IC_{50} value, was determined by solving for log concentration when log [% bound/(100-% bound)] = 0.

RESULTS

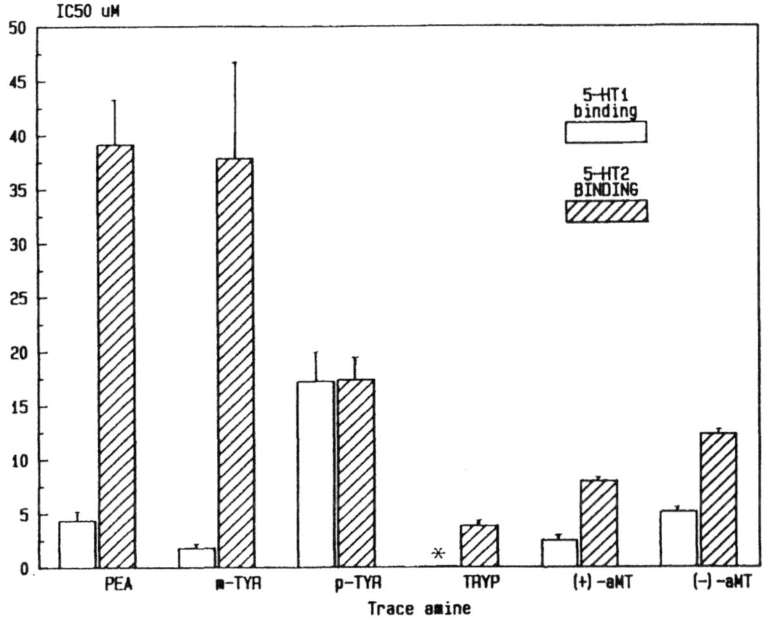

Fig. 1. Inhibition of 5-HT_1 and 5-HT_2 binding by trace amines. The IC_{50} values are the means, ± the standard

error of the means of 4-7 determinations. *The IC_{50} value for tryptamine and the 5-HT_1 receptor is 66.7 $\pm$ 4.8 nM. PEA = 2-phenylethylamine; m-TYR and p-TYR = *m*- and *p*-tyramine; TRYP = tryptamine; (+)-aMT and (-)-aMT = (+)- and (-)-α-methyltryptamine. These data are a graphic representation of results previously reported by Locock *et al.* (1984).

The IC_{50} values for the inhibition of serotonin recognition sites, 5-HT_1 and 5-HT_2, by the trace amines are presented in Fig. 1. The lower the bar the more potent the inhibition of specific binding to serotonin recognition sites.

DISCUSSIONS AND CONCLUSIONS

The results presented in Fig. 1 illustrate the marked differential effects of the trace amines not only between the two serotonin sites but also among the compounds themselves. An exception to this is *p*-tyramine which was equally potent at both the 5-HT_1 and 5-HT_2 binding sites.

Tryptamine was the most potent amine at both sites. The IC_{50} values found at the 5-HT_1 site, 66.7 nM and the 5-HT_2 sites, 3.85 μM, are in reasonable agreement with values reported by Leysen (1982; Ki = 197 nM for the 5-HT_1 site of rat hippocampus and Ki = 5.42 μM for the 5-HT_2 site of rat cortex) and by Nelson *et al.* (1983; 176 nM for the IC_{50} for the 5-HT_1 site of rat cortex).

The IC_{50} values presented in Fig. 1 assume a straight line fit for all the data in the Hill plots. A close examination of the data for 2-phenylethylamine and *m*-tyramine has revealed the inhibition of the 5-HT_2 binding site to be multiphasic, with Hill slopes of approximately 0.90 for both compounds indicating multiple binding sites and a complex interaction of these trace amines with serotonin binding sites.

α-Methyltryptamine has not been identified as an endogenous trace amine in the central nervous system, but it was included in this study as a methyl analogue of tryptamine which is pharmacologically active (Dewhurst and Marley, 1965) and since it can exist as two optically active isomers. Significant differences in the IC_{50} values for (+) and (-) isomers were observed at both the 5-HT_1 and 5-HT_2 binding sites. This observation may indicate that the serotonin recognition sites exhibit some stereospecificity, although the differences observed are not very marked.

The IC_{50} values for serotonin binding sites by most trace amines were in the micromolar range and therefore it may be concluded that in normal circumstances when these compounds are in very low concentrations in the central nervous system a profound effect on serotonin binding sites should not be expected. Since large amounts of trace amines such as 2-phenylethylamine, m- and p-tyramine and tryptamine may accumulate after monoamine oxidase inhibition (Boulton and Juorio, 1982), the differential inhibition by the trace amines of the serotonin specific binding sites suggests that the trace amines may have different effects at specific membrane receptors for serotonin which will be apparent when the trace amines are present in larger than trace concentration. In addition, these interactions may be important in behavioural studies in which large amounts of these trace amines are administered to laboratory animals.

ACKNOWLEGEMENTS

Funds were provided by the Alberta Mental Health Research Fund and the Medical Research Council of Canada. The authors are grateful to Dr. R. G. Micetich and Mr. J. Pyrozko for isolation of the α-methyltryptamine isomers and to Mrs. J. van Muyden for expert technical assistance.

REFERENCES

Boulton A. A. and Juorio A. V. (1982) Brain trace amines, in Handbook of Neurochemistry (A. Lajtha, ed.), vol. 1, pp. 184-222. Plenum Press, New York.

Dewhurst W. G. and Marley E. (1965) The effect of alpha methyl derivatives of noradrenaline, phenylethylamine and tryptamine on the central nervous system of the chicken. Brit. J. Pharmacol. **3**: 682-704.

Glowinski J. and Iversen L. L. (1966) Regional studies of catecholamines in the rat brain. 1. The disposition of [^{3}H] norepinephrine, [^{3}H] dopamine and [^{3}H] dopa in various regions of the brain. J. Neurochem. **13**: 655-669.

Leysen J. E. and Tollenaere J. P. (1982) Biochemical models for serotonin receptors. Ann. Rept. Med. Chem. **17**:

1-10.

Locock R. A., Baker G. B., Coutts R. T. and Dewhurst W. G. (1984) Displacement of serotonin from binding sites in rat cortex: Effects of biogenic "trace" amines. Prog. Neuropsychopharmacol. & Biol. Psychiat. **8:** 701-704.

Martin L. L. and Sanders-Bush E. (1982) Comparison of the pharmacological characteristics of $5\text{-}HT_1$ and $5\text{-}HT_2$ binding sites with those of serotonin autoreceptors which modulate serotonin release. Arch. Pharmacol. **321:** 165-170.

Nelson D. L., Weck B. and Taylor W. (1983) Discrimination of heterogeneous serotonin receptors by indolealkylamines, in CNS Receptors - From Molecular Pharmacology to Behaviour (P. Mandel and F. V. DeFeudis, eds.), pp. 337-347. Raven Press, New York.

Scott J. A. and Crews F. T. (1985) Increases in $serotonin_2$ receptor density in rat cerebral cortex slices by stimulation of beta-adrenergic receptors. Biochem. Pharmacol. **34:** 1585-1588.

EFFECTS OF 2-PHENYLETHYLAMINE AND TYRAMINE ON CENTRAL NORADRENALINE AND DOPAMINE SYSTEMS: AN ELECTROPHYSIOLOGICAL STUDY

Lars Oreland[1], Per-Åke Lundberg[1] and Göran Engberg[2]

[1] Dept. of Pharmacology, University of Uppsala, Sweden

[2] Dept. of Pharmacology, University of Göteborg, Sweden

INTRODUCTION

Monoamine oxidase (MAO, E.C. 1.4.3.4.), which has been shown to exist in two forms (termed A and B), plays an essential role in the metabolism of tyramine and of 2-phenylethylamine (PEA). The two forms of the enzyme are defined according to their sensitivity to clorgyline; the A-form being highly sensitive and the B-form less sensitive. In contrast to clorgyline, FLA 336(+) is a reversible MAO-A inhibitor (1), while (-)deprenyl is an irreversible inhibitor with a high specificity for the MAO-B form (2). In most tissues tyramine is an equally well preferred substrate for both forms of the enzyme, while PEA is mainly a substrate for the B-form (3).

PEA is a sympathomimetic amine, which is produced endogenously from phenylalanine in mammalian tissues (4,5). It has been suggested to exert a neuromodulatory action in brain (6,7) and, mainly based on clinical data about its urinary excretion, PEA has been hypothesized to be involved in the etiology of several psychiatric disorders such as schizophrenia (8,9,10) and depression (11,12,13) as well as migraine (14,15) and stress (16).

In contrast to tyramine and catecholic biogenic amines, PEA is reported to easily penetrate the blood brain barrier (17). In rats, administration of PEA has been reported to induce amphetamine-like behaviour, e.g. increased locomotor activity and stereotypies and these effects are particularly evident after pretreatment with MAO-inhibitors (18,19). No such effects have been reported after administration of tyramine. Biochemical studies have revealed that PEA decreases the content of brain noradrenaline (NA) and dopamine (DA), probably via an amphetamine-like, catecholamine releasing action (20,21,22). A similar mechanism of action, however on peripheral noradrenergic nerves, is suggested to, at least in part, explain the hypertensive crisis as well as headachde and palpitations induced by tyramine if ingested after administration of MAO-inhibitors. Our knowledge about the putative effect of ingested tyramine on the central nervous system is poor.

In order to further examine the effect of PEA and tyramine on central NA and DA mechanisms, we have utilized single unit recording techniques to study, directly, the effects of the drugs on brain NA neurons in the nucleus locus coeruleus (LC), as well as on DA neurons in the substantia nigra (SN).

MATERIALS AND METHODS

Male sprague-Dawley rats, 200-300 g, were anaestetized with chloral hydrate (400 mg/kg, i.p., additional supplements of anaesthetics were given via a lateral tail vein) and placed in a stereotaxic instrument. A burr hole was drilled at the approximate coordinates of the LC (1.2 mm posterior to lambda and 1.1 mm lateral to the midline) and of zona compacta, SN (1.6-2.6 mm anterior to lambda and 1.6-2.7 mm lateral to midline). An electrode was then lowered into the LC or SN by means of a hydraulic microdrive. NA neurons in the LC and DAneurons in the SN were identified in vivo by their locations and by their electrophysiological characteristics (23, 24). Body temperature was maintained at 37° C by a thermostatically controlled heating pad. In all studies when intravenous route of administration was used, only one cell was studied in each animal.

Glass micropipettes filled with 2M NaCl, saturated with Fast Green (impedance 3-6 M Ohms at 135 Hz) were used to record extracellular single unit activity. For the microiontophoretic experiments a five-barrel micropipette was used. The central barrel was filled with 2M NaCl solution saturated with Fast Green and was used for measuring action potentials. One of the four side-barrels was filled with 4M NaCl solution for automatic current balancing and the others contained PEA-HCl (0.5 M, pH 4.0). The in vitro impedances were typically 2 - 4 M Ohms in the central barrel and 40-80 M Ohms in the side barrels. To avoid deadtime artefacts, ejections were performed frequently. During baseline recording, a retaining current of about 5 nA was maintained between ejections to avoid leakage of drugs from the pipett. Ejections of drugs were accomplished by passing of 10-50 nA through the selected barrel.

Following termination of each experiment, the location of the electrode was marked by iontophoretic ejection of Fast Green for at least 15 min. The rats were then perfused through the heart with 10% buffered formaline solution and subjected to conventional histological procedures. The position of the dye was then verified under a light microscope. Only cells within the LC or SN were included in the study.

RESULTS AND DISCUSSION

Locus coeruleus

Systemic injection of low doses of PEA produced an instantaneous and dose-dependent inhibition of firing rate of the LC neurons (ED_{50} = 1.0 mg/kg; Fig. 1). The effect was strongly antagonized by administration of the $alpha_2$-receptor antagonist yohimbine (1 mg/kg, i.v.) or by depletion of endogenous stores of NA by pretreatment with reserpine (10 mg/kg, i.p., 6 h), but unaffected by inhibition of tyrosine hydroxylase with alpha-methyl-p-tyrosine (250 mg/kg, i.p., 30 min; Fig. 2). In contrast, the inhibitory effect of PEA was potentiated by pretreatment with the relatively selective MAO-B inhibitor pargyline (2 mg/kg, 1 h), but also, unexpectedly, since PEA is deaminated by MAO-B (see ref.3), by pretreatment with the selective MAO-A inhibitors clorgyline (2 mg/kg, 1 h) or FLA 336(+) (2 mg/kg, 1 h) (see Fig. 2). This latter finding, which implicates that MAO-A inhibitor potentiate the effect of PEA via some mechansim other than increasing the PEA concentration, makes the clinical use of MAO-A inhibitors reasonable,even if the hypo-

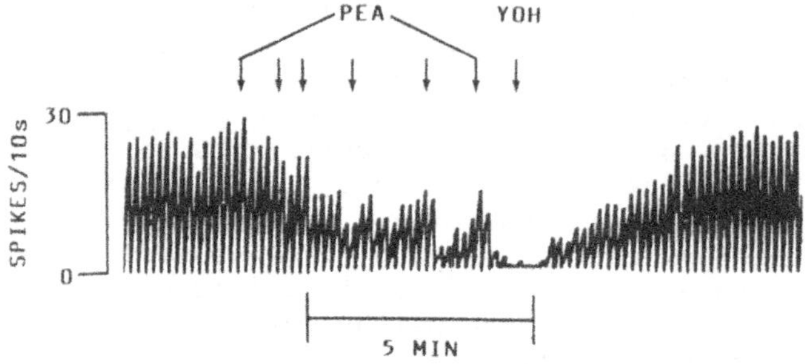

Fig. 1

PEA-induced inhibition of firing of a NA neuron in the LC (PEA, 0.3 + 0.3 + 0.6 + 1.2 + 2.4 + 4.8 mg/kg, i.v., at arrows). The effect is readily reversed by administration of the $alpha_2$-receptor antagonist yohimbine (YOH, 1 mg/kg).

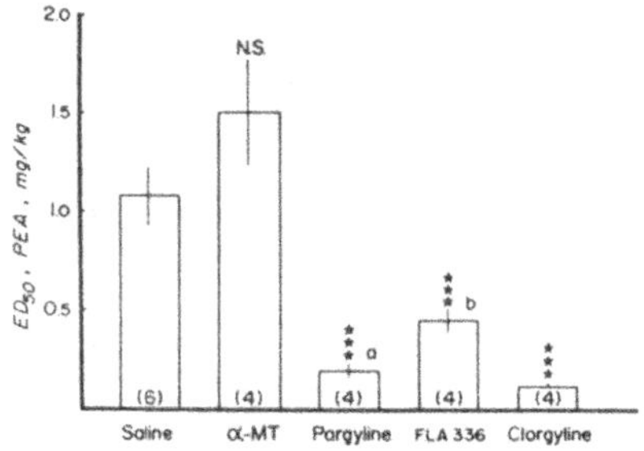

Fig. 2

Mean ED_{50} values (dose required to obtain a 50 per cent decrease in firing rate) for the inhibitory effect of intravenously administered PEA on LC noradrenergic neurons in control rats and in rats pretreated with alpha-methyl-p-tyrosine (250 mg/kg, i.p., 30 min), pargyline (2 mg/kg, i.p., 1 h), FLA 336 (2 mg/kg, i.p., 1 h) and clorgyline (2 mg/kg, i.p., 1 h). Number of cells recorded from are given in parenthesis for each group. For statistical calculations, one-way analysis of variance followed by t-test was used.
*** $p<0.001$, a) N.S. compared to clorgyline, b) $p<0.005$ compared to clorgyline

thesis of the involvement of PEA in the pathogenesis of e.g. depressive states (see Introduction) would be true. When microiontophoretically applied directly onto the LC neurons, PEA produced inhibition of a vast majority of the NA neurons tested (n = 19), an effect that was blocked by the simultaneous administration of the $alpha_2$-receptor antagonist yohimbine (data not shown).

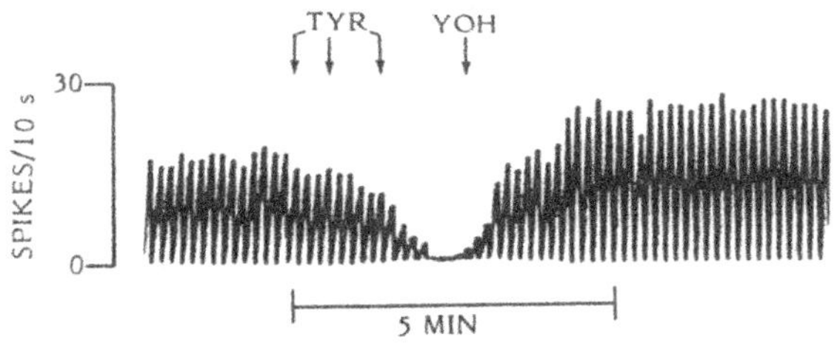

Fig.3

Inhibition of a NA cell in the LC by low doses of tyramine (TYR, 0.5 + 0.5 + 1 mg/kg, i.v., at arrows) following clorgyline pretreatment (10 mg/kg, i.p., 1 h). The effect is reversed by yohimbine (YOH, 1 mg/kg, i.v.)

In control rats, tyramine administration (up to 15 mg/kg, i.v.) caused a negligible inhibition of the LC units. However, in the presence of clorgyline or FLA 336(+), administration of tyramine caused a consistent and dose-dependent inhibition of firing rate of the LC neurons (Fig. 3,4). These results indicate that the ability of tyramine to affect central NA neurons is not mainly limited by its polarity, but rather by its rapid deamination by MAO. The inhibition by tyramine, as well as that produced by PEA, was reversed by yohimbine and prevented by depletion of endogenous stores of central NA (Fig. 3,4). Thus, the action of PEA and tyra-

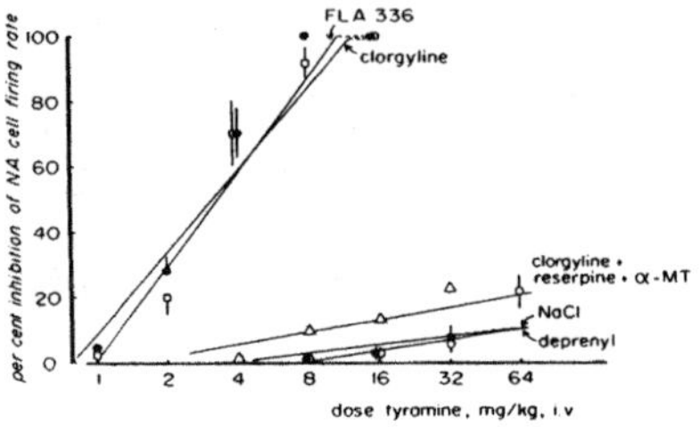

Fig.4

Dose-response curves for the inhibitory effect of intravenously administered tyramine on NA cells in the LC of saline control rats (n=4) and rats pretreated with clorgyline (10 mg/kg, i.p., 1 h) □ (n=4), FLA 336 ● (3 mg/kg, i.p. 1 h) (n=4), deprenyl (10 mg/kg, i.p., 1 h) ■ (n=4) and reserpine (10 mg/kg, i.p., 6 h) + alpha-methyl-p-tyrosine (250 mg/kg, i.p., 30 min) + clorgyline (10 mg/kg, i.p. 1 h) △ (n=4). Lines were plotted using linear regression analysis. Shown are the means +/SEM.

mine on NA neurons in the LC seems to be an indirect effect, requiring availability of a reserpine-sensitive storage pool of NA, and mediated via activation of central $alpha_2$-receptors within the LC.

PEA and tyramine containing foods are known to induce migraine (25,14,15), a disease widely accepted to be mediated via a cerebrovascular reaction (26). Increasing evidence indicate that central NA neurons, functionally and pharmacologically, resemble those of the peripheral nervous system, e.g. LC neurons have been shown to control cerebral blood flow (27) and to be involved in arousal reactions (28,29). Thus, the LC might be a primary target for PEA or tyramine to produce migraine attacks. The hypothesis is strengthened by the fact that propranolol and clonidine, both drugs known to affect LC firing rate following chronic treatment (30,31) is successfully used in the prophylactic treatment of migraine (32,33).

Substantia nigra

Systemic administration of PEA supressed the spontaneous firing rate of the DA containing neurons in a dose-dependent manner (ED_{50} 2.0 mg/kg; Fig.5). The effect was instantaneous in onset and was rapidly reversed by intravenous injection of the dopamine receptor antagonist haloperidol (0.1 mg/kg, i.v.). However, in contrast to the PEA-induced inhibition of central noradrenergic neurons, the dopamine cell inhibition was blocked by pretreatment with alpha-methyl-p-tyrosine (250 mg/kg, i.p., 30 min), but not with reserpine (10 mg/kg, i.p., 6 h). Thus, PEA requires the presence of newly syntetized dopamine rather than that granu-

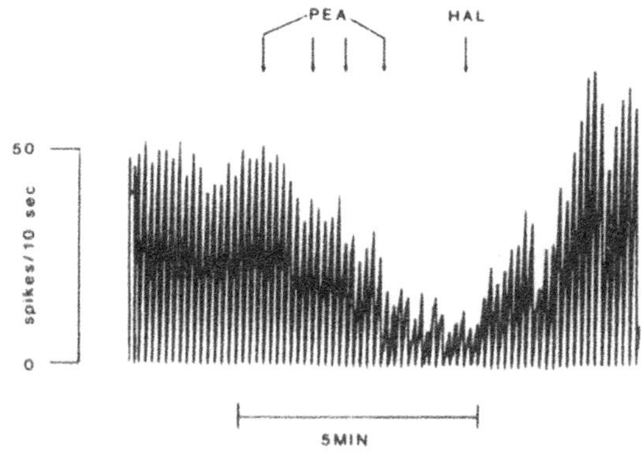

Fig. 5

Typical inhibition of a DA neuron in the SN by PEA (0.4 + 0.4 + 0.8 + 1.6 mg/kg i.v., at arrows). This action is readily antagonised by administration of the DA antagonist haloperidol (HAL, 0.1 mg/kg).

larly stored in order to inhibit the firing rate of dopaminergic neurons. This electrophysiological finding is in consonance with the view that there is a basic difference between dopamine and noradrenaline storage and release processes. In contrast to PEA, tyramine could not, even when the animals were pretreated with a variety of selective MAO-inhibitors, produce any inhibitory effect on nigrostriatal DA neurons. This is not likely to be explained by a low concentration of tyramine in the brain tissue as a consequence of a poor penetration through the blood brain barrier, since after pretreatment with MAO inhibitors, the spontaneous firing of noradrenergic neurons in the locus coeruleus was markedly depressed by tyramine administration (see above). Our results are in excellent agreement with a large body of behavorial evidence, indicating that PEA, but not tyramine, produces increased locomotor activity in rats.

SUMMARY AND GENERAL CONCLUSIONS

In the presence of selective MAO-A inhibitors tyramine suppressed LC neuronal activity, while inhibition of MAO-B by deprenyl did not display this tyramine promoting effect. The results suggest that the ability of tyramine to affect the CNS is not mainly limited by its poor penetration through the blood brain barrier, but rather by its rapid inactivation by MAO.

The inhibitory effect of PEA on the LC neurons was potentiated by various MAO inhibitors (A- as well as B-selective). Furthermore, the inhibition by tyramine and PEA was blocked by yohimbine (an $alpha_2$-receptor blocker) and by reserpine (depletion of endogenous NA). Thus, the inhibitory effect of PEA and tyramine on the LC firing rate seems to be carried out by release of granularly stored NA, and mediated via stimulation of presynaptic, inhibitory LC alpha-2 receptors.

Given morphological and functional evidence for the LC to control cerebral blood flow, we propose that the LC may be a primary site for PEA and tyramine to induce migraine attacks.

Tyramine, even administered in high doses (50 mg/kg) and in

the presence of a variety of MAO-selective inhibitors, did not affect the firing rate of nigrostriatal neurons. In contrast, low doses of PEA caused a marked inhibition of the firing of the DA neurons. These results are in excellent agreement with previous behavioural reports showing that PEA, but not tyramine increases locomotor activity in rats. (see Introduction).

PEA is likely to exert an amphetamine-like, catechol-releasing effect on SN DA-neurons, since it was prevented by alpha-MT and reversed by haloperidol. Moreover, as the effect of PEA on the nigrostriatal DA neurons was not abolished by reserpine pretreatment, it is likely that PEA, like amphetamine, requires the presence of newly syntezised DA rather than that granularly stored. Thus, the present finding is in consonance with the view that there is a basic difference between NA and DA storage and release processes.

REFERENCES

1. Ask, A.-L., Högberg, K., Schmidt, L., Kiessling, H. and Ross, S.B. (1982). 4-dimethylamino-alpha-methylphenylalkylamine (FLA 336), a selective inhibitor of the A form of monoamine oxidase in the rat brain. Biochem. Pharmacol. 31, 1401-1406.

2. Fowler, C.J., Callingham, B.A., Mantle, T.J. and Tipton, K.F. (1978). Monoamine oxidase A and B: a useful concept? Biochem. Pharmacol. 27, 97-101.

3. Fowler, C.J., Oreland, L. and Callingham, B.A. (1981). The acetylenic monoamine oxidase inhibitors clorgyline, deprenyl, pargyline and J-508: their properties and applications. J. Pharmac. Pharm. 33, 341-347.

4. Nakajima, T., Kakimoto, Y. and Sano, I. (1964). Formation of beta-phenyletylamine in mammalian tissue and its effect on motor activity in the mouse. J. Pharmacol. Exp. Ther. 143, 319-325.

5. Borison, R.L., Mosnaim, A.D. and Sabelli, H.C. (1974).

Biosynthesis of brain 2-phenylethylamine: Influence of decarboxylase inhibitors and d-amphetamine. Life Sci. 15, 1837-1848.

6. Sabelli, H.C. and Giardina, W.J. in Chemical Modulation of Brain Function, H. C. Sabelli, Ed. pp 225-259, Raven Press, New York (1973).

7. Boulton, A.A. (1979). Trace amines: Neurohumors (cytosolic, pre- and/or postsynaptic, secondary, indirect?) Behav. and Brain Sci. 2, 418.

8. Potkin, S.G., Karoum, F., Chuang, L.W., Cannon-Spoor, H.E. and Philips, I. (1979). Phenylethylamine in paranoid chronic schizophrenia. Science 206, 470-471.

9. Karoum, F., Potkin, S.G., Murphy, D.L. and Wyatt, R.J.in Non-Catecholic-Phenylethylamines, part 2, A. D. Mosnaim and M. E. Wolf, Eds. pp 177-191, Marcel Dekker, Inc., New York (1980).

10. Reynolds, G.P., Phenylethylamine and schizophrenia - clinical and pharmacological results. in Neurobiology of the Trace Amines, A. A. Boulton, G. B. Baker, W. G. Dewhurst and M. Sandler, Eds. pp 515-523, Humana Press, Clifton, New Jersey (1984).

11. Fischer, E., Heller, B. and Miro, A.N. (1968). Beta-phenylethylamine in human urine. Arzneim. Forsch. 18, 1486.

12. Fischer, E., Spatz, H. and Saavedra, J.M. (1972). Uninary elimination of phenethylamine. Biol. Psychiatry. 5, 139-147.

13. Murphy, D.L., Karoum, F., Alterman, I., Lipper, S. and Wyatt, R.J. Phenylethylamine, tyramine and other trace amines in patients with affective disorders: Association with clinical state and antidepressant drug treatment. in Neurobiology of the Trace Amines, A. A. Boulton, G. B. Baker, W. G. Dewhurst and M. Sandler, Eds. pp 499-514, Humana Press, Clifton, New Jersey (1984).

14. Sandler, M., Youdim, M.B.H. and Hannington, E. (1974). A phenylethylamine oxidising defect in migraine. Nature 250, 335-337.

15. Sandler, M. (1977). Transitory platelet monoamine oxidase deficit in migraine: Some reflectations. Headache 17, 153-158.

16. Paulos, M.A. and Tessel, R.E. (1982). Excretion of beta-phenylethylamine is elevated in humans after profound stress. Science 215, 1127-1129.

17. Oldendorf, W.H. (1971). Brain uptake of radiolabeled amino acids, amines and hexoses after arterial injection. Amer. J. Physiol. 221, 1629-1639.

18. Mantegazza, P. and Riva, M. (1963). Amphetamine-like activity of beta-phenylethylamine after a monoamine oxidase inhibitor in vivo. J. Pharm. Pharmac. 15, 472-478.

19. Moja, E.A., Stoff, D.M. and Gillin, C.J. (1976). Dose-response effects of beta-phenylethylamine on stereotyped behaviour in pargyline-pretreated rats. Biol. Psychiatry 11, 731-742.

20. Carlsson, A., Fuxe, K., Hamberger, B. and Lindqvist, M. (1966). Biochemical and histochemical studies on the effects of imipramin like drugs and amphetamine on central and peripheral catecholamine neurons. Acta Phys. Scand. 67, 481-497.

21. Fuxe, K., Grobecker, H. and Jonsson, G. (1967). The effect of beta-phenylethylamine on central and peripheral monoamine-containing neurons. Eur. J. Pharmacol. 2, 202-207.

22. Jackson, D.M. (1975). Some further observations of the effect of beta-phenylethylamine on locomotoractivity in mice. Arzneim. Forsch. 25, 622-626.

23. Aghajanian, G.K., Cedarbaum, J.M. and Wang, R.Y. (1977). Evidence for norepinephrine-mediated collateral inhibition of locus coeruleus neurons. Brain Res. 136, 570-577.

24. Grace, A. and Bunney, B. (1984). The control of firing pattern in nigral dopamine neurons: Single spike firing. J. Neurosci. 11, 2866-2876 (1984).

25. Hanington, E. (1967). Preliminary report on tyramine headache. Brit. Med. J. 2, 550-551.

26. Bruyn, G.W. The biochemical basis of migraine: A critique. In: Clinical Neuropharmacology, vol. 1, H.L. Klawans (Ed.) Raven Press, New York (1976).

27. Katayama, Y., Ueno, Y., Tsukiyama, T., Tsubokawa, T. (1981). Long-lasting suppression of firing of cortical neurons and decrease in cortical blood flow following train pulse stimulation of the locus coeruleus in the cat. (1981). Brain Res. 216, 173-179.

28. Aston-Jones, G. and Bloom, F.E. (1981a). Activity of norepinephrine-containing locus coeruleus neurons in behaving rats anticipates fluctuations in the sleep-waking cycle. J. Neurosci. 1, 879-886.

29. Aston-Jones, G. and Bloom, F.E. (1981b). Norepinephrine--containing locus coeruleus neurons in behaving rats exhibit pronounced responses to non-noxious environment stimuli. J. Neurosci. 1, 887-900.

30. Dahlöf, C., Engberg, G. and Svensson, T.H. (1981). Effects of beta-adrenoceptor antagonists on the firing rate of noradrenergic neurons in the locus coeruleus of the rat. Naunyn-Schmiedeberg's Arch. Pharmacol. 317, 26-30.

31. Engberg, G., Elam, M. and Svensson, T.H. (1982). Clonidine withdrawal: Activation of brain noradrenergic neurons with specifically reduced $alpha_2$-receptor sensitivity. Life Sci. 30, 235-243.

32. Weber, R.G. and Reinmuth, O.M. (1972). The treatment of migraine with propranolol. Neurology, 22, 366-369.

33. Shafar, J., Tallett, E.R. and Knowlson, P.A. (1972). Eva-

luation of clonidine in prophylaxis of migraine. Double-blind trial and follow-up. Lancet 1, 403-407.

FURTHER INVESTIGATIONS OF THE INTERACTION BETWEEN PHENYLETHYLAMINE AND DOPAMINE IN THE RAT NIGROSTRIATAL SYSTEM

P.S. McQuade, J.W. Richard and M. Thakur

Douglas Hospital Research Centre, Dept. of Psychiatry, McGill University

6875 LaSalle Blvd., Verdun, Québec H4H 1R3 Canada

INTRODUCTION

Dopamine (DA) metabolism proceeds by two parallel pathways - either through the acid metabolite dihydroxyphenylacetic acid (DOPAC) or through the minor amine 3-methoxytyramine (3-MT) (Westerink, 1979). Both of these metabolites are further processed to form homovanillic acid (HVA). This cascade of metabolites can be used in the following model - DA concentrations reflect synthesis, storage and reuptake of the transmitter while DOPAC concentrations reflect at least two processes, synthesis and reuptake. 3-MT concentrations are an index of the amount of DA released into the synaptic cleft (Carlsson and Lindqvist, 1963; Wood et al., 1982) while HVA concentrations represent the ultimate excretion product of DA metabolism (Roffler-Tarlov et al., 1971). The changes produced concomitantly in all three metabolites and any change in the precursor DA concentrations may accurately reflect the physiological status of dopaminergic neurotransmission.

The intraperitoneal administration of phenylethylamine (PE) increases the striatal concentrations of 3-MT and subsequently elevates DOPAC and HVA concentrations (McQuade and Wood, 1983). This pattern of metabolite concentration changes was identical to that observed after electrical stimulation (Wood et al., 1983) thus suggesting that PE acted by releasing dopamine. Electrophysiological studies suggested that PE potentiated the effects of DA in the caudate (Jones and Boulton, 1980). Previous synaptosomal and ventricular perfusion techniques using radiolabelled DA also suggested that a DA and PE interaction existed (Baker et al., 1976). Most recently Philips and Robson (1983) have observed that PE increases the concentrations of endogenous DA appearing in the perfusate from a push-pull cannula implanted in the caudate. All these techniques demonstrate that PE produces an effect on DA metabolism in the rat corpus striatum. Does PE excite the striatonigral neurons? In order to answer this question, we injected PE into both the striatum and the substantia nigra.

METHODS

Male Sprague-Dawley rats (Charles River Breeding Laboratories, St. Constant, Quebec; 170-190 g) were implanted 1 week prior to experimentation with 22 gauge guide cannulas. The striatal implantations were at A, 7.9; L, 2.6; V, 0.5 (Konig and Klippel, 1963). The injection volume was 1.0 microliters.

For the intranigral injections, the following coordinates were used: A, 2.2; L, 2.0; V, 2.2. Injection volume was 0.5 microliters (Wood and Richard, 1982). The animals were injected with 5, 10 and 25 µg of PE and microwaved fifteen min subsequent to the injections. The right and left (injected) caudates were analyzed separately using small tissue punches. Da, DOPAC, HVA and 3-MT concentrations were measured by GC-MS operating in the negative chemical ionization mode (Wood, 1982). The pentafluoropropionyl-hexafluoropropionyl diesters were separated chromatographically and the following masses were monitored (appropriate deuterated internal standard masses listed as well): DOPAC (d_0-463, d_5-468); HVA (d_0-330, d_5-334); DA (d_0-571, d_4-575) and 3-MT (d_0-311, d_4-315).

RESULTS

After fifteen min, the injection of 5 or 10 µg of PE into the corpus striatum of the rat resulted in a very large increase in 3-MT concentrations (to 480% of the control value or 1,180% (Table 1). DOPAC concentrations and HVA concentrations are increased from 36 to 45%. While the highest dose used (25 µg) also increased HVA and 3-MT concentrations these values are significantly decreased when compared to those produced by a dose of 10 µg of PE (Table 1). DA concentrations are also significantly reduced by this dose of PE.

Table 1

Effect of the Injection of PE into the Corpus Striatum on Dopamine Metabolism in the Rat

Left Side (Injected)

Sample	Metabolites (pmoles/mg protein) DOPAC	HVA	DA	3-MT
Controls	110± 8 (6)	110±9 (6)	970± 80 (6)	5±0.7 (6)
Phenylethylamine (15')				
5 µg	150±10* (6)	150±9* (6)	1000± 50 (6)	24±6* (6)
10 µg	157± 5*† (6)	160±4* (6)	1000± 50 (6)	59±8*† (6)
25 µg	130±10 (5)	120±6*† (5)	700±100* (5)	46±6*† (5)

All values are means ± S.E.M. with the number of animals in parentheses.

* $p < 0.05$ Dunnett's t-test.
† $p < 0.05$ Neuman Keul test.

In the contralateral side, only the 10 µg dose affected DA metabolism, increasing DOPAC concentrations (Table 2). The intranigral injection of PE did not release DA or affect DA metabolism fifteen min after injection (Table 3).

Table 2

Effect of the Injection of PE into the Corpus Striatum on Dopamine Metabolism in the Rat

Right Side (Non-Injected)

Sample	DOPAC	HVA	DA	3-MT
	Metabolites (pmoles/mg protein)			
Controls	110±20	93±10	940±100	5.5±0.9
Phenylethylamine (15')				
5 µg	120± 6	110± 9	1000± 50	10±3
10 µg	140± 4*	110± 4	1060± 40	6.1±1
25 µg	130± 8	110± 4	1080± 30	5.8±2

All values are means ± S.E.M.
* $p < 0.05$ Dunnett's t-test.

Table 3

Effect of the Injection of PE into the Substantia Nigra on the Ipsilateral Striatal Dopamine Metabolism

Sample	DOPAC	HVA	DA	3-MT
	Metabolites (pmoles/mg protein)			
Controls	60±3	110±3	390±20	2.7±0.3
Phenylethylamine (15')				
5 µg	27±2	110±4	410±30	3.1±0.3
10 µg	62±3	100±3	390±30	2.5±0.1

All values are means ± S.E.M. of seven animals.

DISCUSSION

The failure of the intranigral injection of phenylethylamine to stimulate the release of dopamine in the caudate nucleus suggests that PE does not excite the nigrostriatal neurons. Previous electrophysiological evidence also suggested that PE failed to stimulate the firing rates of these caudate neurons (Jones and Boulton, 1981). While receptors for PE in whole rat brain have been described, (Hauger et al., 1982) and the electrically stimulated release of PE from rat striatal slices has also been observed, (Niddam et al., 1985), it is possible that PE acts in an indirect manner as previously proposed (Fuxe et al., 1967). PE may act on receptors at the terminals of the nigrostriatal neurons that control the release of dopamine thus acting after release into the extracellular space. Alternatively since PE is a highly lipophillic compound (Oldendorf, 1971), PE may penetrate the cell membrane, enter the storage vesicles and displace dopamine (Philips and Robson, 1983). Another alternative is that PE alters the terminal membranes thus promoting fusion of the storage vesicles and release.

Nielsen et al (1983) described an inhibition of L-DOPA synthesis following a large intraperitoneal dose of 100 mg/kg of PE. As well they observed a decrease in the concentrations of DOPAC measured in their ventricular perfusions. They suggested that the inhibition of DA synthesis is due to blockade of DA reuptake, resulting in longer postsynaptic DA receptor activation. This mechanism may explain the increase in DOPAC concentrations in the contralateral corpus striatum as well. This increase may occur as a result of a compensatory increase in DA synthesis produced via a feedback loop possibly involving the substantia nigra and crossing over to the contralateral corpus striatum.

In an earlier experiment a stimulation of DA synthesis was observed 20 to 30 minutes following PE injection (McQuade and Wood, 1983). Nielsen et al (1983) also observed this stimulation of DA synthesis. They proposed this was due to the depletion of a "specific" intraneuronal DA pool which normally regulates tyrosine hydroxylase by end product inhibition. Rather than invoking two separate mechanisms it is possible that PE produces these dose effects by the same mechanism - a larger dose displaces more DA into this "specific" intraneuronal pool thus inhibiting DA synthesis since monoamine oxidase is overwhelmed by the higher amounts of PE .

In conclusion, PE at short times and low doses represents a "pure" releasing substance (as indicated by the elevated 3-MT concentrations); PE thus appears to regulate DA metabolism at the terminals of the nigrostriatal tract.

ACKNOWLEDGEMENTS

The authors wish to thank Mrs. J. Currie for her assistance in preparing the manuscript. Financial assistance was provided by the Medical Research Council of Canada and the Douglas Hospital Research Centre.

REFERENCES

Baker, G., Raitori, M., Bartollini, A., Del Carmine, R., Keane, P. and Martin, I. (1976) Interaction of β-phenylethylamine with dopamine and noradrenaline in the central nervous system of the rat. J. Pharm. Pharmacol. 28, 456-457.

Carlsson, A. and Lindqvist, M. (1963) Effect of chlorpromazine or haloperidol on formation of 3-methoxytyramine and normetanephrine in mouse brain. Acta Pharmacol. Toxicol. 20, 140-144.

Fuxe, K., Grobecker, H. and Jonoson, J. (1967) The effect of β-phenylethylamine on central and peripheral monoamine containing neurons. Eur. J. Pharmacol. 20, 202-207.

Hauger, R., Skolnick, P. and Paul, S.M. (1982) Specific [^{3}H]β-phenylethylamine binding sites in rat brain. Eur. J. Pharmacol. 83, 147-148.

Jones, R.S.G. and Boulton, A.A. (1980) Interactions between p-tyramine, m-tyramine or β-phenylethylamine and dopamine on single neurons in the cortex and caudate nucleus of the rat. Can. J. Physiol. Pharmacol. 58, 222-227.

Konig, J. and Klippel, R. (1963) The Rat Brain: A Stereotaxic Atlas of the Forebrain and Lower Parts of the Brain Stem. R.E. Krieger Publishing Co., Huntington, New York.

McQuade, P.S. and Wood, P.L. (1983) The effects of β-phenylethylamine on tyramine and dopamine metabolism. Prog. Neuro-Psychopharmacol. & Biol. Psychiat. 7, 755-759.

Niddam, R., Arbilla, S., Baud, P. and Langer, S.Z. (1985) [^{3}H]β-Phenylethylamine but not [^{3}H](+) amphetamine is released by electrical stimulation from perfused rat striatal slices. Eur. J. Pharmacol. 110, 121-124.
Nielsen, J.A., Chapin, D.S. and Moore, K.E. (1983) Differential effects of d-amphetamine, β-phenylethylamine, cocaine and methylphenidate on the rate of dopamine synthesis in terminals of nigrostriatal and mesolimbic neurons and on the efflux of dopamine metabolites into cerebroventricular perfusates of rats. Life Sci. 33, 1899-1907.
Oldendorf, H. (1971) Brain uptake of radiolabelled amino acids, amines and hexoses after arterial injection. Am. J. Physiol. 22, 1629.
Philips, S.R. and Robson, A.M. (1983) In vivo release of endogenous dopamine from rat caudate nucleus by phenylethylamine. Neuropharmacol. 22, 1297-1301.
Roffer-Tarlov, S., Sharman, D.F. and Tegerdine, P. (1971) 3,4-Dihydroxyphenylacetic acid and 4-hydroxy-3-methoxyphenylacetic acid in the mouse striatum: A reflection of intra- and extra- neuronal metabolism of dopamine? Br. J. Pharmacol. 42, 343-351.
Westerink, B. (1979) Further studies on the sequence of dopamine metabolism in the rat brain. Eur. J. Pharmacol. 56, 312-322.
Wood, P. (1982) A selected ion monitoring assay for dopamine and its metabolites using negative chemical ionization. Biomed. Mass Spectrom. 9, 302-306.
Wood, P.L. and Richard, J.W. (1982) Morphine and nigrostriatal function in the rat and mouse: Role of nigral and striatal opiate receptors. Neuropharmacol. 21, 1305-1310.
Wood, P., Nair, N.P.V. and Bozarth, M. (1982) Striatal 3-methoxytyramine as an index of dopamine release: Effects of electrical stimulation. Neurosci. Lett. 32, 291-294.

TRANYLCYPROMINE AND AN N-CYANOETHYL ANALOGUE: EFFECTS ON BRAIN LEVELS OF THE TRACE AMINES β-PHENYLETHYLAMINE AND TRYPTAMINE

Adil J. Nazarali, Glen B. Baker, Ronald T. Coutts, David R. Hampson, Tse W. Hall and Ronald G. Micetich.

Neurochemical Research Unit and PMHAC Research Unit, Department of Psychiatry and Faculty of Pharmacy and Pharmaceutical Sciences, University of Alberta, Edmonton, Alberta, Canada T6G 2G3

INTRODUCTION

Imbalances in the neuronal homeostasis of the trace amines β-phenylethylamine (PE) and tryptamine (T) have been implicated in the aetiology of certain neuropsychiatric disorders, including depression (Dewhurst, 1984 for review). Tranylcypromine (TCP), a monoamine oxidase (MAO) inhibitor used clinically as an antidepressant, has been shown to elevate substantially brain concentrations of these trace amines (Philips and Boulton, 1979). A number of N-alkylated derivatives of TCP have been synthesized in our laboratories to test them as potential prodrugs of TCP. The prodrug N-(2-cyanoethyl)tranylcypromine (CE-TCP) was found to be a potent MAO inhibitor in its own right (Baker et al., 1984). A neurochemical study is now reported in which the effects of TCP and CE-TCP on brain levels of the trace amines PE and T are compared.

MATERIALS AND METHODS

Equimolar doses (0.1 mmol/kg i.p.) of TCP and CE-TCP were administered to male Sprague-Dawley rats (220-250 g). Groups of rats were killed at 5, 15, 30, 60, 120 and 240 min after drug administration and the brains rapidly dissected out (the pineal gland and meninges removed). The brains were immediately frozen solid in isopentane on solid carbon dioxide and stored at -20°C until time of analysis.

Brain levels of PE were analysed after derivatization with trichloroacetic anhydride in aqueous medium (Baker, Nazarali and Coutts, submitted) or after derivatization with acetic anhydride in aqueous medium followed by anhydrous pentafluorobenzoylation (Hampson *et al.*, 1984). Brain levels of T were analysed after derivatization with acetic anhydride in aqueous medium followed by anydrous pentafluoropropionylation (Baker *et al.*, 1980). Analysis of PE and T was performed on a gas-liquid chromatograph equipped with a silica fused capillary column and an electron-capture detector.

RESULTS

Brain concentrations of PE and T after administration of TCP and CE-TCP are illustrated in Table 1.

β-Phenylethylamine

Brain levels of PE after administration of TCP are elevated rapidly, reaching a concentration of 84.2 ± 6.3 ng/g by 5 min. Brain PE levels continued to rise, with levels at 60 min significantly higher than the levels at 30 min ($p<0.01$, Student's *t*-test). The PE levels reached a maximum (191.8 ± 20.1 ng/g) at 120 min. At 240 min the brain PE levels had fallen to 91.6 ± 16.4 ng/g but were still significantly higher than the control value ($p<0.0001$).

Brain levels of PE after administration of CE-TCP showed a steady increase up to 30 min, the levels of PE being elevated significantly above control values at 5 min $p<0.001$). There was a marked elevation of PE levels between 30 and 60 min ($p<0.02$), followed by a plateau effect with no significant difference between values at 60 and 240 min.

Tryptamine

Brain levels of T also increased rapidly after administration of TCP. At 240 min T levels remained at a high of 87.5 ± 7.9 ng/g, representing an increase of approximately 175 times the control value, assuming a control level of 0.5 ng/g as found by Philips *et al.* (1974) employing high resolution mass spectrometry.

Brain levels of T also showed an immediate increase after administration of CE-TCP, with levels at 5 min significantly higher than those in the TCP-treated rats ($p<0.05$). Thereafter, however, the increases of T were not as large as those observed after treatment with TCP.

TABLE 1 Brain concentrations (ng/g) of PE and T in the rat after administration of TCP or CE-TCP.

CONTROLS	TCP	5 MIN	15 MIN	30 MIN	60 MIN	120 MIN	240 MIN
2.2 ± 0.2	PE	84.2 ± 6.3	84.6 ± 3.7	102.4 ± 8.6	169.4 ± 20.9	191.8 ± 20.1	91.6 ± 16.4
<1	T	4.6 ± 0.66	7.8 ± 1.6	26.0 ± 2.7	40.2 ± 4.4	68.1 ± 5.3	87.5 ± 7.9

CONTROLS	CE-TCP	5 MIN	15 MIN	30 MIN	60 MIN	120 MIN	240 MIN
2.2 ± 0.2	PE	11.6 ± 2.0	38.9 ± 7.0	50.0 ± 12.2	209.9 ± 66.9	217.4 ± 14.7	214.1 ± 29
<1	T	9.9 ± 0.9	8.1 ± 0.3	7.5 ± 0.9	17.4 ± 2.2	16.1 ± 1.1	20.1 ± 2.9

Results represent mean ± S.E.M. (N = 4-6). Drug dosage was 0.1 mmol/kg i.p.

DISCUSSION

The brain levels of PE and T (Table 1) were approximately 45 and 175 times the control values respectively at 240 min after administration of TCP. However, the concentration of PE had reached a maximum at 120 min (191.8 $\pm$ 20.1 ng/g) and its level had declined to 91.6 $\pm$ 16.4 ng/g, mean $\pm$ S.E.M. at 240 min. Similar findings have also been reported by Philips and Boulton (1979), but an explanation for the differential effect was not clarified. The reasons for the decline in PEA concentrations are not obvious. Maitre et al. (1976) have postulated the existence of two pools of MAO, the "bulk" mitochondrial MAO with a turnover of about 12 days (Goridis and Neff, 1971) and a smaller pool of MAO with a rapid turnover rate (12 to 18 hours). This smaller MAO pool has been claimed to be resistant to inhibition. The trend towards the decline of brain PEA would tend to support the hypothesis for the existence of a resistant pool of MAO. Since only PEA seems to be affected this leads one to postulate that perhaps this resistant pool of MAO is primarily MAO-B. T is a substrate for both MAO-A and MAO-B but its levels remain significantly above control at 240 min and do not show a decline with the passage of time (Table 1). It has been reported that administration of TCP causes an increase in brain concentrations of tryptophan (Grahame-Smith, 1971). Tryptophan is a precursor of T and may well significantly contribute to the elevated levels of T in the presence of inhibited MAO. The mechanism of this increase in brain tryptophan is unknown.

An interesting observation was made after administration of the prodrug CE-TCP. The decline of PE between 120 and 240 min observed with TCP was not apparent with this drug. Both PE and T remained elevated at 240 min, representing an increase of approximately 100 and 40 times the control values respectively. Although it is too early to offer a conclusive explanation of these observations (a longer time study needs to be conducted), the present study reveals some interesting data. It is possible that the small pool of MAO (Maitre et al., 1976) is inhibited by CE-TCP (unlike TCP), thereby preventing the decline of PE. A second possibility may be that since CE-TCP forms substantial amounts of TCP in brain (Nazarali et al., 1984), the two MAO inhibitors could have a synergistic effect. Finally, the presence of novel metabolites of CE-TCP with a different MAO inhibitory pattern cannot be excluded. Recently, Hampson (1984) identified the presence of *p*-

hydroxytranylcypromine (p-OH-TCP) in rat brain and urine after administration of TCP. This metabolite was found to possess MAO inhibitory activity although it differs somewhat from TCP with regard to its specificity towards the A and B forms of MAO. It is possible that high levels of p-OH-TCP and 4-hydroxy-N-(2-cyanoethyl)tranylcypromine (p-OHCE-TCP) could occur in brain after administration of CE-TCP. Since the prodrug CE-TCP did not have as large an effect as TCP on the levels of T, it is possible that CE-TCP has less effect on tryptophan levels or on amino acid decarboxylase activity than does TCP.

The above findings suggest that CE-TCP may have potential as an antidepressant drug and indicate that effects on trace amines should be considered in any future investigations of prodrugs of this type.

ACKNOWLEDGEMENTS

Financial support form the Alberta Heritage Foundation for Medical Research, the Medical Research Council of Canada, and the Provincial Mental Health Advisory Council (PMHAC) is gratefully acknowledged. The authors acknowledge the skilled typing of this manuscript by Miss Heather Schmidt.

REFERENCES

Baker G. B., Martin I. L., Coutts R. T., and Benderly A. (1980) Determination of 5-hydroxytryptamine in rat brain regions by gas-liquid chromatography with electron-capture detection. J. Pharmacol. Methods **3**: 173-179.

Baker G. B., Nazarali A. J., Coutts R. T., Micetich R. G. and Hall T. W. (1984) Brain levels of 5-hydroxytryptamine, tryptamine and 2-phenlethylamine in the rat after administration of N-cyanoethyltranylcypromine. Prog. Neuropsychopharamcol. & Biol. Psychiat. **8**: 657-660.

Dewhurst W. G. (1984) Trace Amines: The Early Years, in Neurobiology of the Trace Amines (A. A. Boulton, G. B. Baker, W. G. Dewhurst, and M. Sandler, eds.), pp. 3-12. Humana Press, Clifton, N.J.

Goridis C., and Neff N. H. (1971) Monoamine oxidase: an approximation of turnover rates. J. Neurochem. **18:** 1673-1682.

Grahame-Smith D. G. (1971) Studies in vivo on the relationship between brain tryptophan, brain 5-HT synthesis and hyperactivity in rats treated with a monoamine oxidase inhibitor and L-tryptophan. J. Neurochem. **18:** 1053-1066.

Hampson D. R. (1984) The monoamine oxidase inhibitor tranylcypromine: a comprehensive neurochemical study. Ph.D. Thesis. Univ. of Alberta, Edmonton, Alberta.

Hampson D. R., Baker G. B., and Coutts R. T. (1984) A rapid and sensitive method for quantitation of 2-phenylethylamine in brain tissue and urine. Res. Commun. Chem. Pathol. Pharmacol. **43:** 169-172.

Maitre L., Delini-Stula A., and Waldmeir P. C. (1976) Relationship between the degree of monoamine oxidase inhibition and some psychopharmacological responses to monoamine oxidase inhibitors in rats, in Monamine Oxidase and Its Inhibition, Ciba Foundation Symposium 39, pp. 247-270. Elsevier/Excerpta Medica/North-Holland, Amsterdam.

Nazarali A J. (1984) Analysis of bioactive arylalkylamines and their prodrugs in the mammalian central nervous system. Ph.D. Thesis. Univ. of Alberta, Edmonton, Alberta.

Philips S. R., and Boulton A. A. (1979) The effect of monoamine oxidase inhibitors on some arylalkylamines in rat striatum. J. Neurochem. **33:** 159-167.

Philips S. R., Durden D. A., and Boulton A. A. (1974) Identification and distribution of tryptamine in the rat. Can J. Biochem. **52:** 447-451.

IN VIVO RELEASE OF ENDOGENOUS DOPAMINE FROM RAT CAUDATE NUCLEUS BY PHENYLETHYLAMINE

S. R. Philips
Psychiatric Research Division
Cancer and Medical Research Building
University of Saskatchewan
Saskatoon, Saskatchewan
Canada S7N OWO

INTRODUCTION

Phenylethylamine (PE) is a normal constituent of rat and human brain (1,2). It is interesting because of its structural similarity to amphetamine, and because of its ability to induce stereotypic behaviors similar to those evoked by amphetamine (3). Although normally present in tissues in very small amounts, PE has a rapid turnover rate, and hence its concentration may be greatly increased by blocking its oxidative degradation with monoamine oxidase (MAO) inhibitors (4). PE is able to release dopamine (DA) from rat striatal prisms *in vitro* (5). When administered to animals in high doses, it depletes whole brain DA (6) and increases cerebrospinal fluid (CSF) levels of DA in the awake primate (7). In rats treated intraperitoneally with 50 mg/kg PE, striatal concentrations of the amine briefly rise to at least 200 µM (8), an amount known to increase the release of endogenous DA *in vivo* (9).

PE has generally been considered to exert its behavioral effects through DA-mediated mechanisms, although recent evidence indicates that some of its effects are mediated by serotonergic mechanisms (10).

The purpose of the experiments described here is to show that DA release can be significantly increased by amounts of PE which are present in the rat striatum under certain physiological conditions. The results provide direct evidence that some of the behavioral effects of PE may be mediated by DA. The effect of α,α-dideuterated PE on DA release is also examined and compared to that of the non-deuterated compound.

METHODS

Male Wistar rats (240-300 g) were anesthetized with urethane (1.20 g/kg, i.p.) and placed in a stereotaxic frame. A small hole 2 mm in diameter was drilled in the exposed skull and a push-pull cannula was inserted into the medial portion of the caudate nucleus (11). Two series of experiments were performed. In one, the tissue was perfused for 45 min with artificial CSF containing 10 μM pargyline to establish the unstimulated DA release rate; three 15-min samples were collected. The tissue was then perfused with PE (5, 0.5 or 0.05 μM) in the presence of 10 μM pargyline, and three 15-min samples were collected. Finally, the tissue was again perfused with CSF containing only 10 μM pargyline, and another three 15-min samples were obtained.

In the second set of experiments, nine 15-min collections were obtained from tissue perfused with artificial CSF alone. At the end of the third collection period, PE hydrochloride (5, 12.5, 25 or 50 mg/kg) or α,α-dideuterated PE hydrochloride (5, 12.5 or 25 mg/kg) was injected intracardially. In both series of experiments, tissue perfusates were collected into 0.1 ml of 0.6 M perchloric acid, and DA was determined radioenzymatically (11).

RESULTS

In the first series of experiments, DA release was increased to significantly greater than the resting rate (defined as the release rate measured in sample 3) by 5 μM, and briefly by 0.5 μM and 0.05 μM PE in the presence of 10 μM pargyline (Fig. 1). The release was also significantly increased by all except the lowest intracardially administered dose of PE. The effect lasted for 15 min following 12.5 mg/kg PE, and for 30 min following 25 or 50 mg/kg PE (Fig. 2). α,α-Dideuterated PE significantly stimulated DA release at all doses tested; following 5 or 12.5 mg/kg, the effect lasted for only 15 min, but after 25 mg/kg the

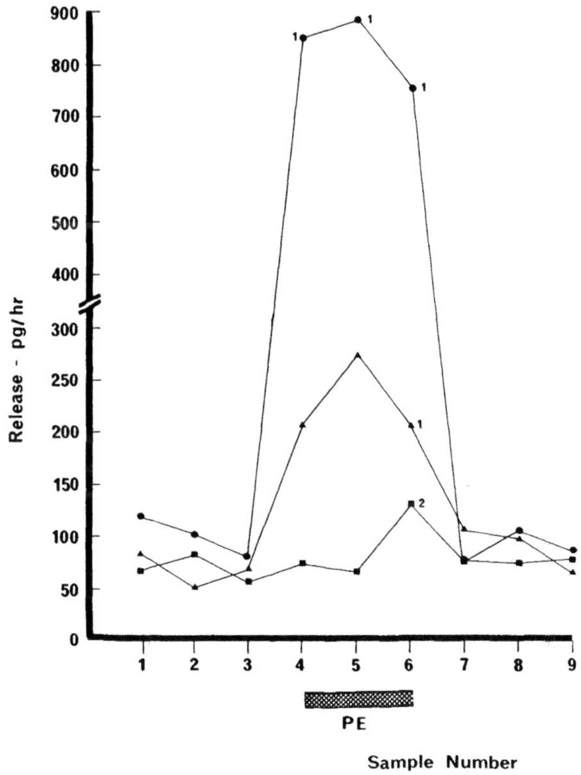

Figure 1. In vivo release of DA by PE at concentrations of 5μM (●——●), 0.5 μM (▲——▲) and 0.05 μM (■——■) in the presence of 10 μM pargyline. Points represent the mean value of 8-9 samples. Statistically significant differences from the unstimulated release rate are denoted by numbers: [1] $p < 0.05$; [2] $p < 0.01$ (Student's t-test).

release rate remained significantly above its resting rate for 45 min (data not shown). Although deuterated PE stimulated DA release to a greater extent than did the non-deuterated amine at all doses tested, the difference was statistically significant only at the 5 mg/kg dose ($p < 0.05$; Student's t-test).

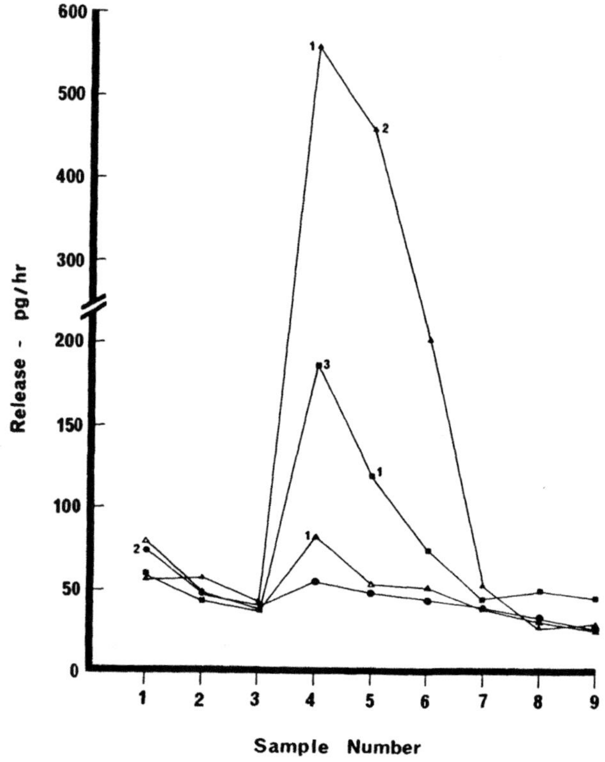

Figure 2. In vivo release of DA by intracardial injection of PE hydrochloride: 5 mg/kg (●——●); 12.5 mg/kg (△——△); 25 mg/kg (■——■); 50 mg/kg (▲——▲). Points represent the mean value of 10-13 samples. Statistically significant differences from the unstimulated release rate are denoted by numbers: [1] $p < 0.05$; [2] $p < 0.01$; [3] $p < 0.001$ (Student's t-test).

DISCUSSION

Assuming that the specific gravity of striatal tissue is unity, the PE concentrations used in the first series of experiments (5, 0.5 and 0.05 μM) correspond to tissue levels of about 600, 60 and 6 ng/g, respectively. Since amounts up to 200 ng/g are known to be present in the striata of rats treated with low to moderate doses of MAO inhibitors (12), it is likely that under certain conditions, physiologically occurring amounts of PE are able to cause a significant increase in DA release.

In the second series of experiments, the dose of PE required to stimulate the release of DA agreed closely with that required to elicit changes in behavior (13). Furthermore, the duration of the two effects was similar (13). Deuteration of the alkyl side chain in the PE molecule enhanced the ability of PE to stimulate DA release and to evoke behavioral effects (14), presumably by diminishing its rate of oxidation (15), and thereby effectively increasing its tissue concentration. The results of these experiments are consistent with the view that PE-induced behavioral changes may be mediated, at least in part, by an increase in DA release.

ACKNOWLEDGEMENTS

I thank Dr. A.A. Boulton for his interest in this work, Mr. H. Miyashita and Miss C.B. Leier for expert technical assistance, and the Medical Research Council of Canada and Saskatchewan Health, Province of Saskatchewan, for continuing financial support.

REFERENCES

1. Durden D.A., Philips S.R. and Boulton, A.A. (1973). Identification and distribution of β-phenylethylamine in the rat. Can. J. Biochem. 51, 995-1002.
2. Philips S.R., Rozdilsky B. and Boulton A.A. (1978). Evidence for the presence of *m*-tyramine, *p*-tyramine, tryptamine and phenylethylamine in the rat brain and several areas of the human brain. Biol. Psychiat. 13, 51-57.
3. Mantegazza P. and Riva M. (1963). Amphetamine-like activity of β-phenylethylamine after a monoamine oxidase inhibitor *in vivo*. J. Pharm. Pharmacol. 15, 472-478.
4. Durden D.A. and Philips S.R. (1980). Kinetic measurements of the turnover rates of phenylethylamine and tryptamine *in vivo* in the rat brain. J. Neurochem. 34, 1725-1732.

5. Baker G.B. and Yasensky D.L. (1981). Interactions of trace amines with dopamine in rat striatum. Prog. Neuro-Psychopharmacol. 5, 577-580.
6. Fuxe K., Grobecker H. and Jonsson J. (1967). The effect of β-phenylethylamine on central and peripheral monoamine containing neurons. Eur. J. Pharmacol. 2, 202-207.
7. Perlow M.J., Chiueh C.C., Lake C.R. and Wyatt R.J. (1980). Increased dopamine and norepinephrine concentrations in primate CSF following amphetamine and phenylethylamine administration. Brain Res. 186, 469-473.
8. Dyck L.E. (1984). The behavioral effects of phenelzine and phenylethylamine may be due to amine release. Brain Res. Bull. 12, 23-28.
9. Philips S.R. and Robson A.M. (1983). In vivo release of endogenous dopamine from rat caudate nucleus by phenylethylamine. Neuropharmacology 22, 1297-1301.
10. Dourish C.T. (1981). Behavioral effects of acute and chronic β-phenylethylamine administration in the rat: evidence for the involvement of 5-hydroxytryptamine. Neuropharmacology 20, 1067-1072.
11. Philips S.R., Robson A.M. and Boulton A.A. (1982). Unstimulated and amphetamine-stimulated release of endogenous noradrenaline and dopamine from rat brain in vivo. J. Neurochem. 38, 1106-1110.
12. Philips S.R. and Boulton A.A. (1979). The effect of monoamine oxidase inhibitors on some arylalkylamines in rat striatum. J. Neurochem. 33, 159-167.
13. Dourish C.T. (1984). Studies on the mechanisms of action of β-phenylethylamine stereotypy in rodents: implications for a β-phenylethylamine animal model of schizophrenia. In, Neurobiology of the Trace Amines (Boulton A.A., Baker G.B., Dewhurst W.G. and Sandler M., eds), pp. 389-411. Humana Press, Clifton, N.J.
14. Dourish C.T., Greenshaw A.J. and Boulton A.A. (1983). Deuterium substitution enhances the effects of β-phenylethylamine on spontaneous motor activity in the rat. Pharmacol. Biochem. Behav. 19, 471-475.
15. Yu P.H., Barclay S., Davis B.A. and Boulton A.A. (1981). Deuterium isotope effects on the enzymatic oxidative deamination of trace amines. Biochem. Pharmacol. 30, 3089-3094.

PHENYLETHYLAMINE AND 5-HYDROXYTRYPTAMINE FUNCTION

GAVIN P. REYNOLDS

Department of Pathology,
University of Nottingham Medical School,
Queen's Medical Centre, Nottingham NG7 2UH, U.K.

ABSTRACT

Phenylethylamine (PE) is implicated in a range of neurological and psychiatric disorders although in most cases the evidence for its involvement is at best circumstantial. However, PE does have profound behavioural effects on administration to experimental animals. Pharmacological studies indicate that the monoamine neurotransmitter systems may mediate these effects; recently 5-hydroxytryptamine (5-HT) has been proposed to have a particular role in this respect. PE, but not its methylated derivative amphetamine, appears to exhibit a direct agonist action on 5-HT systems.

The effects of PE on the displacement of ligand binding to $5\text{-}HT_1$ and $5\text{-}HT_2$ receptors using human brain tissue preparations are reported, and compared with the effects of amphetamines and some similar neuroactive drugs.

INTRODUCTION

Pharmacological investigations of PE have shown the amine to have many similarities to amphetamine, particularly in presynaptic effects on monoamine neurotransmitter function. There are differences, however. The more obvious difference is that PE is a good substrate for monoamine oxidase (MAO) and thereby quickly metabolised, whereas amphetamine is not. This necessitates the use of different dosage levels or the use of MAO inhibitors leading to difficulties in interpreting in vivo pharmacological data.

Past studies have suggested that PE (unlike amphetamine) has a direct effect on post-synaptic dopamine receptors (Jackson, 1975) while more recently an agonist action at 5-HT receptors has been proposed (Sloviter et al., 1980). Certainly a rigorous investigation of the various behavioural effects of PE on mice (Dourish, 1982) indicated all three monoamine neurotransmitters to be involved, 5-HT being particularly implicated in many of these behaviours. The reported selectivity of clozapine as an antagonist for certain PE (but not amphetamine) induced stereotypies has been proposed as providing an improved animal model for schizophrenia (Borison et al., 1977) but has been explained on the basis of clozapine's high affinity for 5-HT_2 receptors (Dourish and Boulton, 1982).

In the light of these reports, it was decided to study the effects of PE and some related compounds including the amphetamines on radioligand binding to 5-HT_1 and 5-HT_2 receptor sites. These sites are throught to reflect physiological receptors to 5-HT, the 5-HT_2 site being primarily post-synaptic and mediating the "5-HT behavioural syndrome", while the 5-HT_1 site may be presynaptic.

METHODS

Binding studies were carried out using human post-mortem frontal cortex tissue from which a crude membrane fraction was prepared by a simple homogenisation and centrifugation procedure in the appropriate buffer. 5-HT_1 sites were defined by (^{3}H)5-HT binding (displaced with cold 5-HT at 1 µM) at 4nM in 0.05M Tris Buffer at pH 7.7, containing 10mM $CaCl_2$. 5-HT_2 sites were defined by (^{3}H)ketanserin binding (displaced with 1µM LSD) at 1nM in 0.05M Tris at pH 7.4. In each case incubations were made at 37°C for 15 minutes. IC_{50} values were obtained as the mean of 2-5 experiments each performed in triplicate.

Table 1

Inhibition of specific (^{3}H) 5-HT binding

	IC_{50} (μM)
Phenylethylamine	14
(+) Amphetamine	195
(-) Amphetamine	52
Phenylacetate	> 1000
Propranolol	5.8
(+) Tranylcypromine	74
(-) Tranylcypromine	5.4
(+) Deprenyl	97
(-) Deprenyl	19
Clorgyline	32

Table 2

Inhibition of specific (^{3}H) ketanserin binding

	IC_{50} (μM)
Phenylethylamine	32
(+) Amphetamine	130
(-) Amphetamine	100
Phenylacetate	> 1000

RESULTS AND DISCUSSION

PE has a high affinity, relative to the amphetamines, in binding to both $5\text{-}HT_1$ and $5\text{-}HT_2$ receptors as defined by the binding of radioligands described above (tables 1 and 2). The concentrations in which PE is generally administered to animals (typically 10-100 mg/kg) is likely to produce brain concentrations of PE high enough to bind to both 5-HT receptor sub-type. The lack of any effect with phenylacetic acid, the major metabolite of PE, indicates that this compound is unlikely to contribute to any direct 5-HT receptor mediated action of PE *in vivo*.

The results for the action of the amphetamines and the related MAO inhibitors at the $5\text{-}HT_1$ site show clear isomeric differences with, in each case, a more powerful effect for the (-) isomer. Several of these compounds, including PE, approach the activity of propranolol which is one of the most active antagonists of 5-HT in terms of both *in vivo* and *in vitro* action. (-) Tranylcypromine in particular may well reach concentrations in the brain after administration of the racemate, which are adequate to affect $5\text{-}HT_1$ receptor function. This effect may contribute to the antidepressant effect of the racemic drug.

These results do not distinguish agonist and antagonist effects. Assessment of second messenger function (phosphatidyl inositol for $5\text{-}HT_2$, adenylate cyclase for $5\text{-}HT_1$) may provide a biochemical identification of possible 5-HT agonist effects of these compounds.

REFERENCES

Borison R. L., Havdala H. S. and Diamond I. (1977) Chronic phenylethylamine stereotypy in rats: a new animal model for schizophrenia. Life Sci. 21, 117-122.

Dourish C. T. (1982) A pharmacological analysis of the hyperactivity syndrome induced by β-phenylethylamine in the mouse. Brit. J. Pharmacol. 77, 129-139.

Dourish C. T. and Boulton A. A. (1982) Putative PEA receptors? Psychiatr. Res. 7, 387.

Jackson D. M. (1975) β-phenylethylamine and locomotor activity in mice. Arzneim. Forsch. 25, 622-626.

Sloviter R. S., Connor J. D. and Drust E. G. (1980) Serotonergic properties of β-phenylethylamine in rats. Neuropharmacol. 19, 1071-1074.

DISTRIBUTION OF CNS SITES SENSITIVE TO TRYPTAMINE AND SEROTONIN IN PAIN PROCESSING

Alice A. Larson

University of Minnesota
Department of Veterinary Biology
295 AnSci/Vet Med Building
St. Paul, Minnesota 55108 U.S.A.

ABSTRACT

Areas sensitive to the antinociceptive effect of intrathecally (IT) administered serotonin (5HT) were compared to sites most sensitive to the hyperalgesic effect of IT tryptamine (TA). Antinociception and hyperalgesia were determined by increases and decreases, respectively, in the latency of the tail-flick response to a radiant heat source. Drugs were injected via permanently indwelling IT cannulas. In general, the injection of TA alone into the thoracic level of the spinal cord produced the most intense hyperalgesia, while injection of 5HT produced the most pronounced antinociception in the lumbosacral area. Methysergide, which had no effect on pain-thresholds by itself, antagonized the antinociceptive effect of 5HT and potentiated TA-induced hyperalgesia in areas sensitive to 5HT-induced antinociception. The apparent hyperalgesic effect of TA is evident when applied to a variety of sites in the CNS, while the antinociceptive effect of 5HT is most effective in the area directly involved in the reflex. This suggests different mechanisms of interaction of these indoleamines with the neuronal substrates involved in nociception.

INTRODUCTION

Intrathecal injection of TA i.e. into the spinal subarachnoid space, causes an apparent hyperalgesia in rats [10]. In contrast, IT injections of 5HT cause a dose-related antinociception in the rat [9, 25], cat and primate [23], consistent with the hypothesis that manipulations which enhance serotonergic activity tend to potentiate analgesia [16].

In contrast to the antinociceptive effect of 5HT, which is sensitive to inhibition by 5HT antagonists [25], the decrease in latency of response to pain after TA not only fails to be inhibited, but is potentiated by pretreatment with methysergide [10]. Although antinociception can be elicited by injection of high doses of TA, this response readily converted to hyperalgesia when preceeded by methysergide. Thus it would appear that the antinociception produced by TA results from an interaction of TA with serotonergic receptors, while TA-induced decreases in tail flick latency, which will be referred to here as hyperalgesia, result from an action on a population of receptors distinct from those producing analgesia [10].

The existence of high affinity binding sites for (^{3}H)-TA, which are distinct from 5HT receptor binding sites, have been demonstrated in the rat CNS [2, 8, 15]. These data are consistent with the postulate that a tryptaminergic system exists in the CNS [13, 14, 19]. The ability of exogenously administered TA and 5HT to effect pain-reflex activity [10] would further suggest an involvement of these two indoleamines in the modulation of pain-pathways.

The present study was designed to investigate the anatomical distribution of sensitivities to TA and 5HT as modulators of pain-perception. The degree of hyperalgesia induced by TA in various areas of the brain and spinal cord were compared to the antinociceptive effects of 5HT and to TA's effect after pretreatment with methysergide.

METHODS

Male Sprague-Dawley albino rats (375-425 g) were fed ad libidum. Rats were anesthetized with ether during cannulation and then allowed a minimum of 7 days recovery from surgery. Animals were chronically implanted with either IT cannulas, by the method of Yaksh and Rudy [24], or intracerebroventricular (ICV) cannulas. IT cannulas were constructed of various lengths of PE10 tubing and inserted into the spinal subarachnoid space via the cisterna magna. ICV cannulas, constructed of PE20 tubing with stainless steel plugs, were placed to allow injections into the lateral ventricle and thus inserted to a depth of 3 mm through a hole in the skull which was drilled 1.5 mm lateral to the midline and 0.5 mm caudal with respect to bregma.

After each cannulation, an injection of saline was made to clear the catheter of accumulated debris.

Drug solutions were prepared fresh daily. Drugs administered IT were injected in a maximum volume of 20 μl, followed by a 12 μl flush with sterile water. Tryptamine HCl and serotonin oxalate were both purchased from Sigma (St. Louis, MO). Methysergide bimaleate was a gift from Sandoz Pharmaceuticals (Hanover, NJ).

Analgesia and hyperalgesia were measured using the tail-flick assay of D'Amour and Smith [4], using a 10 sec cutoff (heat lamp set at 50V AC). Tail-flick times were determined 5 min prior to drug injections and again at 2, 5, 10 and 30 min after injections. The average post-drug latency was then expressed as a percent of the group's average pre-drug control latency. Values significantly greater than 100% control values reflect analgesia and those less than 100% reflect hyperalgesia. Significance was determined using Student's t-test for dependent measures ($P < 0.05$). IT injection of water, equal to the largest volume of drug solution used IT, produced no change in reaction times measured at 2, 5, 10 or 30 min compared to the control values.

RESULTS

As reported previously for the caudal spinal area [10], IT injections of 200 μg/rat of TA onto various regions of spinal cord produced an immediate and marked decrease in the latency of response to a noxious thermal stimulus. The most intense hyperalgesia was seen 5 min after injection of TA via a cannula 2.5 cm in length (Fig. 1B) which corresponds to the upper thoracic level. Of those sites tested, using cannulas ranging in length from 1.0 to 8.6 cm, the rostral segments of the spinal cord generally appeared to be more sensitive to the hyperalgesic effect of TA, as indicated by decreased reaction times after injection via 1.0 to 5.5 cm long cannulas. In contrast, the lumbosacral area of the spinal cord was least sensitive to TA-induced hyperalgesia.

IT injections of 200 μg/rat of 5HT increased the tail flick latency at 5 min (Fig. 1A). When injected various distances from the cisterna magna, the distribution of sensitivities to 5HT-induced analgesia was distinct from that of TA-induced hyperalgesia. The lumbosacral area was most sensitive to the antinociceptive effect of 5HT, as indicated by an increase in the reaction time 5 min after 5HT was injected via 7.0 cm long cannulas. More rostral por-

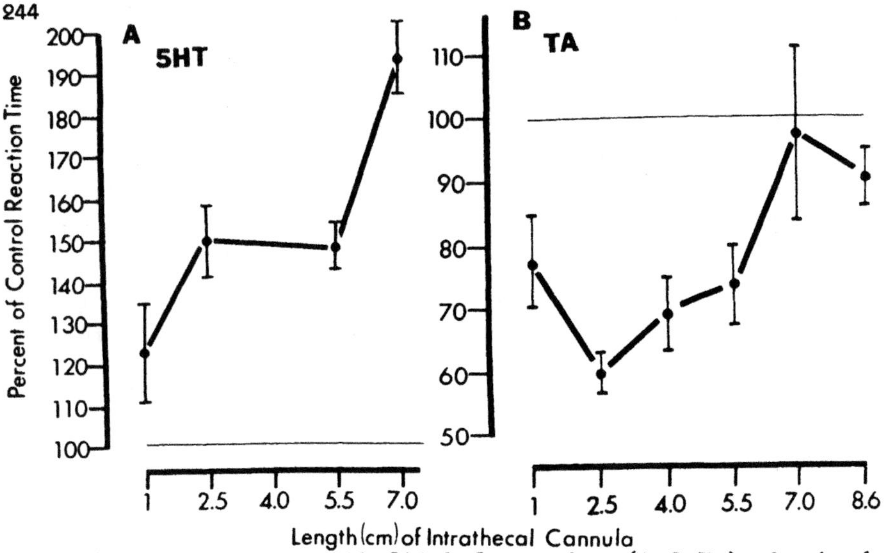

Fig. 1. The average tail flick latencies (± S.E.) obtained from groups of 5 to 9 rats 5 min after IT injection of 200 μg/rat of 5HT various distances from the cisterna magna, shown in A or TA shown in B. Latencies are expressed as the percent of each group's average control (pre-drug) reaction time. Hyperalgesia is defined as values significantly less than 100% ($p < 0.05$) while analgesia is defined as values significantly greater than 100%.

tions of the cord were only moderately effected by 5HT, such that injections 2.5 and 5.5 cm from the cisterna magna elicited an analgesic effect one half the magnitude of that obtained at 7.0 cm, while injections of 5HT just 1.0 cm from the cisterna magna failed to produce any significant antinociception.

IT injections of methysergide in the lumbosacral area not only fail to inhibit the hyperalgesic effect of TA, but potentiate its hyperalgesic activity. The present study shows that the degree of hyperalgesia produced by TA varies, depending on the level in the spinal cord into which it is injected (Fig. 1B).

The time course of TA's hyperalgesic effect after injection of 200 μg/rat at three different areas of the cord is shown in figure 2. In spite of the variability from one spinal area to another in the degree of hyperalgesia after TA alone, 5 min after 50 μg/rat of methysergide IT, TA produced very similar effects on reaction times in each area of the cord. Progressively greater degrees of potentiation of hyperalgesia were achieved in more caudal areas of

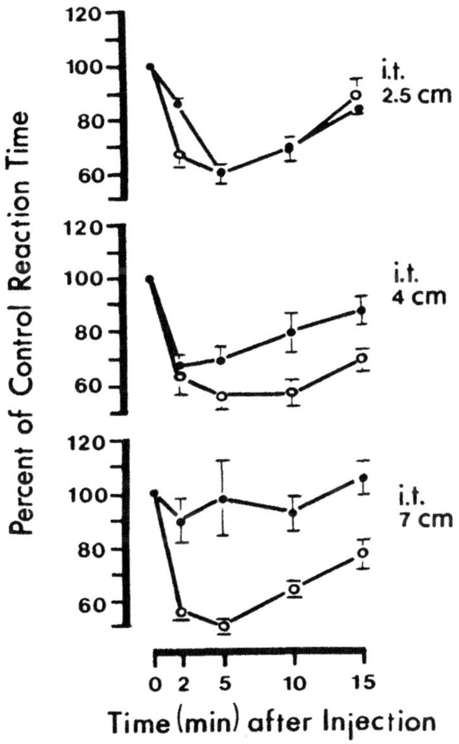

Fig. 2. Each point represents the average tail-flick latency, expressed as a percent of the pre-drug reaction time (± S.E.), for groups of 8 to 10 rats at various times after injection of 200 µg/rat of TA alone (closed circles) and after identical injections 5 min after 50 µg/rat of methysergide IT (open circles) at three different levels of the spinal cord.

the spinal cord where TA alone was least effective but 5HT alone was maximally antinociceptive.

While 5HT causes analgesia when administered ICV [21], our results indicate that 200 µg TA/rat injected ICV decrease the tail-flick latency at 5 min after its injection (Fig. 3). Pretreatment with 50 µg of methysergide ICV did not significantly alter the magnitude of this TA-induced hyperalgesia. Thus the effect of TA supraspinally, either alone or after methysergide, appears to be similar to that elicited from the thoracic area of the spinal cord.

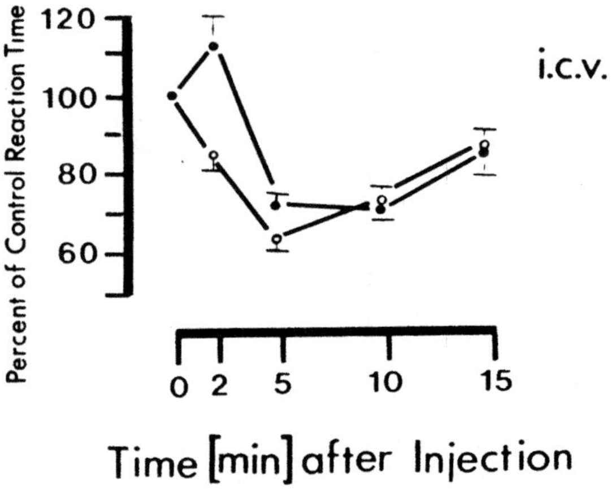

Fig. 3. Each point represents the average tail-flick latency, expressed as a percent of the pre-drug reaction time (± S.E.), for groups of 8 and 9 rats at 5, 10 and 15 min after an i.c.v. injection of 200 µg/rat of TA into the lateral ventricle (closed circles) and after identical injection of TA 5 min after 50 µg/rat of methysergide i.c.v. (open circles).

DISCUSSION

Applied IT onto the lumbosacral area of the spinal cord, 5HT produces analgesia [9, 23, 25], consistent with the majority of evidence describing 5HT's role in pain and analgesia [16]. Although TA differs from 5HT by only one hydroxyl group, low doses of TA applied IT onto the lumbosacral level of the spinal cord, result in modest hyperalgesia as measured by both tail-flick and hot plate assays [10]. This is consistent with the TA-induced enhancement of the C-fiber reflex [1], an electrophysiological reflection of activity along spinal pain-pathways.

The present study demonstrates that the apparent hyperalgesia produced from IT injections of TA differs significantly in intensity, depending on the site of injection along the spinal cord (Fig. 1). The high sensitivity of the more rostral levels of the cord lends a greater physiologic significance to the low concentration of TA found endogenously [18].

Given ICV, 5HT causes antinociception when measured by a procedure which elicits vocalization [21], but not when measured by a foot-pinch technique [20]. In contrast, ICV injection of TA produces a hyperalgesic effect, as measured by the tail-flick assay (Fig. 3). Our previous results indicate that IT TA caused hyperalgesia in the hot plate assay [10] which requires supraspinal circuitry for its behavioral cutoff. These data indicate that TA-induced hyperalgesia, in contrast to 5HT-induced analgesia, does not appear to be limited to direct actions on pain-pathways at the spinal cord level, that TA increases pain-perception rather than merely increasing spinal reflex activity [11, 12, 17, 22], and that TA can effect pain processing by an action in a variety of areas within the CNS.

It is possible that TA produces greater hyperalgesia when administered more rostrally than when administered more caudally due to the differential distribution of TA-binding in CNS [15]. However, after methysergide, TA's hyperalgesic effect did not vary throughout the spinal cord, indicating that TA interacts with serotonergic as well as tryptaminergic receptors, attenuating the hyperalgesic effect of TA. Methysergide simply unmasks tryptamine's full hyperalgesic effect.

In conclusion, just as the effects of TA and 5HT differ in other systems [3, 6], their effects on nociception also appear to differ. Areas in the CNS most sensitive to each of these indoleamines, with respect to pain-perception, are not identical. This correlates well with studies indicating a difference in the high affinity binding sites with which these two compounds interact [8], in the distribution of their binding sites in the CNS [2, 15], and in the distribution of TA and 5HT themselves [5, 7].

ACKNOWLEDGEMENTS
This work was supported by a PMAF Starter Grant and NIH Grant NS17407.

REFERENCES

1. Bell, J.A. and Martin, W.R. (1974) Studies of tryptamine and lysergic acid diethylamide (LSD) on cutaneous C-fiber and polysynaptic reflexes in the cat, J. Pharm. exp. Ther. 190, 492-500.
2. Cascio, C.S. and Kellar, K.J. (1983) Characterization of [^{3}H]tryptamine binding sites in brain, Eurp. J. Pharm. 95, 31-39.
3. Cox, B., Lee, T.F. and Martin, D. (1981) Different hypothalamic receptors mediate 5-hydroxytryptamine- and tryptamine-induced core temperature changes in the rat, Brit. J. Pharm. 72, 477-482.

4. D'Amour, F.E. and Smith, D.L. (1941) A method for determining loss of pain sensation, J. Pharm. exp. Ther. 72, 74-79.
5. Hadjiconstantinou, M., Paunla, P., Lackovic, Z. and Neff, N.H. (1984) Spinal cord serotonin: A Biochemical and immunohistochemical study following transection, Brain Res. 322, 245-254.
6. Jones, R.S.G. (1981) A comparison of cortical neurone and 5-hydroxytryptamine (5-HT), Neuroscience 7, 256.
7. Juorio, A.V. and Durden, D.A. (1984) The distribution and turnover of tryptamine in the brain and spinal cord, Neurochemical Res. 9, 1283-1293.
8. Kellar, K.J. and Cascio, C.S. (1982) [^{3}H]tryptamine: high affinity binding sites in rat brain, Eurp. J. Pharmacol. 78, 475-478.
9. Larson, A.A. and Armstrong, M.J. (1980) Morphine analgesia after intrathecal administration of a narcotic agonist, chloroxymorphine and antagonist, chlornaltrexamine, Eurp. J. Pharmacol. 68, 25-31.
10. Larson, A.A. (1983) Hyperalgesia produced by the intrathecal administration of tryptamine to rats, Brain Res. 265, 109-117.
11. Marley, E. and Vane, J.R. (1967) Tryptamines and spinal cord reflexes in cats, Brit. J. Pharmacol. 31, 447-465.
12. Martin, W.R. and Eades, C.G. (1970) The action of tryptamine on the dog spinal cord and its relationship to the agonistic actions of LSD-like psychotogens, Psychopharmacolgia (Berl.) 17, 242-257.
13. Martin, W.R., Sloan, J.W., Buchward, W.F. and Clements, T.H. (1975) Neurochemical evidence for tryptaminergic ascending and descending pathways in the spinal cord of the dog, Psychopharmacologia (Berl.) 43, 131-134.
14. Martin, W.R., Thompson, J.A. and Nozaki, M. (1976) Physiologic evidence for descending tryptaminergic pathways in the spinal cord, Life Sci. 19, 1383-1386.
15. McCormack, J.K., Beitz, A.J. and Larson, A.A. (1983) ^{3}H-tryptamine binding in the brain and spinal cord: in vitro autoradiography. Neuroscience 9, 333.
16. Messing, R.B. and Lytle, L.G. (1977) Serotonin-containing neurons: their possible role in pain and analgesia, Pain 4 (1977) 1.
17. Nozaki, M., Bell, J.A., Vaupel, D.B. and Martin, W.R. (1977) Responses of the flexor reflex to LSD, TA, 5-hydroxytryptamine, methoxamine and d-amphetamine, Psychopharmacologia (Berl.) 55, 13.

18. Philips, S.R., Durden, D.A. and Boulton, A.A. (1974) Identification and distribution of tryptamine in the rat, Can. J. Biochem. 52, 447-451.
19. Sloan, J.W., Martin, W.R., Clements, T.H., Buchwald, W.F. and Bridges, S.R. (1975) Factors influencing brain and tissue levels of tryptamine: species, drugs and lesions, J. Neurochem. 24, 523-532.
20. Sparkes, C.G. and Spencer, P.S.J. (1971) Antinociceptive activity of morphine after injection of biogenic amines in the cerebral ventricles of the conscious rat, Brit. J. Pharmacol. 42, 230-241.
21. Tamayo, L. and Contreras, E. (1965) Effect of 5-hydroxytryptophan on the analgesic actions of morphine in rats treated with morphine, guanethidine and tolazoline, Arch. Biol. Med. Expts. 2, 70.
22. Vane, J.R., Collier, H.O.J., Corne, S.J., Marley, E. and Bradley, P.B. (1961) Tryptamine receptors in the central nervous system, Nature 191, 1068-1069.
23. Yaksh, T.L. (1978) Analgesic actions of intrathecal opiates in cat and primate, Brain Res. 153, 205-210.
24. Yaksh, T.L. and Rudy, T.A. (1976) Chronic catherization of the spinal subarachnoid space, Physiol. Behav. 17, 1031-1036.
25. Yaksh, T.L. and Wilson, P.R. (1979) Spinal serotonin terminal system mediates antinociception, J. Pharm. exp. Ther. 208, 446-453.

TYROSINE, TYRAMINE AND DOPAMINE CONCENTRATIONS AFTER ELECTROLYTIC LESIONS OF MIDBRAIN RAPHÉ NUCLEI

A. V. Juorio and A. J. Greenshaw
Psychiatric Research Division
Cancer and Medical Research Building
University of Saskatchewan
Saskatoon, Saskatchewan
Canada S7N OWO

The existence of ascending 5-hydroxytryptamine (5-HT) containing cell bodies that project from the raphé nuclei to the striatum has been established for a number of years (Dahlström and Fuxe, 1964). More recently it has been shown that ascending 5-HT containing neurons exert a negative tonic effect on striatal dopamine (DA) turnover (Dray, Gonye, Oakley and Tanner, 1976). This pathway possibly forms part of a regulatory system for dopaminergic neurotransmission; in accord with this view, it has been reported that unilateral lesions of the dorsal raphé result in a reduction in ipsilateral striatal 5-HT and a concomitant increase in 3,4-dihydroxyphenylacetic acid (DOPAC) and homovanillic acid (HVA) that were observed thirty days after lesioning (Nicolaou, Garcia-Munoz, Arbuthnott and Eccleston, 1979). In addition, pharmacological evidence supports the existence of a reciprocal relationship between striatal DA turnover and *p*- and *m*-tyramine (*p*-TA and *m*-TA) concentrations; the direction of these changes are such that increases in DA turnover are accompanied by a reduction in *p*-TA and an increase in *m*-TA and changes in the opposite direction occur if DA turnover is reduced (see Juorio, 1982, for a review). These data suggest that if an increase in DA turnover occurs in response to raphé lesioning, it should be accompanied by a reduction in *p*-TA and no change or an increase in *m*-TA.

The present experiments were directed to an analysis of the effects of combined dorsal and median raphé nuclei lesions on striatal *p*-tyrosine, *p*-TA, *m*-TA, DA, DOPAC and HVA.

METHODS

In order to test the above hypothesis we investigated the effects of electrolytic lesions (1 mA, 10s, anodal) of the median and dorsal raphé nuclei of Wistar rats (Juorio and Greenshaw, 1985), on the striatal concentrations of DA, DOPAC, HVA, *p*-tyrosine, 5-HT and 5-hydroxyindoleacetic acid (5-HIAA), *p*-TA and *m*-TA; the assays were performed 1 and 7 days after lesioning. The concentrations of *p*-tyrosine, DA, DOPAC, HVA, 5-HT and 5-HIAA were measured by high performance liquid chromatography coupled with electrochemical detection (Saller and Salama, 1984) as already described (Juorio and Greenshaw, 1985). The *p*- and *m*-isomers of TA were estimated by the high resolution mass spectrometric selected ion monitoring (integrated ion current) technique using deuterated *p*- and *m*-TA as an internal standard. Complete details concerning these procedures have been described (Philips, Durden and Boulton, 1974; Philips, Davis, Durden and Boulton, 1975).

RESULTS

Electrolytic lesions of the median and dorsal raphé nuclei induced a statistically significant reduction in rat striatal *p*-TA (to 64% of controls), an effect that was observed one day after the lesion was made (Table 1). At this time striatal concentrations of *m*-TA, DOPAC and HVA were significantly increased (to 153%, 118% and 132% of their respective controls; Table 1). One week later the concentrations of the *p*- and *m*-isomers of tyramine, DOPAC and HVA had returned to their control values (Table 1).

The rat striatal concentrations of 5-HT were reduced at 1 or 7 days following the dorsal and median raphé nuclei lesions (to 69% or 58% of their respective controls) while the concentrations of 5-HIAA were reduced at 7 days (to 36% of controls) (Table 1). The raphé nuclei lesions resulted in no significant changes in rat striatal *p*-tyrosine (Table 1).

Table 1

The effects of dorsal and median raphé nuclei (RN) lesions on rat striatal *p*-tyrosine (*p*-TYR), *p*-tyramine (*p*-TA), *m*-tyramine (*m*-TA), dopamine (DA), 3,4-dihydroxyphenylacetic acid (DOPAC), homovanillic acid (HVA), 5-hydroxytryptamine (5-HT) and 5-hydroxyindoleacetic acid (5-HIAA)

	1 day		7 day	
	Sham lesion $ng\ g^{-1}$	RN lesion $ng\ g^{-1}$	Sham lesion $ng\ g^{-1}$	RN lesion $ng\ g^{-1}$
p-TYR	14,410	11,711	12,270	12,477
p-TA	11.1	7.10*	9.90	11.4
m-TA	3.40	5.20*	4.00	3.90
DA	10,390	10,107	9,728	9,980
DOPAC	1,050	1,240*	950	1,000
HVA	760	1,000*	610	660
5-HT	690	480*	720	350*
5-HIAA	850	722	800	290*

Values are means in $ng\ g^{-1}$ of fresh tissue. The star indicates statistical significance of the results with respect to the corresponding sham operated controls.

DISCUSSION

The present results further confirm that lesions of the midbrain raphé nuclei result in increased DA turnover; the effects, however, were confined to the first day after the lesions were made and subsided after a week (Table 1). Other researchers have reported that unilateral lesions of the median raphé nucleus result in a significant increase in striatal DOPAC and HVA that may be observed at 30 days after lesioning (Nicolaou, Garcia-Munoz, Arbuthnott and

Eccleston, 1979). Furthermore, increases in striatal DA turnover have been observed at 7-14 days after lesioning the median raphé nucleus (Dray, Gonye, Oakley and Tanner, 1976). The differences between the results of the present experiments (Table 1) and those reported by other groups (Nicolaou, Garcia-Munoz, Arbuthnott and Eccleston, 1979; Dray, Gonye, Oakley and Tanner, 1976) may be attributed to the differential extent of lesions of the raphé nuclei.

The increases in DA turnover induced by raphé nuclei lesions were accompanied by the predicted changes in p- and m-tyramine concentrations and supported further the existence of a reciprocal relationship between their concentrations and DA turnover (Juorio, 1982). Since these effects were observed within a day after the lesions were made and had recovered by 7 days, the existence of a so far unexplained compensatory mechanism is suggested. In addition, the fact that the p- and m-TA concentrations are not reduced at 7 days after the present lesion indicate that the striatal tyramine-containing neurons do not originate in or pass through the dorsal or median raphé nuclei. This latter observation is particularly interesting in relation to previous reports of alterations in tyramine levels after lesions of the substantia nigra pars compacta (Juorio and Jones, 1981).

ACKNOWLEDGEMENTS

We thank Dr. A.A. Boulton for his advice; Dr. D.A. Durden for supervising the mass spectrometric analyses, Dr. B.A. Davis for synthesizing the deuterated tryptamine, J. Broadbent for staining the histological material, M. Mizuno and E.P. Zarycki for expert technical assistance. Dr. T.B. Wishart kindly lent us the lesion maker. Saskatchewan Health provided continuing financial support. A.J. Greenshaw is the recipient of a Saskatchewan Health Research Board Fellowship.

REFERENCES

Dahlström A. and Fuxe K. (1964) Evidence for the existence of monoamine-containing neurons in the central nervous system. Acta Physiol. Scand. 62, Suppl. 232, 1-55.

Dray A., Gonye T.J., Oakley N.R. and Tanner T. (1976) Evidence for the existence of a raphé projection to the substantia nigra in rat. Brain Res. 113, 45-57.
Juorio A.V. (1982a) A possible role for tyramines in brain function and some mental disorders. Gen. Pharmacol. 13, 181-183.
Juorio A.V. and Greenshaw A.J. (1985) Tryptamine concentrations in areas of 5-hydroxytryptamine terminal innervation after electrolytic lesions of midbrain raphé nuclei. J. Neurochem. (in press).
Juorio A.V. and Jones R.S.G. (1981) The effect of mesencephalic lesons on tyramine and dopamine in the caudate nucleus of the rat. J. Neurochem. 36, 1898-1903.
Nicolaou N.M., Garcia-Munoz M. and Arbuthnott G.W. (1979) Interactions between serotonergic and dopaminergic systems in rat brain demonstrated by small unilateral lesions of the raphé nuclei. Europ. J. Pharmacol. 57, 295-305.
Philips S.R., Durden D.A. and Boulton A.A. (1974) Identification and distribution of p-tyramine in the rat. Can. J. Biochem. 52, 366-373.
Philips S.R., Davis B.A., Durden D.A. and Boulton A.A. (1975) Identification and distribution of m-tyramine in the rat. Can. J. Biochem. 53, 65-69.
Saller C.F. and Salama A.I. (1984) Rapid automated analysis of biogenic amines and their metabolites using reverse phase HPLC with electrochemical detection. J. Chromatog. 309, 287-298.

PHARMACOLOGICAL CHARACTERISATION OF OCTOPAMINE-SENSITIVE ADENYLATE CYCLASE IN *PERIPLANETA AMERICANA*

R.G.H. Downer, J.W.D. Gole and G.L. Orr

Dept. of Biology, University of Waterloo

Waterloo, Ontario, Canada, N2L 3G1.

Octopamine (OA), the monohydroxyphenolic analogue of norepinephrine, occurs in high concentrations in nervous tissues and hemolymph of many invertebrates and in sympathetically innervated tissues of vertebrates (Hicks, 1977; Bailey *et al.*, 1984). In insects, OA functions as a neurotransmitter, neuromodulator and neurohormone (Orchard, 1982) and it has been proposed as the sympathomimetic effector of physiological responses to excitation (Hoyle, 1975; Downer, 1979). The role of OA in vertebrates is uncertain although it has been suggested that specific receptors for OA are present in the rat brain and effects of OA on neuronal firing and potentiation of norepinephrine-mediated responses have been reported (Axelrod and Saavedra, 1977; Jones, 1984). The relatively low concentration of OA in vertebrate tissue and, by inference, the small population of specific OA-receptors have prevented investigation of the pharmacological properties of OA-receptors in vertebrates. By contrast, some insect tissues are highly sensitive to octopamine and, therefore, offer an excellent model system with which to study the pharmacology of OA-receptors.

At least some of the physiological effects of OA in insects are mediated through interaction of the amine with a membrane receptor that is coupled to an adenylate cyclase complex (Nathanson and Greengard, 1973; Nathanson, 1976; Bodnaryk, 1982; Gole *et al.*, 1982, 1983; Orchard *et* al., 1983; Downer et al., 1984). Thus, the interaction of various agonists and antagonists with OA-sensitive receptors can be determined by monitoring the effects of these compounds on cyclic AMP production in OA-sensitive cells. OA-sensitive adenylate cyclase has been identified in several tissues of the american cockroach, *Periplaneta americana* including hemocytes (Orr et al., 1985), nerve cord (Gole et al., 1983) and the corpus cardiacum-corpus allatum complex (Downer et al., 1984). Additivity studies indicate that the nerve cord and corpus-cardiacum have separate dopamine- and 5-hydroxytryptamine-sensitive adenylate cyclases in addition to the OA-sensitive enzyme complex (Gole et al., 1983; Downer et al., 1984) whereas washed membrane fragments from disrupted hemocytes appear to be free of other amine-sensitive receptors (Orr et al, 1985).

The current study examines the effects of a variety of compounds that are known to inhibit or mimic aminergic responses in vertebrate nervous tissues on the expression of OA-sensitive adenylate cyclase activity in the nerve cord, corpus cardiacum and hemocyte membrane fragments of *P. americana* and, on this basis, presents a preliminary pharmacological profile of the OA-sensitive receptor(s) in this insect.

MATERIALS AND METHODS

Procedures for maintaining insects (Downer, 1972), preparing washed membrane fragments of hemocytes (Orr et al., 1985) and determining cyclic AMP production in the three model systems have been described previously. (Gole et al., 1983; Downer et al., 1984; Orr et al., 1985). Ki values for the inhibition of amine-mediated elevation of cyclic AMP production were calculated according to the formula, $Ki = IC_{50}/(1 + S/Ka)$ where IC_{50} is the concentration of antagonist at which the agonistic response is inhibited by 50%, S is the concentration of agonist and Ka is the concentration of agonist required to effect 50% of the maximal activation.

RESULTS AND DISCUSSION

The ability of various aminergic antagonists to inhibit OA-mediated elevation of cyclic AMP production in the hemocyte membrane preparation, intact corpora cardiaca and nerve cord homogenates of P. americana is indicated in Table 1. The most potent antagonist of the OA-response in each of the preparations tested is mianserin which, in the vertebrate nervous system, is classified as antihistaminic and an inhibitor of 5-hydroxytryptamine. Cyproheptadine, another blocker of histamine and 5-hydroxytryptamine, shows pronounced inhibition of the OA-response in nerve cord homogenates but is a less potent inhibitor in the corpora cardiaca and hemocyte preparations. The α-adrenergic blocker, phentolamine and the serotoninergic-blocker, gramine also show marked inhibition of the

Table 1. Effects of various antagonists on octopamine-mediated elevation of cyclic AMP production in hemocyte membrane preparations, intact corpora cardiaca and nerve cord homogenates of *Periplaneta americana*.

ANTAGONIST	Ki (M)*		
	HEMOCYTE MEMBRANE	CORPORA CARDIACA	NERVE CORD HOMOGENATE
MIANSERIN	1.99×10^{-8}	3.68×10^{-7}	2.10×10^{-8}
PROMETHAZINE	1.99×10^{-7}	8.65×10^{-6}	1.74×10^{-7}
PHENTOLAMINE	2.29×10^{-7}	4.13×10^{-6}	5.11×10^{-8}
GRAMINE	5.60×10^{-7}	2.32×10^{-6}	6.30×10^{-8}
CIS-FLUPENTHIXOL	8.90×10^{-7}	1.47×10^{-5}	8.92×10^{-7}
CYPROHEPTADINE	1.99×10^{-6}	5.70×10^{-6}	2.64×10^{-8}
PROPRANOLOL	2.51×10^{-5}	9.89×10^{-5}	6.60×10^{-4}

*$Ki = IC_{50}/(1 + S/Ka)$

OA-response in the three preparations with the nerve cord displaying greater sensitivity to these compounds than the hemocyte preparation which, in turn, is more sensitive than the corpus cardiacum. However, the relative sensitivities of the response to the vertebrate neuroleptic compound, cis-flupenthixol, are similar within each tissue. The antihistaminic compound, promethazine caused marked inhibition of the OA-response in the hemocyte preparation and nerve cord homogenates but was less effective as an inhibitor in the corpus cardiacum. It is possible that the use of intact corpora cardiaca rather than isolated membrane preparations or homogenates has resulted in apparent pharmacological differences that may be attributed to the inability of some antagonists to penetrate to the binding site. An alternative explanation is that there are several types of OA-receptor and that each demonstrates a distinctive pharmacological profile. Such a diversity of OA receptors has been proposed by Evans (1981) on the strength of observations in locust leg muscle and the present study lends support to this proposal. The lack of sensitivity of the three preparations to the β-adrenergic blocking agent, propranolol, is consistent with the findings of other workers (Nathanson and Greengard, 1973; Harmer and Horn, 1977, Morton, 1984), who also support the conclusion of the present study that the most effective blockers of OA-mediated processes are those that affect α-adrenergic, histamine or serotoninergic receptors in vertebrates.

The nature of the inhibition of the OA-mediated stimulation of cyclic AMP production by mianserin, promethazine and gramine has been investigated in the hemocyte preparation and shown to be competitive (Orr et al., 1985). Competitive inhibition of the OA-response has been demonstrated also in the cockroach nervous system for cyproheptadine (Harmar and Horn, 1977) and phentolamine (Gole et al., 1983).

The sensitivity of the OA-mediated response to inhibition by classical serotoninergic and histamine blockers contrasts with the relatively poor agonism obtained with histamine and 5-hydroxytryptamine against the hemocyte preparation. The effect of a variety of potential agonists on cyclic AMP production in the hemocyte membrane preparation is indicated in Table 2.

Table 2. Effect of potential agonists on cyclic AMP production in hemocyte membrane preparations of *Periplaneta americana**.

Agonists (1×10^{-5} M)	Change in cyclic AMP production relative to control incubations (pmoles cAMP/min/mg protein)**
Octopamine	12.59 ± 1.16 (18)+
Synephrine	6.34 ± 1.37 (4)+
Tyramine	4.62 ± 1.09 (9)+
Dopamine	1.63 ± 0.61 (9)
5-Hydroxytryptamine	-0.08 ± 0.42 (5)
Norepinephrine	1.52 ± 0.35 (5)
Phenylephrine	2.29 ± 0.91 (5)
Histamine	0.62 ± 0.48 (5)
Clonidine	2.03 ± 0.27 (5)+
Ephedrine	0.17 ± 0.15 (4)
Ergonavine maleate	1.02 ± 0.23 (4)
Norphenylephrine	-0.79 ± 0.28 (4)
Norephedrine	-0.99 ± 0.37 (4)
Phenylethanolamine	0.80 ± 0.40 (4)
Tolazoline	-0.06 ± 0.41 (4)
Isoproterenol	-0.01 ± 0.36 (4)
Naphazoline	4.35 ± 0.96 (4)+
Apomorphine	0.87 ± 0.40 (4)

* From Orr *et al.* (1985).

** Values indicate mean ± standard error of the mean for the number of determinations indicated in parenthesis; values represent increase or decrease (-) in cyclic AMP production relative to control incubation (7.54 ± 1.41(7) pmoles cAMP/min/mg protein).

\+ Statistically significant (< 0.05, Dunnetts Test).

The results show that histamine and 5-hydroxytryptamine cause no statistically significant increase in cyclic AMP production whereas synephrine, tyramine and the α-adrenoreceptor agonists naphazoline and clonidine stimulate OA-sensitive adenylate cyclase. These findings are consistent with those of previous investigations on the pharmacology of OA-receptors in insects (Evans, 1981).

It is apparent from the results of the present investigation that the pharmacological properties of the OA-receptor differ from those of any recognised aminergic receptor of vertebrates. The OA-receptor must be considered as belonging to a separate receptor class and, therefore, an independent pharmacological classification of OA-receptors is required. It is hoped that the present study will contribute to the development of such a classification.

ACKNOWLEDGEMENTS

The study was supported by the grants from the American Cyanamid Company and the Natural Sciences and Engineering Research Council of Canada. The counsel of Dr. Kent Jennings and his colleagues at American Cyanamid is gratefully acknowledged.

REFERENCES

Axelrod J. and Saavedra J.M. (1977) Octopamine. Nature 265, 501-504.

Bailey B.A. Martin R.J. and Downer R.G.H. (1984) Haemolymph octopamine levels during and following flight in the american cockroach, Periplaneta americana. Can. J. Zool. 62, 19-22.

Bodnaryk R.P. (1982) Biogenic amine-sensitive adenylate cyclases in insects. Insect Biochem. 12, 1-6.

Downer R.G.H. (1972) Interspecificity of lipid regulating factors from insect corpus cardiacum. Can. J. Zool. 50, 63-65.

Downer R.G.H. (1979) Induction of hypertrehalosemia by excitation in *Periplaneta americana*. J. Insect Physiol. 29, 55-63.

Downer R.G.H., Orr G.L., Gole J.W.D. and Orchard I. (1984) The role of octopamine and cyclic AMP in regulating hormone release from corpora cardiaca of the american cockroach. J. Insect Physiol. 30, 457-462.

Evans P.D. (1981) Multiple receptor types for octopamine in the locust. J. Physiol. 318, 99-122.

Gole J.W.D., Downer R.G.H. and Sohi S.S. (1982) Octopamine-sensitive adenylate cyclase in haemocytes of the forest tent caterpillar *Malacosoma disstria* Hubner (Lepidoptera: Lasiocampidae). Can. J. Zool. 60, 825-829.

Gole J.W.D., Orr G.L. and Downer R.G.H. (1983) Interaction of formamidines with octopamine sensitive adenylate cyclase receptors in the nerve cord of *Periplaneta americana*. Life Sci. 32, 2939-2947.

Harmar A.J. and Horn A.S. (1977) Octopamine-sensitive adenylate cyclase in cockroach brain : effects of agonists, antagonists and guanylyl nucleotides. Molec. Pharmacol. 13, 512-520.

Hicks T.P. (1977) The possible role of octopamine as a synaptic transmitter : a review. Can. J. Physiol. Pharmacol. 55, 137-155.

Hoyle G. (1975) Evidence that insect dorsal unpaired median (DUM) neurons are octopaminergic. J exp. Zool. 193, 425-431.

Jones R.S.G. (1984) Electrophysiological studies of the possible role of trace amines in synaptic function, in *Neurobiology of the trace amines* (Boulton A.A., Baker G.B., Dewhurst W.G. and Sandler M. eds.), pp 205-223. Humana Press, Clifton, N.J.

Morton, D.B. (1984) Pharmacology of the octopamine-stimulated adenylate cyclase of the locust and tick CNS. Comp. Biochem. Physiol. 78C, 153-158.

Nathanson J.A. (1976) Octopamine-sensitive adenylate cyclase and its possible relationship to the octopamine receptor, in Trace amines and the brain (Usdin E. and Sandler M. eds.) pp. 161-190. Marcel Dekker, N.Y.

Nathanson J.A. and Greengard P. (1973) Octopamine - sensitive adenylate cyclase : evidence for a biological role of octopamine in nervous tissue. Science 180, 308-310.

Orchard I. (1982) Octopamine in insects : neurotransmitter, neurohormone and neuromodulator. Can. J. Zool. 60, 659-669.

Orchard I., Gole J.W.D. and Downer R.G.H. (1983) Pharmacology of aminergic receptors mediating an elevation in cyclic AMP and release of hormone from locust neurosecretory cells. Brain Res. 288, 249-353.

Orr G.L., Gole J.W.D. and Downer R.G.H. (1985) Characterisation of an octopamine-sensitive adenylate cyclase in haemocyte membrane fragments of the American cockroach, Periplaneta americana. Insect Biochem. (in press).

EFFECTS OF DRUGS ON OCTOPAMINE METABOLISM IN THE RAT

David J. Edwards and M. L. Sedlock

Dept. Pharmacology-Physiology, Univ. Pittsburgh

546 Salk Hall, Pittsburgh, PA 15261

There is substantial evidence to suggest that octopamine (OCT) is a neurotransmitter in invertebrates, but its role in the neural function of higher organisms is not yet established. Although the presence of OCT in mammalian brain has been demonstrated by a variety of techniques in several laboratories, its presence alone is not sufficient to prove a physiological function.

OCT is structurally like norepinephrine (NE) but lacks the ring 3-OH. Its biosynthesis uses the same metabolic pathway as NE, except NE requires an additional first step, tyrosine hydroxylase (TH), which converts tyrosine to DOPA. The enzymes involved in the next two steps of NE synthesis, aromatic amino acid decarboxylase (AAAD) and dopamine-β-hydroxylase (DBH), are relatively nonspecific. Consequently, these same enzymes convert DOPA → dopamine (DA) → NE and also tyrosine → tyramine → OCT. Although AAAD has an approximately 30-fold higher Km for tyrosine than for DOPA (Lovenberg et al., 1962), cellular concentrations of tyrosine far exceed those of DOPA. Thus, significant amounts of OCT would be expected wherever NE is made.

In contrast to the nervous systems of many invertebrates that contain OCT but not TH and NE, the mammalian brain contains both NE and OCT, with the concentration of OCT less than 3% of NE. This poses the exceedingly difficult problem of determining whether OCT is simply a by-product of NE synthesis or whether it has a role in synaptic transmission, as a neuromodulator of NE function or as a neurotransmitter in specific OCT neurons.

One approach to solving this problem is to correlate the effects of drugs on OCT metabolism with clinical or behavioral effects. In such studies, it is important to measure levels not only of the amine but also of its major metabolites or other indices of turnover, since the function of any brain amine is more likely to be reflected by changes in turnover than in its steady-state level.

We have focused on the neutral metabolite, p-hydroxyphenylglycol (pHPG) as a marker of OCT turnover, since pHPG rather than the acidic metabolite, p-hydroxymandelic acid (pHMA), should be the major metabolite of OCT in brain for the same reasons that other β-hydroxyphenethylamines are converted mainly to glycol rather than to acidic metabolites (e.g. to 3-methoxy-4-hydroxyphenylglycol (MHPG) and not vanillylmandelic acid (VMA) in the case of NE) (Sandler et al., 1976). On the other hand, pHMA has been detected in rat brain in a somewhat greater concentration (3-5 ng/g) (Karoum et al., 1975a,b) than that of pHPG (Edwards et al., 1985; Sedlock et al., 1985). Man excretes large amounts of pHMA, about 542 μg/hr (Tyce et al., 1983) or 4 times the amount of VMA. However, it is doubtful that most of the pHMA in brain and urine derives from OCT. A portion probably derives from dietary sources, such as p-sympatol (N-methyl-OCT), which is found in oranges (Gjessing and Armstrong, 1963). Part of the pHMA may be artifactual, since p-hydroxyphenylpyruvic acid decomposes to pHMA (Gjessing et al., 1965). A report that pHMA excretion was unaltered by either an MAO inhibitor or a DBH inhibitor (Gjessing et al., 1965) (although in a study by Hoeldtke and Wurtman (1974) it was reduced by the MAO inhibitor iproniazid) adds further to the uncertainties about the origin of pHMA. We believe, therefore, that pHPG is a more suitable marker of OCT turnover.

In this chapter we will present the results of studies on the effects of drugs on OCT metabolism. Studies on antidepressants (ADs) may lead to a better understanding of not only the role of OCT in brain function but also of the mechanism of action of ADs and of the biochemical abnormalities underlying affective disorders. Clues to the role of OCT in neural function may also be provided by studies with drugs having known actions on catecholamines (CAs).

METHODS

The neutral metabolites were analyzed in brain and urine samples using gas chromatography-chemical ionization-mass spectrometry (GC-CI-MS), as described in detail elsewhere (Edwards et al., 1985; Edwards and Rizk, 1981; Edwards et al., 1981). OCT was also determined by GC-CI-MS (Sedlock et al., 1985). Brain concentrations of CAs and tyrosine were determined by high-performance liquid chromatography with electrochemical detection (Edwards et al.,1985).

RESULTS

Effects of IMI and other ADs on OCT metabolism

As shown in Table 1, the acute administration of IMI (20 mg/kg) to rats reduced the brain concentration of pHPG to about 50% of control when measured 24 hr later. The 24-hr urinary excretion of pHPG was lowered to about 30%. In contrast, MHPG levels in brain and urine were decreased only to about 85 and 60%, respectively. Neither brain OCT nor NE were changed (data not shown).

In another experiment, animals were killed either 3 or 6 hr after IMI. Brain pHPG was not lowered at 3 hr (1.40 ±.29 vs. 1.50±.14 ng/g for control group) but had dropped to 33% (0.5±.05 ng/g; $p<.001$) by 6 hr. Brain OCT levels were 81 and 75% of control at 3 and 6 hr, respectively, but these decreases were nonsignificant.

Following 14 days of IMI treatment, pHPG excretion was reduced to 29% of control ($p<.02$), which is similar to the decrease to 24% observed after one injection (Expt. 4, Table 1). However, brain pHPG had returned to 97% of control level after repeated IMI (Fig. 1).

Iprindole (IPR), an atypical AD that unlike IMI does not inhibit amine uptake, resembled IMI in its effects on pHPG. One injection of IPR reduced brain pHPG to 63% of control 24 hr later, as compared to 62% for IMI (Expt. 3, Table 1). In another experiment, IPR lowered brain pHPG to 69% of control ($p<.005$) 6 hr after injection. The 24-hr excretion of total pHPG nonsignificantly decreased to 50%, but free pHPG excretion was significantly reduced to 70% by IPR. The percent of total pHPG excreted in the free form (4-5%) was unchanged by either IMI or IPR.

Table 1. Effects of IMI on pHPG and MHPG in 4 experiments

EXPT	GROUP	WEIGHT (g)	BRAIN (ng/g) pHPG	BRAIN (ng/g) MHPG	URINE (μg/24h) pHPG	URINE (μg/24h) MHPG
1	CONTROL	178.3 ±3.8	1.2 ±0.5	73.7 ±4.8	1.0 ±0.3	19.8 ±3.8
	IMI	179.3 ±11.2	0.5 ±0.1	61.0 ±6.1	0.5 ±0.2	12.9 ±4.8
			42%	83%	50%	65%
2	CONTROL	307.2 ±7.5	5.0 ±0.5	59.6 ±3.9	3.1 ±0.4	61.2 ±10.6
	IMI	307.5 ±8.2	2.3 ±0.7	52.8 ±3.8	1.0 ±0.1	38.6 ±5.1
			46%[b]	88%	32%[c]	63%
3	CONTROL	245.8 ±3.0	1.9 ±0.2	53.1 ±1.6	1.6 ±0.4	20.8 ±3.8
	IMI	245.5 ±4.4	1.2 ±0.2	50.7 ±3.3	0.4 ±0.1	12.9 ±2.1
			62%[a]	95%	25%[a]	62%
4	CONTROL	348.2 ±4.0	1.8 ±0.2	60.4 ±2.2	2.3 ±0.5	51.4 ±6.0
	IMI	347.8 ±2.8	1.0 ±0.2	51.3 ±2.1	0.7 ±0.1	26.5 ±2.5
			56%[a]	85%[b]	30%[b]	52%[c]

All values are the mean ± S.E. (n=4-6). Male Sprague-Dawley rats were weight-matched and received either IMI (20 mg/kg, i.p.) or saline 24 hours before they were killed. In Expt. 4, both groups also received saline daily for 13 days before the final injection, while a third group received IMI for all 14 days (see Fig. 1).
[a]$p<.05$; [b]$p<.02$; [c]$p<.005$, compared by two-tailed t-tests.

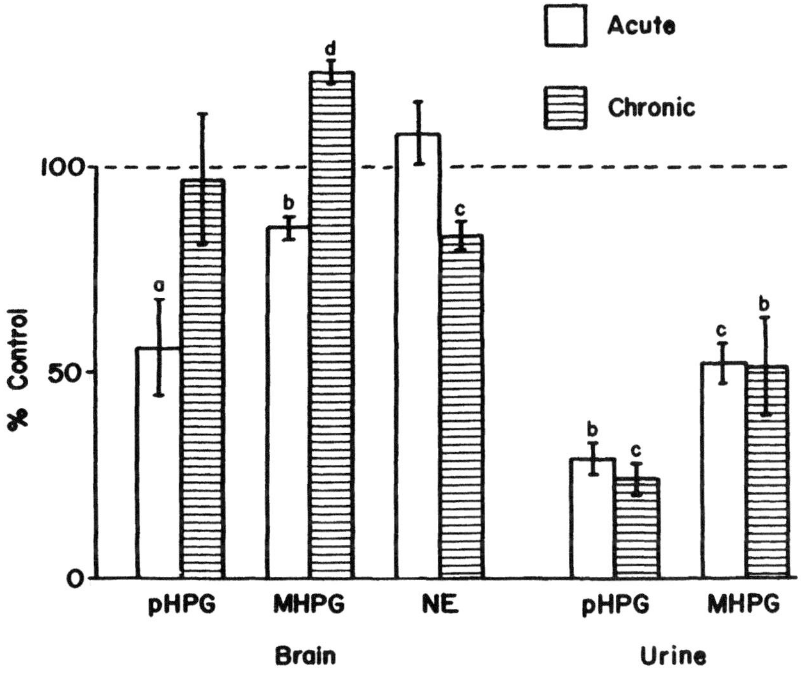

Figure 1. Acute and chronic effects of IMI. Rats in the acute group (n=6) were given saline once daily for 13 days and IMI (20 mg/kg) on day 14; rats in the chronic group(n=8) received the same dose of IMI once daily for 14 days. Control rats (n=6) received saline injections for all 14 days. Control values are given in Expt. 4, Table 1. [a]p<.05; [b]p<.02; [c]p<.005; [d]p<.001, vs. control.

Preliminary studies of other atypical ADs indicated that amoxapine, mianserin, trazodone and zimelidine (20 mg/kg) nonsignificantly decreased brain pHPG after one injection. Bupropion was the only exception, but 50 mg/kg of the drug (which has a half-life of about 1 hr in the rat) did reduce brain pHPG to 74% 6 hr after injection. All of the above drugs except mianserin lowered urine pHPG levels, but the decreases were significant only for amoxapine (to 32%, p<.05) and zimelidine (to 30%, p<.001). Chlorpromazine (CPZ), a phenothiazine with antipsychotic properties, lowered brain pHPG to 45% (p<.02) 24 hr after 20 mg/kg and 24-hr excretion to 19% (p<.001).

Effects of CA-altering drugs on OCT metabolism

6-Hydroxydopamine (6-OHDA)

This neurotoxin injected intraventricularly into rats (2x250 μg) (Edwards et al., 1985) depleted brain pHPG levels 3 weeks later to 30% of control (.23±.02 vs. .77±.13 ng/g, p<.005). By comparison, brain levels of the NE metabolites, MHPG and 3,4-dihydroxyphenylglycol (DHPG), were reduced to 15 and 6.5%, respectively (p<.001). OCT levels were unaffected.

Yohimbine

This α_2-adrenoceptor antagonist (10 mg/kg) raised brain MHPG levels to 261, 279 and 282% of control at 2, 4 and 6 hr, respectively (p<.001, control = 98.0±3.8 ng/g). Concomitantly, NE levels fell to 54 (p<.001), 63 (p<.001) and 80% (p<.025), indicating that synthesis was unable to keep up with the enhanced NE release. In contrast, brain pHPG was not elevated, but instead was decreased to 75% at 6 hr (control = 1.36±.09 ng/g). OCT levels were unaffected (control = 1.54±.44; 2 hr = 1.67±.44; 4 hr = 1.74±.56; 6 hr = 1.63±.51 ng/g). Tyrosine levels were decreased to 72% at 6 hr (p<.025, control = 12.3±1.2 μg/g).

DISCUSSION

Although Harmar and Horn (1976) reported 4- to 5-fold increases in brain OCT in mice treated with ADs, in the present study in the rat no effect was found. In contrast, the decrease in pHPG produced by many ADs suggests they decrease OCT turnover. These results raise the possibility that at least some ADs exert their clinical effects via an interference with OCT metabolism. Although the time course for the effect on pHPG, which was observed after acute but not chronic IMI, does not match that of the clinical response, which does not appear until 1-2 weeks after treatment is begun, this does not rule out the possibility that OCT triggers other biochemical changes, such as the modification of receptors, that lead to the eventual clinical response. The finding that CPZ also lowered pHPG levels does not necessarily argue against this as a mechanism of AD action, since CPZ may have AD properties and it down regulates the NE-sensitive adenylate cyclase system, as do the ADs (Vetulani and Sulser, 1975).

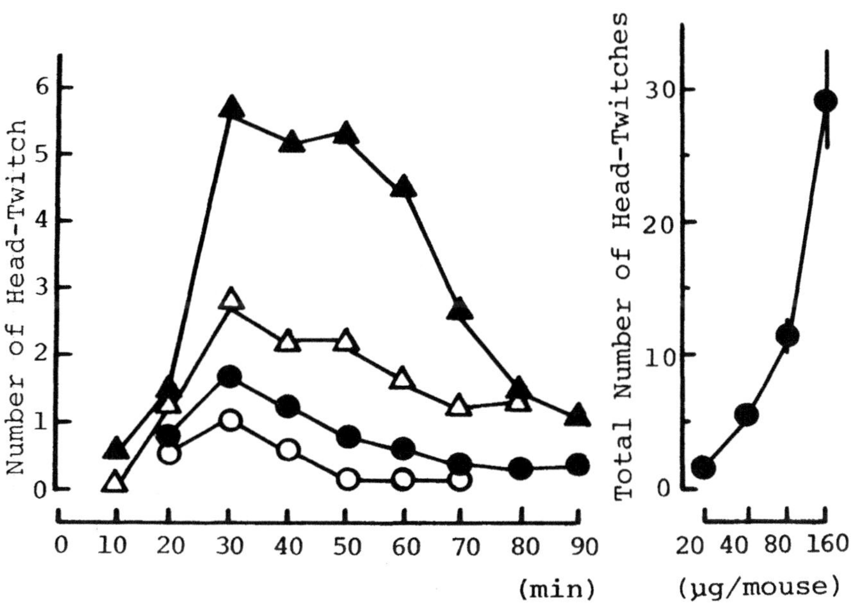

Fig.1. Effect of intracerebroventricular injection p-hydroxyamphetamine in a dose of 20 (-○-; N=10), 40 (-●-; N=10), 80 (-△-; N=10) and 160 (-▲-; N=20) µg/mouse on head-twitch response in mice.

The i.c.v. administration of p-OHA (20-160 µg/mouse) increased spontaneous motor activity, piloerection, salivation, weeping and the number of head-twitches in mice. HTR induced by p-OHA continued for 20-80 min and the peak time of the head-twitch activity was approximately 30 min after i.c.v. administration of p-OHA. Peak number were 7 twitches/2 min for a dose of 20 µg/mouse, 8 for a dose of 40 µg/mouse, 10 for 80 µg/mouse and 30 for 160 µg/mouse. Thus, i.c.v. administration of p-OHA dose-dependently increases the number of head-twitches (Fig. 1.).Pretreatment with reserpine (1 mg/kg/day i.p. for 4 days), a serotonergic and catecholaminergic denervator, markedly potentiated HTR induced by p-OHA. The total number of head-twitches in mice preadministered (i.p.) with reserpine (in doses of 1 mg/kg/day for 3 or 4 days) was approximately 4.5-12 times that of the mice not treated with reserpine.

Effect of serotonergic drugs on HTR

The i.c.v. administration of serotonin alone caused HTR in mice, and its intensity and duration were more marked compared to p-OHA alone. A dose of serotonin (10 µg/mouse, i.c.v.) which did not elicite HTR, was used to see the combined effect of serotonin plus p-OHA. Simultaneous injection of serotonin and p-OHA increases in the number of head-twitches. The total number of HTR in this co-administration caused 2.2-2.6 fold of increase compared to non-serotonin controls. Pretreatment with p-chlorophenylalanine (p-CPA), an inhibitor of serotonin biosynthesis, resulted in dose dependent reduction of HTR induced by p-OHA. The total number of head-twitches induced by p-OHA was completely blocked by three consecutive injections of p-CPA (200 mg/kg). Pretreatment with p-CPA methylester (500 µg/mouse, i.c.v.) also significantly decreased the total number of head-twitches. Several other drugs that block serotonergic receptors in the brain were also tested for their ability to change the intensity of HTR evoked by p-CPA. Of those drugs tested, pretreatment with either cyproheptadine, dimethothiazine or morphine markedly reduced the total number of head-twitches induced by p-OHA in a dose-dependent manner.

DISCUSSION

It has been widely accepted that amphetamine elicites abnormal behavior (Randrup and Munkvad, 1966), and also induces changes in regional brain catecholamine levels (Lewander, 1971). The abnormal behavior induced by amphetamine has been reported to be related to catecholamine release and inhibition of re-uptake (Moore, 1964; Ross and Renyi, 1966). After administration of amphetamine, one of its metabolites, p-OHA, is highly accumulated in the hypothalamus and striatum (Jori and Caccia, 1974). Furthermore, Randrup and Munkvad (1970) have reported that p-OHA, injected bilaterally into the rat caudate nuclei, induces amphetamine-like stereotyped behavior.

In the present study, the i.c.v. administration of p-OHA, one of the metabolites of amphetamine and of methamphetamine, induced HTR which has been proposed as an experimental model for hallucination (Keller and Umbreit, 1956; Corne and Pickering, 1967), since hallucinogens strongly induce this response in mice (Corne and Pickering, 1967; Teresa et al., 1975). Therefore, it is assumed that p-OHA-induced HTR may be a symptoms of hallucination. As well as a wide range of hallucinogens, 5-HTP (Corne et al ., 1963; Sakurada, 1975), serotonin (Nakamura and Fukushima, 1978; Orikasa et al., 1980), also induce HTR in mice. HTR evoked by these drugs may involve the serotonergic system in the central nervous system, since effects caused by these drugs are preferentially blocked by antiserotonergic drugs such as dimethothiazine (Collier et al ., 1974) and methysergide (Corne et al., 1963), markedly enhanced by i.c.v. injection of serotonin (Orikasa, 1980). In this study, p-CPA, a serotonin biosynthesis inhibitor, or the serotonin receptor blocking agents, dimethothiazine and cyproheptadine, decreased the number of head-twitches induced by p-OHA in a dose dependent manner. Moreover, i.c.v. infusion of serotonin in a dose which itself does not influence this response, markedly potentiated the number of head-twitches. The HTR elicited in the present experiment appeared to be similar to those produced by serotonin, tyramine and various hallucinogens and may involve the serotonergic system of the brain. Our data together with these related findings led us to conclude that the response induced by p-OHA may be mediated via the brain serotonergic system.

REFERENCES

Brittain R. T. and Handley S.L. (1967) Temperature changes produced by the injection of catecholamines and 5-hydroxytryptamine into the cerebral ventricles of the conscious mouse. J. Physiol. 192, 805-813.

Collier H.O.J., Hammond M.D. and Schneider C. (1974) Biogenic amines and head twitches in mice during ethanol withdrawal. Br. J. Pharmacol. 51, 310-311.

Corne S.J., Pickering R.W. and Warner B.T. (1963) A method for assessing the effects of drugs on the central actions of 5-hydroxytryptamine. Br. J. Pharmacol. 20, 106-120.

Corne S.J. and Pickering R.W. (1967) A possible correlation between drug-induced hallucinations in man and a behavioural response in mice. Psychopharmacol. 11, 65-78.

Jori A. and Caccia S. (1974) Distribution of amphetamine and its hydroxylated metabolites in various areas of the rat brain. J. Pharm. Pharmacol. 26, 746-748.

Keller D.L. and Umbreit W.W. (1956) "Permanent" Alteration of behavior in mice by chemical and psychological means. Science 124, 723-724.

Lewander T. (1971) Displacement of brain and heart noradrenaline by p-hydroxynorephedrine after administration of p-hydroxyamphetamine. Acta Pharmacol. Toxicol. 29, 20-32.

Malick J.B., Doren E. and Barnett A. (1977) Quipazine-induced head-twitch in mice. Pharmacol. Biochem. Behavior 6, 325-329.

Moore K.E. (1964) The role of endogenous norepinephrine in the toxicity of d-amphetamine in aggregated mice. J. Pharmacol. Exp. Ther. 144, 45-51.

Nakamura M. and Fukushima H. (1978) Effects of reserpine, para-chlorophenylalanine, 5,6-dihydroxytryptamine and fludiazepam on the head twitches induced by 5-hydroxytryptamine or 5-methoxytryptamine in mice. J. Pharm. Pharmacol. 30, 254-256.

Orikasa S., Sakurada S. and Kisara K. (1980) Head-twitch response induced by tyramine. Psychopharmacol. 67, 53-59.

Randrup A. and Munkvad I. (1966) Role of catecholamines in the amphetamine excitatory response. Nature 211, 540.

Randrup A. and Munkvad I. (1970) Biochemical, anatomical and psychological investigations of stereotyped behavior induced by amphetamines. In: International symposium on amphatamines and related compounds. Ed: E. Costa and S. Garattini. Raven Press. New York.

Ross S.B. and Renyi A. (1966) In vivo inhibition of ^{3}H-noradrenaline uptake by mouse brain slices in vitro. J. Pharm. Pharmacol. 18, 322-323.

Sakurada S. (1975) Central action of β-phenylethylamine derivatives (6) Head-twitches induced by intracerebrally administered tyramine in isocarboxazide pretreated mice. Folia. Pharmacol. Japon 71, 779-787.

Teresa M., Silva A. and Calil H.M. (1975) Screenig hallucinogenic drugs: Systematic study of three behavioral tests. Psychopharmacol. 42, 163-171.

CATECHOLAMINE, SEROTONIN AND TETRAHYDROBIOPTERIN LEVELS IN BRAIN REGIONS OF MALE, C57Bl/6J MICE TREATED WITH MPTP (1-METHYL-4-PHENYL-1,2,3,6-TETRAHYDROPYRIDINE)

John F. Reinhard, Jr., John Y. Chao and Charles A. Nichol
The Wellcome Research Laboratories, Burroughs Wellcome Co. 3030 Cornwallis Rd., Research Triangle Park NC 27709, USA.

The neurotoxin MPTP produces secondary Parkinson's disease (PD) in humans (1), a syndrome resembling parkinsonism in monkeys (2), and striatal dopamine (DA) depletions in mice (3) and dogs (4). In contrast, MPTP has little effect on DA levels in rats (5) or in guinea pigs (6). While there is biochemical, behavorial and histological evidence for loss of dopaminergic cells in the monkey (2), the corresponding effects of MPTP in mice are controversial (3,7,8). Thus, the validity of the MPTP-treated mouse as a small animal model of PD is uncertain. As part of a research program on the role of tetrahydrobiopterin (BH_4), the cofactor for tyrosine hydroxylase, in the regulation of dopamine synthesis, we have examined the levels of catechols, indoles and BH_4 in different brain regions of mice treated with MPTP, and find marked decreases in striatal DA and BH_4 with partial recovery, as contrasted to other brain regions where little or no effect was observed.

MATERIALS AND METHODS

Male, C57Bl/6J mice were obtained from the Jackson Laboratories (Bar Harbor ME) housed in wire-mesh cages, allowed <u>ad libitum</u> access to food and water and were illuminated from 0800-2000 hours daily. The MPTP hydrochloride salt (Research Biochemicals Inc., Wayland, MA) was dissolved in saline and injected i.p. with 10 ml/kg body weight (36 mg/kg; 0.17 mmol/kg) daily for 5 days. The mice were killed by cervical dislocation at the times indicated following the last dose of MPTP, and the tissues were rapidly frozen in liquid nitrogen and stored it -65 degrees C. until analysis, within a few days.

Tissue catechols and indoles were assayed as described (9), using liquid chromatography with electrochemical detection, as modified (10). Tissue biopterin levels were determined by liquid chromatography with fluorescence detection after acid and alkaline iodine oxidation (11,12). Locomotor activity was assessed with 6-beam photocell cages by Dr. G. Pollard (Dept. of Pharmacology, The Wellcome Research Labs). Statistical comparisons among 3 or more groups were by 1-way analysis of variance with differences among groups tested by the Neuman-Keuls test. Tests of two sample means were accomplished with a two-tailed t-test.

RESULTS AND DISCUSSION

Since the presence of BH_4 is essential for the formation of catecholamines from tyrosine, treatments which change DA levels in tissue may also affect BH_4 biosynthesis and cellular content. The selective effects of MPTP in the striatum are reflected by changes in the levels of both dopamine and BH_4 the levels of BH_4 were also decreased, but to a lesser extent than those of DA. In contrast, neither noradrenaline (NA) nor serotonin levels in brainstem were decreased while the serotonin and 5-hydroxyindoleacetic acid (5-HIAA) contents of brainstem were somewhat increased, two weeks after MPTP (Figure 1). It had previously been observed that MPTP increased the 5-HIAA level of ventricular cerebrospinal fluid of monkeys treated with MPTP (2). The reduction of DA and BH_4 in mouse striatum suggests that the hydroxylase cofactor is largely contained within DA neurons, as was suggested earlier based on 6-hydroxydopamine lesioning in rats (13). That the BH_4 reduction was of lesser magnitude than that of DA, and localized to the striatum, argues against a primary effect of MPTP on BH_4 regeneration, as was speculated earlier on the basis of biochemical studies on the inhibition of dihydropteridine reductase (14).

The effects of MPTP, studied at three different times after MPTP, are shown in Figure 2. Since only the

FIGURE 1 THE EFFECTS OF MPTP ON CATECHOLS, INDOLES, AND BH_4 IN MOUSE BRAIN REGIONS.

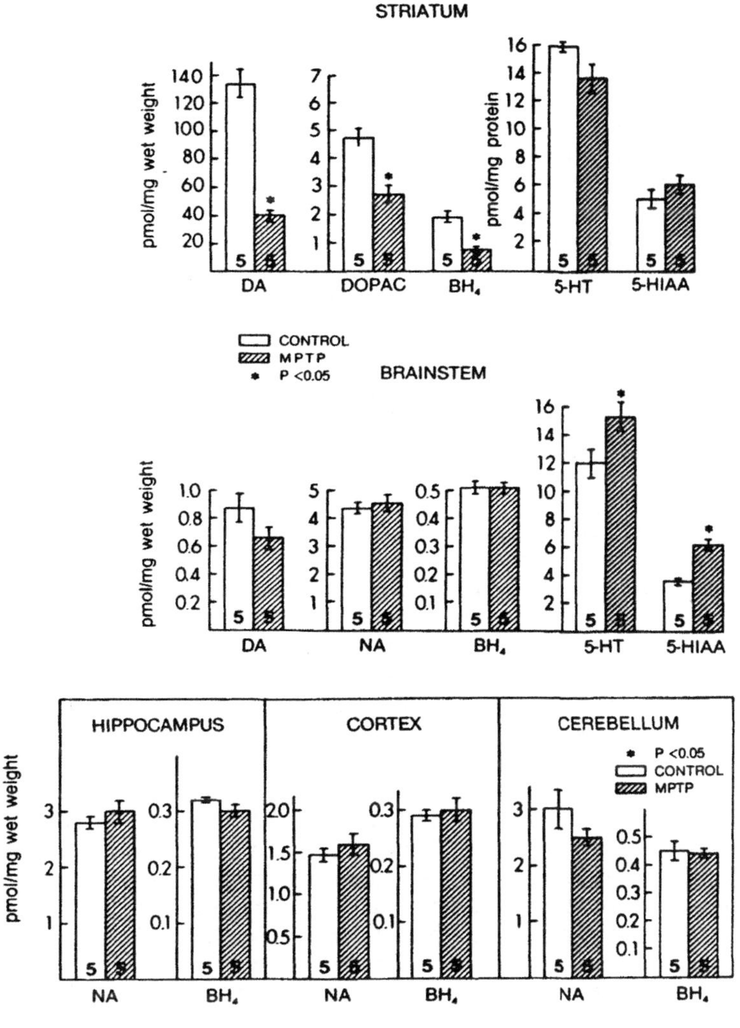

Data are the means ± SEM. Five animals were used per point. Asterisks denote significant differences from the control means ($p<0.05$).

striatum and vas deferens showed decreases in their catechols only those are presented. Significant dopamine depletions occur at the earliest (13 day) point, are greater 21 days after MPTP, and recover to about 50% by 212 days after MPTP. The reduced levels of DA, 7 months after MPTP suggests both neurotoxicity and recovery. The depletion is for a much longer time than that which would be expected from tyrosine hydroxylase inhibition. Moreover the synthesis rate of the striatal dopamine 3 weeks after MPTP is actually increased (15) Histological analysis of MPTP-treated mouse brain tissues reveal a decrease (3), or no change (7,8) in the number of dopamine cells in the substantia nigra, and a decrease in both the number of tyrosine hydroxylase-immunoreactive and formaldehyde-induced fluorescent nerve terminals in the caudate (7). The long duration of the DA depletion, the decrease in the number of dopaminergic nerve terminals and the decreased levels of BH_4 are consistent with neurotoxicity. Assuming a loss of DA neurons produced by MPTP in the mouse, the partial recovery of dopamine could be due to increased levels of DA in the remaining DA terminals, or to reinnervation of the caudate. Given the long projection distance of the nigro-striatal projection, this latter explanation would seem unlikely. As a peripheral structure heavily innervated by noradrenergic fibers of the sympathetic nervous system, the vas deferens was examined. Similarly, the content of dopamine and NA in the vas deferens decreased by approximately 70%, two to three weeks after MPTP. However, the NA levels recovered completely seven months after MPTP while the DA levels remained depleted by 31%. The high ratio of DA to NA in this structure (approximately 20%), and the partial recovery of DA suggest the presence of dopaminergic fibers in the vas deferens. Were this explanation accurate, the DA fibers of this structure might be affected by MPTP in much the same way as their striatal counterparts.

Lastly, the depletion of dopamine elicited by MPTP was without effect on locomotor behavior in mice. MPTP treated mice, in photocell cages generated activity counts which were 118 $\pm$ 11 % of the control animals (100 $\pm$ 7%) in the 15 minute intervals studied. The effect of MPTP on locomotor behavior is controversial (7,8), however the Weaver mutant mouse also shows

FIGURE 2. TIME-COURSE EFFECTS OF MPTP IN STRIATUM AND VAS DEFERENS.

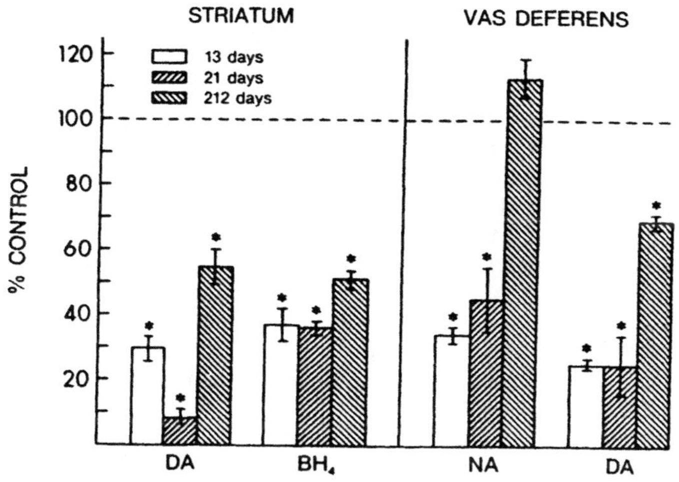

Time-course effects of MPTP on catechols and BH_4 in striatum and vas deferens. The data are the percent of control means ± their s.e.m. of 5 animals per point. Control values for striatal DA are: 13 days; 133.1 ± 9.08; 21 days 88.5 ± 2.0; 212 days, 110.2 ± 3.2. Control values for striatal BH_4 are: 13 days, 1.94 ± 0.19; 21 days, 1.52 ± 0.55; 21 days, 1.63 ± 0.06. Vas deferens noradrenaline: 13 days, 314.8 ± 17.5 pmol/mg protein; 21 days, 39.1 ± 0.1; 212 days, 36.5 ± 3.3. Vas deferens dopamine: 13 days, 52.9 ± 1.8 pmol/mg protein; 21 days, 9.0 ± 4.4; 212 days, 5.3 ± 6.7. Control means are presented as pmol/mg wet tissue weight, except where indicated. Asterisks denote significant differences from the respective control means.

dramatically reduced levels of DA in the striatum and similarly does not show a deficit in locomotor behavior (16).

While the depletion of dopamine in mouse striatum may be a consequence of neurotoxicity, there is clearly no parallel diminution in motor activity as is seen in humans and monkeys exposed to MPTP. Thus, even though MPTP toxicity is associated with substantial depletion of striatal dopamine and BH_4, the lack of effect on locomotor activity in these mice indicates that this is not suitable as a model of human Parkinson's disease.

REFERENCES

1. Langston J.W., Ballard P., Tetrud W. and Irwin I., Science 219:979-980, 1983.
2. Burns R.S., Chiueh C.C, Markey S.P., Ebert M.H., Jacobowitz D.M. and Kopin I.J., Proc. Nat. Acad. Sci. 80:4546-4550, 1983.
3. Heikkila R.E., Hess A. and Duvosin R.C., Science 224:1451-1453, 1984.
4. Burns R.S., Jacobowitz D.M., Chuieh C.C., Phillips J.M., Ebert M.H. and Kopin I.J., Soc. Neurosci. Abstr. 101, part 2, 880, 1984.
5. Boyce S., Kelly E., Reavill C., Jenner P. and Marsden C.D., Biochem. Pharmacol. 33:1747-1752, 1984.
6. Perry T.L., Voon W.Y., Ito M., Jones K., Wall R.A., Foulks J.G., Wright J.M. and Kish S.J., Life Sci. 36:1233-1238, 1985.
7. Hallman H., Lange J., Olson E.L., Strombert I. and Jonsson G., J. Neurochem. 44:117-127, 1985.
8. Chiueh C.C. Johannessen J.N. Chesselet M.-F. and Markey S.P., Fed. Proc. 44(4):893, 1985.
9. Reinhard J.F., Jr. and Roth R.H., J. Pharmacol. Exp. Therap. 221:541-546, 1981.
10. Reinhard J.F., Jr. and Perry J.A., J. Liq. Chrom. 7:1211-1220, 1984.
11. Fukushima T. and Nixon J.C., Anal. Biochem. 102:176-188, 1980.
12. Woolf J.H., Nichol C.A. and Duch D.S., J. Chromatog. 274:398-402, 1983.
13. Levine R.A., Miller L.P. and Lovenberg W., Science 214:919-921, 1981.
14. Abell C.W., Shen R.-S, Gessner W and Brass A., Science 224:405-407, 1984.
15. Reinhard J.F., Jr. and Nichol C.A., J. Neurochem. 40 :S189A, 1985.
16. Schmidt M.J., Sawyer B.D., Perry K.W., Fuller R.W., Foreman M.M. and Ghetti B.J., Neuroscience 2:376-380, 1982.

THE EFFECTS OF PARGYLINE, CLORGYLINE, DEPRENYL AND THEIR METABOLITES ON RAT PERIPHERAL AND CENTRAL BIOGENIC AMINES: A COMPARISON BETWEEN CHANGES IN URINE EXCRETION AND BRAIN CONCENTRATIONS

Farouk Karoum

Neuropsychiatry Branch, National Institute of Mental Health, St. Elizabeths Hosp., Washington, D.C. 20032

ABSTRACT

The chronic effects of pargyline, clorgyline, deprenyl and their metabolites on central and peripheral biogenic amines were evaluated in rats. The amines studies are phenylethylamine (PEA), p-tyramine (Ty), norepinephrine (NE) and tryptamine (TP). All analyses were carried out by mass fragmentography. Pargyline gave rise to three dealkylated products: benzylamine, N-methylbenzylamine and N-propargylbenzylamine (NPB). Of these three metabolites only NPB had in vivo monoamine oxidase (MAO) inhibitory properties. NPB like pargyline markedly elevated urine and brain PEA while only pargyline reduced peripheral and brain 5HT metabolism, suggesting that the metabolism of pargyline to NPB may be largely responsible for MAO type B inhibition. Pargyline, clorgyline, deprenyl, and NPB reduced NE and DA metabolism as reflected upon urine and brain concentrations of the metabolites. These effects remained evident 24 hours after termination of treatment. In contrast to their expected effects on urine and brain PEA and Ty, none of these above four drugs reduced 5HT and TP metabolism as would be expected from their in vitro properties. Furthermore, the relatively mild reductions in 5-hydroxyindole acetic acid (5HIAA) observed after pargyline and clorgyline disappeared 24 hours after drug treatment signifying paradoxically a transient and short lived inhibition of peripheral 5HT metabolism. The effects of D-and L-amphetamine on biogenic amines were also studied. These amines are metabolites of deprenyl in vivo. Both isomers produced peripheral and central changes that are similar, thus adding support to the notion that the in vivo effects of deprenyl on PEA is at least partially related to its metabolism to amphetamine and methamphetamine. Further, in view of the similarities in their effects on the amines studies here, it is unlikely that the behavioral differences between D- and L-amphetamine are directly mediated via changes in perhipheral or central catecholamines, PEA or p-Ty.

INTRODUCTION

The use of monoamine oxidase (MAO) inhibitors (MAOI) to reduce biogenic amine metabolism and hence to increase their availability both peripherally and centrally is a strategy exploited in the treatement of a number of psychiatric disorders that are believed to be associated with deficient monoamine availability especially the catecholamines. Recent trends in the clinical application of MAOIs is towards the employment of the so called selective MAOI; drugs that at low doses preferentially block one of the two major types of MAO (A- and B-types). Serotonin and phenylethylamine are the preferred substrates for the A and B types of MAO respectively. Unfortunately, the therapeutic benefits of these MAOIs are frequently attributed to MAO inhibition by the parent compound with little or no consideration of other possible pharmacological effects. Among these effects one may consider blockade of biogenic amine release and uptake, direct drug effects on biogenic amine receptors and above all the effects of the drugs' metabolites. Indeed, we have recently demonstrated that chronic deprenyl (an MAO type B inhibitor) treatment to humans produced changes in urinary excretion of biogenic amines metabolites that cannot wholely be attributed to MAO inhibition (Karoum et al., 1982) and have suggested that the metabolites of deprenyl (Reynolds et al., 1978) (amphetamine and methamphetamine) may play equally prominent roles as their parent compounds. There is therefore a need for a better understanding of the in vivo pharmacological and biochemical properties of MAOIs commonly employed in research and medicine. Such an evaluation should ideally consider a wide range of relevant biogenic amines and should also consider the effects of the drugs' metabolites.

As part of an ongoing effort to correlate changes in urinary amine excretion in man with central changes employing rats as a model, we have developed a system in which we expose rats to the same treatments as those used in humans and then assay urine and brain contents of a number of biogenic amines and their metabolites. Although, correlations between urine amine changes in rats and those induced centrally may not faithfully mimic the situation occurring in humans, they nevertheless offer an insight to how these two environments respond to systemic administration of drugs. In this communication I report on the chronic effects of pargyline, clorgyline, deprenyl and their metabolites on urinary excretion and brain concentrations of a number of biogenic amines and their metabolites in rats. The metabolites ot pargyline employed are benzylamine, N-methyl benzylamine and N-propargyl benzylamine (NPB). L-amphetamine is the deprenyl metabolite studied. For comparison we have also evaluated the effects ot D-amphetamine. As will be shown, changes in urinary metabolites of biogenic amines closely resemble brain changes. The results obtained imply the existence of more than one variant of MAO type A and B in rats.

MATERIAL AND METHODS

Urine was collected (usually for four hours) from rats placed in "Econo Metabolic Units" (American Scientific Products, Springfield, Va.). A fused silica capillary column, 0.32 mm i.d. bonded coated with a mixture of SP-2250 (methyl-phenylpropyl silicone gum), SPB 5 (methyl phenyl-vinyl-silicone gum) and SP 2401 (fluoropropyl silicone) was used tor the mass fragmentography of phenylethylamine and tyramine (custom made by Supelco Inc., Pa.). For the mass fragmentography of catecholamines and their metabolites a 10 ft. 1/8" o.d. stainless steel column packed with 3% SE 54 was used. Indoleacetic acid (IAA) and 5-hydroxyindoleacetic acid (5IHAA) were chromatographed on a 5 ft. 3% SE 54 pack column. A model 4000 Finnigan gas chromatograph-quadrupole mass spectrometer (Finnigam Corp., Sunnyvale, Ca.) was used. All chemicals and reagents used were of the highest grades commercially available.

Catecholamines in urine and brain were measured as previously reported (Karoum, 1983). Brain and urine free PEA, amphetamine, methamphetamine, and urine m-and p-tyramine were extracted and prepared for mass fragmentography as previously reported (Karoum, 1983) but a fused silica capillary column was used instead of a packed column.

RESULTS AND DISCUSSION

As shown in Figure 1, pargyline is metabolized to benzylamine, N-methylbenzylamine and N-propargyl-benzylamine (NPB). At an acute dose of 30 mg/kg, only NPB among these three metabolites increased PEA excretion. For this reason, only NPB was studies.

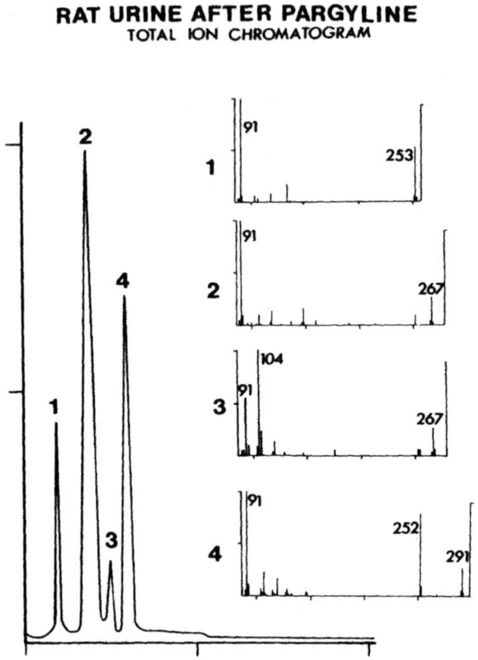

FIGURE I

The total ion chromatograms of an ethyl acetate extract (see method) of rat[a] urine collected for a duration of four hours after treatment with 50 mg/kg pargyline. 1=Benzylamine; 2=N-methylbenzylamine; 3=Phenylethylamine and 4=N-propargylbenzylamine (NPB).

The effects of chronic pargyline (35 μmol/kg), NPB (35 μmol/kg), clorgyline (20 μmol/kg) and deprenyl (20 μmol/kg) on PEA, Ty, MHPG and HVA excretions are summarized in Figure 2.

As can be deduced from this figure all drugs except clorgyline showed good MAO type B inhibition. The ability of pargyline, clorgyline and NPB to increase p-Ty excretion but not deprenyl is interesting and may have some bearing on the failure of this drug to induce the so-called "cheese effects" by other MAOI. The marked reduction in MHPG and HVA excretion during treatment and 24 hours after termination of treatment indicated a

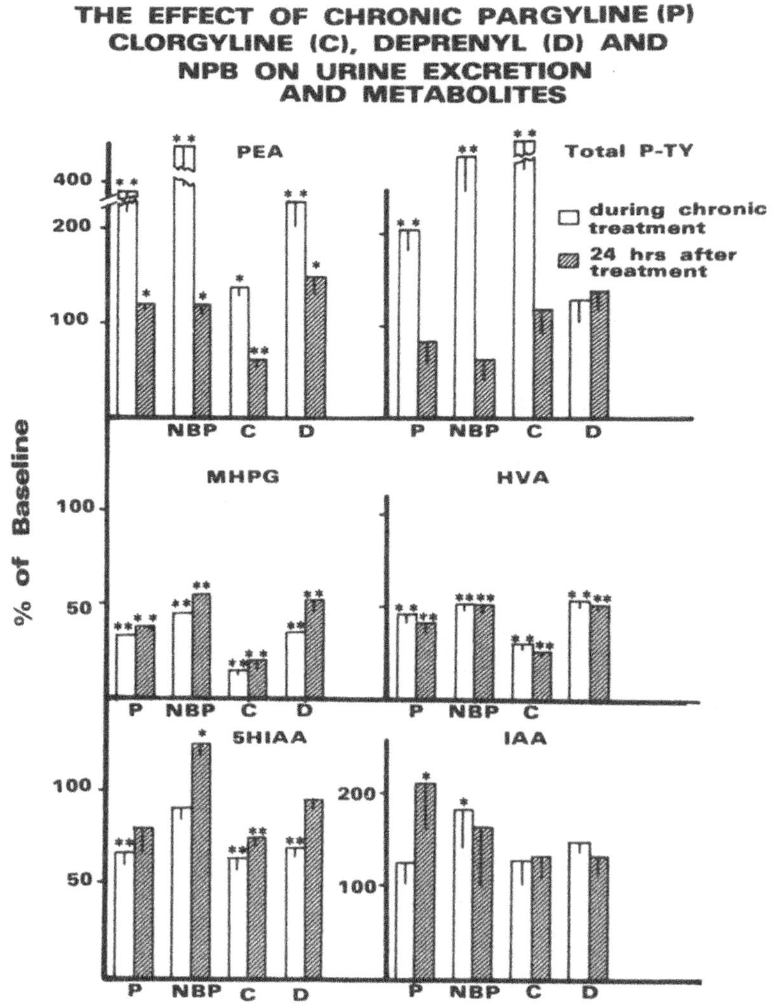

FIGURE 2

Pargyline (35 μmol/kg; 5 mg/kg), NPB (35 μmol/kg; 6.8 mg/kg), clorgyline (20 μmol/kg; 6.1 mg/kg) and deprenyl (20 μmol/kg; 4.5 mg/kg) were administered I.P. daily for 14 days and urine collected for four hours immediately after treatment on the 13th day (urine collection during treatment) and 24 hours after the last treatment. The results are expressed (mean ± SEM) as the percent of appropriate baseline values of untreated rats.

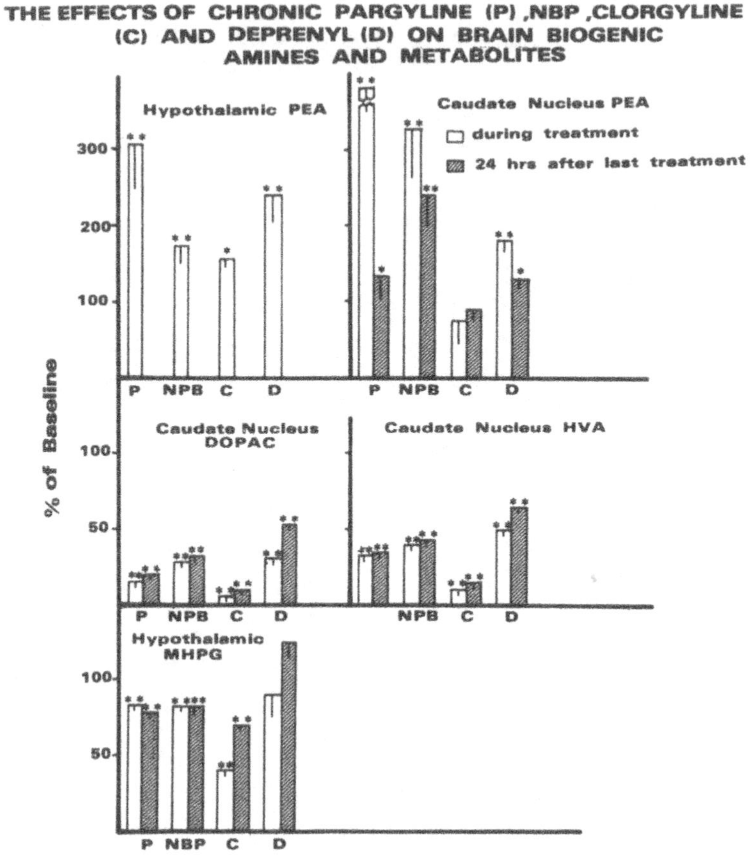

FIGURE 3

Two separate groups ot rats (5 in each group) were treated with MAOI as shown in Figure 2. One group was dacapitated on the 14th day and two hours after the last treatment and the other 24 hours atter the last treatment. Results are expreseed (mean $\pm$ S.E.M.) as the percent ot the appropriate baseline values of untreated rats.

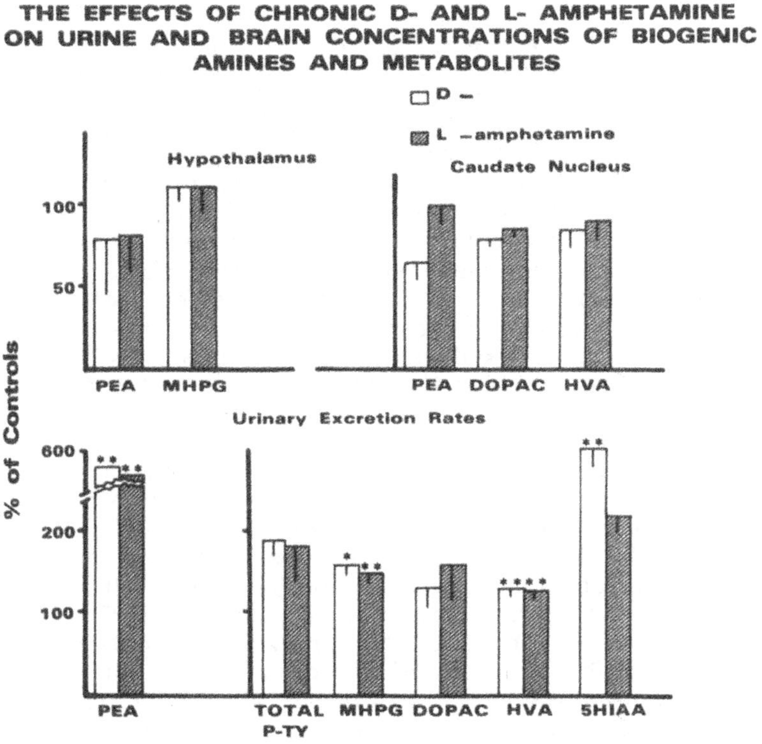

FIGURE 4

Groups of rats (five rats in each group) were treated for 14 days with daily I.P. doses of D- and L-amphetamine (10 μmol/kg; 3.7 mg/kg). Urines were collected as described in Figure 2 and brains obtained as described in Figure 3. Results are expressed (mean $\pm$ S.E.M.) as the percent of the appropriate baseline values of untreated rats.

potent inhibition of MAO type A by all four drugs. In spite of this latter, 5HT metabolism was only mildly and transiently inhibited. These data cannot be logically explained without proposing the existence ot more than one varient of MAO type A and B. Some of these enzymes may show strong in vivo preference for certain amines and thus may exhibit different in vivo sensitivities to MAOI when the deamination to these preferred amines are considered.

The effects of the above four MAOI on brain PEA and catecholamine metabolism are comparable to their effects on the urinary excretion of these substances. Thus it appears that evaluation of urine biogenic amines and their metabolites offers an useful approach to the assessment ot the central effects of MAOI in rats and possibly humans. The central effects of these tour MAOI's are summarized in Figure 3. Here again, the data cannot clearly be explained without speculating a complex MAO inhibition.

A comparison between the effects of chronic treatments with D- and L-amphetamine (10 μ mol/kg) on the urinary output ot a number of biogenic amines and their metabolites revealed no difference between these two drugs. Thus, as shown in Figure 4 both isomers of amphetamine elevated PEA excretion and reduced Ty. These two isomers also reduced NE and DA metabolism. Similar effects ot these isomers was also observed in the brain, hence, again establishing a good correlation between the peripheral effects ot these drugs and those induced in the brain. These results clearly suggest that none of the amines investgated here are directly responsible for the behavioral influences of D- and L-amphetamine and therefore, the direct effects of amphetamine isomers on behavior and probably specific receptors should be considered as the most likely causes. Furthermore, the results summarized in Figure 4 indicate that although PEA crosses easily the blood brain barrier as deduced from PEA ingestion experiments, stimulation of endogenously produced peripheral PEA does not easily lead to increased brain content of PEA.

It is concluded that while the in vitro affinity of MAO to a group of amines (preferred substrates) may be similar, the in vivo inhibition ot the metabolism of these amines by MAOI may not parallel their expected in vitro characteristics. By the same token, the in vivo affinity of a particular type of MAO towards the same amine in the periphery and the brain may not always be similar. Further, the results ot the present study suggest the existence of functionally multiple forms of MAO both in the brain and the periphery where they apparently are specialized in the metabolism ot a narrow range ot amines. As has been demonstrated for pargyline and deprenyl, the metabolites of MAOI should ideally be considered whenever the parent substances are investigated.

REFERENCES

Karoum F. (1983) Mass fragmentography in the analysis of biogenic amines: a clinical, physiological and pharmacological evaluation. In: Method of Biogenic Amine Research, (Parvez, S., Nagatsu, T., Nagatsu, I. and Parves, H., eds.). Elsevier Science Publishers, New York, pp. 237-255.

Karoum, F., Chuang, L-W., Eisler, T., Calne, D.B., Liebowitz, M.R., Quitkin, F.M., Klein, D.F. and Wyatt, R.J. (1982) Metabolism of (-) deprenyl to amphetamine and methamphetamine may be responsible for deprenyl's therapeutic benefit: A biochemical assessment. Neurology 32, 503-509.

Reynolds G.P., Elsworth, J.D., Blau, K., Sandler, M., Lee, A.J. and Stern, G.M. (1978) Deprenyl is metabolized to methamphetamine and amphetamine in man. Br. J. Clin. Pharmacol., 6, 542-544.

DIRECT AND INDIRECT EFFECTS OF MAO INHIBITORS AND TRACE AMINES ON SEROTONERGIC NEUROTRANSMISSION

N. Weiner, T. Ziegenfuß, M. Petersen, and

W. Wesemann, Abt. Neurochemie, Physiol.-Chem.

Institut II, Philipps-Univ., D-3550 Marburg

According to the "dopamine hypothesis of schizophrenia" an increase of dopamine (DA), especially in the mesolimbic system, plays an important role in the pathogenesis of this central disorder (Ackenheil et al., 1978). Since increased DA levels might reflect a deficiency of MAO [monoamine: O_2 oxidoreductase, deaminating (EC 1.4.3.4)] besides DA also the concentrations of other endogenous and exogenous amines could be enhanced in schizophrenia. On the basis of the assumption that decreased MAO activity reflects a genetic defect, the activity of this enzyme should be affected not only in the brain but also in other tissues. As a peripheral marker of synaptic function and metabolism, blood platelets are frequently used. Many studies were published on the MAO activity in platelets of patients with schizophrenia using different substrates. The results obtained are disappointing and controversial: if MAO deficiency plays a role in the pathogenesis of this disease then probably only in different subtypes of chronic schizophrenia. - Besides the "DA hypothesis" O- and N-methylated derivatives of DA and serotonin (5-HT) are discussed as a biochemical basis of schizophrenia ("transmethylation hypothesis"), since some of these derivatives induce experimental psychosis (Krüger, 1981).

The present paper describes two sets of experiments: (1) Four substrates - tyramine, tryptamine, 2-phenylethylamine, benzylamine - are used to measure the kinetics of

Table 1: MAO activity in healthy controls (n=26) and in patients with schizophrenia (n=20)

Substrate	Control K_M	Control V_{max}	Schizophrenia K_M	Schizophrenia V_{max}
Tyramine	143	57	183[a)]	60
Tryptamine	18	12	17	12
2-Phenylethylamine	15	53	11[b)]	48
Benzylamine	110	55	117	51

K_M: [µM], V_{max}: [nmol/mg protein x h]
Difference from control a) $p < 0.01$, b) $p < 0.08$

platelet MAO in patients and healthy control persons to get some information on possible shifts in the concentration of these amines in schizophrenia. (2) The effect of trace amines on 5-HT binding to nerve ending membranes is assayed since in schizophrenia these amines possibly accumulate as a consequence of a deficiency in MAO activity.

KINETICS OF BLOOD PLATELET MAO

The MAO activity in human blood platelets was determined by a modification of the radioenzymatic assay described by McCaman et al. (1965) using [^{14}C] labeled amines as substrate (Tab. 1). The reaction products were extracted with ethyl acetate and measured by liquid scintillation spectrometry. Table 1 shows that platelet MAO exhibits a relatively high affinity towards tryptamine and 2-phenylethylamine as compared with tyramine and benzylamine. In order to elucidate if MAO activity towards the substrates used is differently affected by endogenous and exogenous factors, the effect of schizophrenia, medication, smoking, and age on the kinetics was tested in a pilot study.

The MAO kinetics of 20 patients with schizophrenia were compared with the kinetic constants of control persons (7 patients with hebephrenia; International Classification of Diseases, Injuries and Causes of Death: ICD 295.1; Research Diagnostic Criteria: RDC 1.4.2; 13 patients with paranoic schizophrenia; ICD 295.3; RDC 1.4.1). The K_M value with tyramine as substrate was found to be significantly

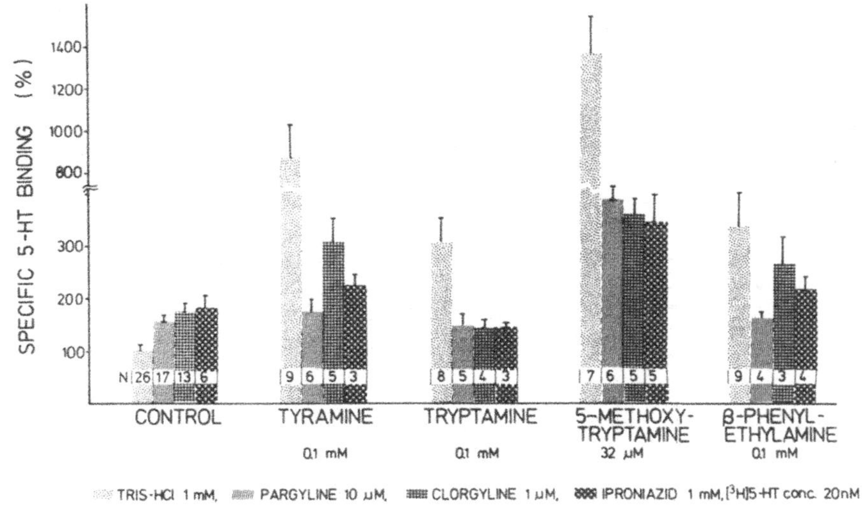

Fig. 1 Effect of MAO inhibitors and biogenic amines on specific 5-HT binding to crude membrane fractions isolated from rat brain. The activation induced by the MAO inhibitors and the amines was expressed as percentage of the control. Number of experiments: N.

enhanced in platelets from schizophrenic patients; the K_M for 2-phenylethylamine showed a tendency to lower values (Tab. 1). No significant differences of the kinetic parameters were found between patients with the diagnosis hebephrenic and paranoic schizophrenia. Though smoking did not significantly influence MAO activity in the controls, a significant decrease of the K_M and V_{max} values was found in the patient group using 2-phenylethylamine as substrate. With tyramine as substrate, in smoking patients the K_M and in nonsmoking patients the V_{max} were significantly enhanced as compared with the smoking and nonsmoking controls, respectively. An age-related significant decrease in the K_M value with tyramine as substrate was observed in the control group (volunteers $<$ 35 years / volunteers $>$ 35 years). No age relationship was found in the group of patients. The patients under investigation were treated with phenothiazines, clozapine, butyrophenones, benzodiazepines, and anticholinergics. Phenothiazines decreased the V_{max} values if tyramine or 2-phenylethylamine were used as substrates. Clozapine, benzodiazepines, and butyrophenones had no effect. Anticholinergics significantly decreased K_M and V_{max} but only if tryptamine was used as substrate.

Applying multi-variant analysis to correct for the effect of smoking, age, and medication, a significant relationship between schizophrenia and MAO activity could not be confirmed in this pilot study. Further studies are necessary with larger groups of patients selected according to the above mentioned parameters. However, the results obtained indicate that these parameters change platelet MAO activity differently depending on the substrate used.

MODIFICATION OF 5-HT BINDING BY TRACE AMINES AND MAO INHIBITORS

The individual biogenic amines are metabolized by MAO with different kinetics and the katabolism is differently affected by endogenous and exogenous factors. Hence, MAO inhibition either by a genetically determined defect or by inhibitors may interfere with 5-HT mediated neurotransmission in a direct and indirect manner: (1) the 5-HT concentration might be enhanced. (2) The trace amine levels might be increased. As a consequence 5-HT binding sites might be modulated.

To test possible indirect effects of MAO inhibition, the influence of 5-methoxytryptamine and of 3 trace amines on specific 5-HT binding to crude membrane fractions isolated from rat brain (Weiner and Wesemann, 1982) was studied. In the presence of 0.1 mM concentrations of the trace amines tyramine, tryptamine, 2-phenylethylamine, and 32 µM solutions of 5-methoxytryptamine specific 5-HT binding was increased by factor 12-14 (Fig. 1). Log dose response curves indicated that tyramine or 2-phenylethylamine concentrations as low as 10 nM enhance 5-HT binding (Weiner and Wesemann, 1982). Though trace amines may enhance 5-HT concentration by MAO inhibition, it is doubtful that the high increase of specific 5-HT binding observed in the presence of these amines can be attributed only to the competitive inhibition of MAO activity: (1) the K_M value of 95 µM reported for 5-HT metabolization by brain mitochondria differs by factor 5,000-20,000 from the K_D values of 5 and 20 nM found for 5-HT binding to synaptic membranes (White and Wu, 1975; Weiner and Wesemann, 1982). (2) Tyramine and tryptamine inhibit 5-HT oxidation by MAO competitively with K_i values of 66 and 4.3 μM contrasting with 2-phenylethylamine which exhibits no inhibitory action on 5-HT metabolization in brain mitochondria (White and Wu, 1975). All three substances, however, increased 5-HT binding to the crude membrane preparation.

MAO inhibitors are frequently used in the treatment of brain disorders, e.g. in depression or M. Parkinson. Since the possibility cannot be excluded that MAO inhibitors act directly on 5-HT binding sites, specific 5-HT binding to crude membrane fractions was measured in the presence or absence of the MAO inhibitors pargyline, clorgyline, and iproniazid. Figure 1 shows that, depending on the drug applied, 5-HT binding was increased up to about 200% by the MAO inhibitors. Since these drugs were applied in concentrations used in receptor binding experiments, this effect should be taken into consideration if studies performed in the presence or absence of MAO inhibitors are compared with each other. The results obtained for 5-HT binding in the presence of MAO inhibitors can be interpreted only partly by the assumption that these substances will maintain higher 5-HT levels in the incubation medium as compared with the untreated controls. In fact, low concentrations of clorgyline inhibit specifically metabolization of 5-HT by MAO A. But it is questionable if the increase in 5-HT binding can be explained merely by the interaction of clorgyline with the active site of the enzyme, since previous experiments showed that 5-HT binding is also increased by deprenyl which inhibits selectively the metabolism of MAO B substrates like 2-phenylethylamine (Weiner and Wesemann, 1982).

The effect of MAO inhibitors and trace amines on specific 5-HT binding was not additive (Fig. 1). The tyramine, tryptamine, 5-methoxytryptamine, 2-phenylethylamine induced activation of 5-HT binding was decreased in the presence of MAO inhibitors.

Iproniazid was used in the treatment of tuberculosis. It induced psychotic disorders in some patients. This effect was explained on the basis of MAO inhibition. However, this interpretation is not acceptable anymore since isoniazid (Neoteben®) which is now used in the treatment of tuberculosis, also induces psychotic disturbances though it lacks MAO inhibitory activity. The clinical observations and the structural similarities of iproniazid and isoniazid suggested to study the effect of isoniazid on specific 5-HT binding in the absence and presence of trace amines. As compared with iproniazid the tuberculostatic agent isoniazid exhibited qualitatively the same effect, namely an increase in specific 5-HT binding and a decrease of the activation induced by the trace amines (Weiner and Wesemann, 1982).

As far as MAO inhibitors are concerned, our working hypothesis to explain the effect of MAO inhibition must be extended: (1) MAO inhibitors may act directly on the serotonergic transmission by enhancing the 5-HT concentration and by sensitizing 5-HT binding. (2) Our experiments substantiate the assumption that varying trace amine levels may modulate 5-HT binding sites. This hypothesis would be in accordance with the side effects observed sometimes after treatment with MAO inhibitors.

REFERENCES

Ackenheil M., Hippius H., and Matussek N. (1978) Ergebnisse der biochemischen Forschung auf dem Schizophreniegebiet. Nervenarzt 49, 634-649.

Krüger H. (1981) Biologische Grundlagen. In: Krüger, H. (Hrsg.): Die Schizophrenien. Enke-Verlag, Stuttgart (1981) 32-61.

McCaman R.E., McCaman M.W., Hunt J.M., and Smith M.S. (1965) Microdetermination of monoamine oxidase and 5-hydroxytryptophan-decarboxylase activities in nervous tissues. J. Neurochem. 12, 15-23.

Weiner N. and Wesemann W. (1982) In vitro studies on the modification of 5-HT receptor sensitivity in the CNS of the rat. In: Biologische Psychiatrie (Beckmann, H., ed.) pp. 247-252. Thieme, Stuttgart and New York.

White H.L. and Wu J.Ch. (1975) Multiple binding sites of human brain monoamine oxidase as indicated by substrate competition. J. Neurochem. 25, 21-26.

SOME ASPECTS OF THE OXIDATIVE DEAMINATION OF TRYPTAMINE BY MAO-A AND MAO-B IN DIFFERENT TISSUES

P.H. Yu

Psychiatric Research Division
Cancer and Medical Research Building
University of Saskatchewan
Saskatoon, Saskatchewan
Canada S7N 0W0

INTRODUCTION

Tryptamine has been found in mammalian brain (Philips et al., 1974); it possesses a high turnover rate (Durden and Philips, 1980). Oxidative deamination catalyzed by monoamine oxidase is its major catabolic pathway (Weissback et al., 1961; Wu and Boulton, 1973). It was, at first, reported to be oxidized solely by MAO-B (Houslay and Tipton, 1974), then by both MAO-A and MAO-B (White and Tansik, 1979) but most recently predominantly by MAO-A in the human (Bieck et al., 1984). This present study was attempted in order to try to understand how MAO-A and MAO-B are involved in the deamination of tryptamine. Mitochondrial MAO-A and MAO-B was prepared from different tissues; a kinetic analysis on the deamination of tryptamine was then carried out.

MATERIALS AND METHODS

Male Wistar rats (200 g) were used in this study. Bovine and hog livers were kindly supplied by Intercontinental Packers Ltd., Saskatoon; post-mortem human brain and liver were provided by Dr. Rozdilsky, Department of Pathology, human placenta was obtained from the Department of Obstetrics and Gynecology, University of Saskatchewan; p-[1-^{14}C]-tyramine, β-[ethyl-^{14}C]-phenylethylamine, 5-[2-^{14}C]-hydroxytryptamine, [3-^{14}C]-tryptamine and Omnifluor were purchased from New England Nuclear Corp. (Boston);

depreny1 (phenylisopropylmethylpropynylamine) was a gift from Prof. J. Knoll, Semmelweis University (Budapest), and clorgyline (2,4-dichlorophenoxypropyl-N-methylpropynylamine hydrochloride) from May and Baker, Ltd., (Pagenharn, UK). All other chemicals were of an analytical grade.

The washed mitochondrial membrane fragments were prepared by differential centrifugation as previously described (Yu, 1979).

A radiometric method was utilised using ^{14}C-labelled tryptamine as substrate. A typical reaction mixture contained 0.05 M phosphate buffer (pH 7.5), 2 x 10^{-4} M tryptamine (containing 0.1 µl, ^{14}C-labelled tryptamine), and washed mitochondrial membrane fragments (15-50 µg) in a total of 200 µl. Blank values were obtained either by including 1×10^{-4} M pargyline or excluding enzyme preparation in the incubation mixture. The enzyme reactions were carried out at 37°C for 30 min. and were stopped by addition of 250 µl 2M citric acid. The reaction products were extracted into 1 ml toluene and 600 µl of the organic phase transferred to counting vials containing 10 ml omnifluor fluid, and then subjected to liquid scintillation counting. In the case of determination of MAO activity towards p-tyramine, β-phenylethylamine and serotonin, the extraction was conducted by using toluene:ethylacetate (1:1).

RESULTS AND DISCUSSION

Effect of pH on MAO Activity Towards Tryptamine

Mitochondrial MAO was obtained from human liver, placenta and brain, rat heart, liver and brain, as well as from hog and bovine liver. A pH optimum of 7.5-8.0 in the range 5.8-8.2 in 0.05 M phosphate for tryptamine deamination with respect to all the above MAO's was observed.

Apparent Kinetic Parameters of MAO Isolated from Different Tissues Utilising Tryptamine as Substrate

As can be seen from Table 1, the apparent maximal velocities (V_{max}) and Michaelis-Menten constants (K_m) for MAO isolated from different tissues differs slightly. MAO from human and rat tissues clearly exhibits higher affinity (i.e. lower K_m) for tryptamine in comparison to the MAO from bovine or hog liver.

Table 1

Comparison of the apparent V_{max} and Michaelis-Menten constants (K_m) for different monoamine oxidase with respect to tryptamine

Tissues	MAO-A V_{max}	MAO-A K_m	MAO-B V_{max}	MAO-B K_m
Human Brain	0.70	0.90	0.62	1.3
Liver	8.26	3.50	3.21	5.4
Placenta	3.28	1.50	--	--
Platelet	--	--	1.10	2.0
Rat liver	1.38	0.8	0.76	1.4
Hog liver	--	--	4.76	21.0
Bovine liver	--	--	2.92	22.0

*V_{max} (nmoles/min/mg protein) and K_m (1×10^{-5} M) were obtained from Lineweaver-Burke double reciprocal plots.

Effects of the Specific MAO-A Inhibitor Clorgyline and MAO-B Inhibitor Deprenyl on MAO Catalyzed Deamination of Tryptamine and Other Amines

In Fig. 1, the inhibition by clorgyline and deprenyl on human liver MAO activity towards tryptamine (1×10^{-4} M) is compared with that of the oxidation of the typical MAO-A substrate serotonin (1×10^{-4} M), the MAO-B substrate β-phenylethylamine (1×10^{-5} M) and the mixed type substrate *p*-tyramine (1×10^{-4} M) as catalysed by the same enzyme. Tryptamine appears to be a mixed type substrate from its biphasic inhibition manner. Approximately 70% of tryptamine is metabolized by MAO-A and 30% by MAO-B, while *p*-tyramine, another mixed type substrate, both MAO-A and MAO-B contribute about 50% each to the deamination.

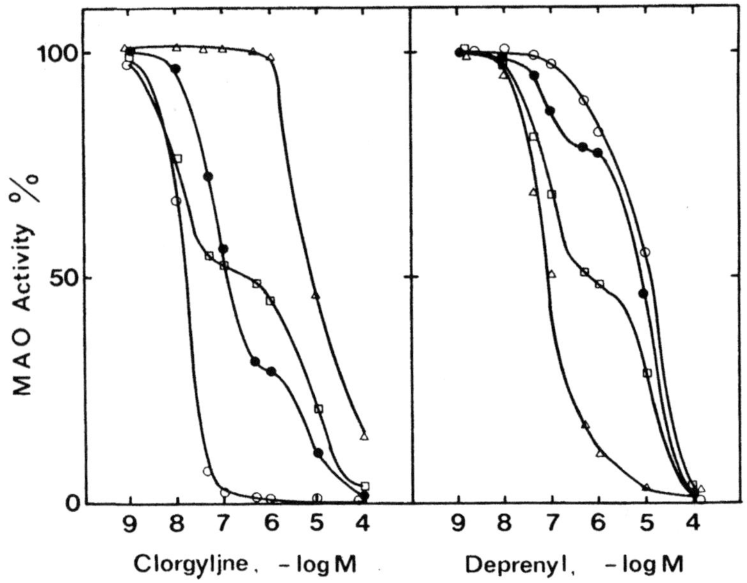

Fig. 1. Inhibition of the deamination of 100 μM tryptamine (●), serotonin (o) and p-tyramine (□) and 10 μM β-phenylethylamine (△) by clorgyline and deprenyl, catalysed by human liver MAO.

Tryptamine is oxidized exclusively by MAO-B in bovine or hog liver and human platelets and predominantly by MAO-A in human placenta MAO. In rat brain and liver, tryptamine is oxidized by both MAO-A and MAO-B. The proportion of oxidation catabolized by MAO-A or MAO-B is dependent upon the substrate concentration applied.

Inhibition of human liver MAO activity towards tryptamine at different concentrations by clorgyline and deprenyl

As shown in Fig. 2b, the single-sigmoid clorgyline inhibition curve obtained at higher concentrations of tryptamine (1×10^{-3} M) changed into a double-sigmoid curve when the tryptamine concentrations were decreased. The proportion of tryptamine oxidised by human liver MAO-A decreased with decreasing tryptamine concentration. This phenomenon can also be demonstrated using deprenyl to inhibit MAO-B selectively (result not shown).

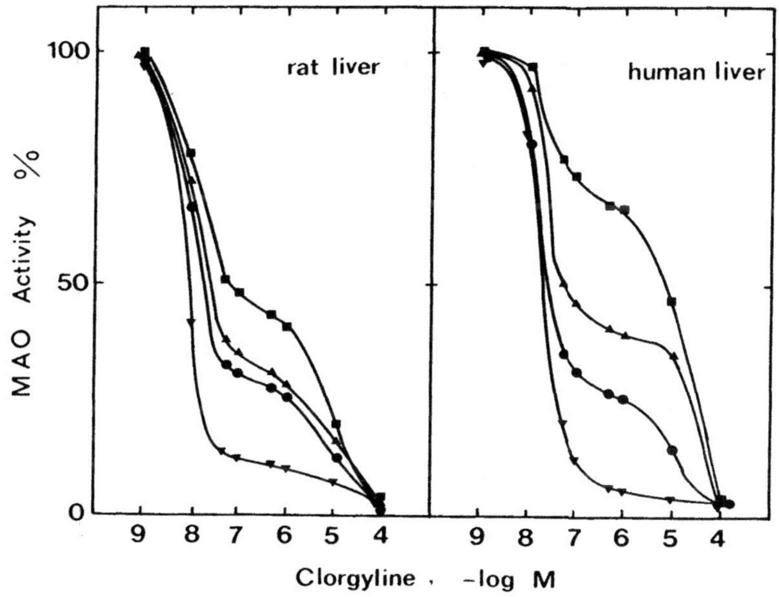

Fig. 2a and 2b. Inhibition of rat and human liver MAO activity towards tryptamine at different concentrations by various concentrations of clorgyline. The concentrations of tryptamine are: $5x10^{-5}$ M (■), $1x10^{-4}$ M (▲), $2x10^{-4}$ M (●), and $1x10^{-3}$ M (▼).

In rat liver, tryptamine was also solely deaminated by MAO-A at 1 mM, up to 50% was catabolized by MAO-B at low concentration (0.05 mM) (Fig. 2a).

Effect of Tranylcypromine on MAO Activities
Towards Tryptamine and Other Substrates

Tranylcypromine has been found to be the most effective MAO inhibitor in elevating tryptamine levels in rat brain (Philips and Boulton, 1979) and in human urinary excretion (Bieck et al., 1984). It is, therefore, interesting to know whether tranylcypromine can preferentially affect MAO activity towards tryptamine. As indicated in Fig. 3, tranylcypromine exhibits rather non-selective MAO inhibitory activity towards different substrates (including tryptamine). Also there are no distinct differences between MAO's from the rat and from the human liver in response to tranylcypromine inhibitory activity.

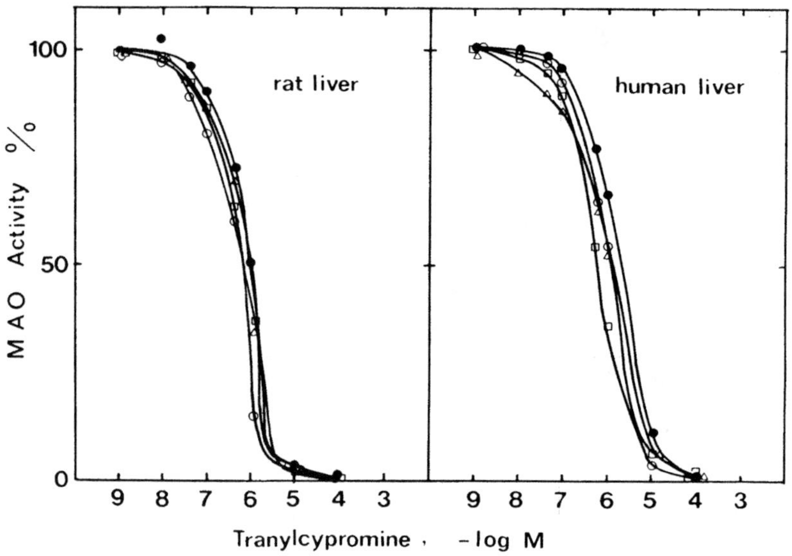

Fig. 3. Effect of tranylcypromine on rat (a) and human liver (b) mitochondrial MAO activities towards different substrates. The substrates used are 5×10^{-4} M serotonin (○), p-tyramine (□), tryptamine (●) and 1×10^{-5} M β-phenylethylamine (△).

The increases in tryptamine levels in the rat brain after i.p. administration of the MAO-B inhibitor, deprenyl, are more pronounced than those that occur with clorgyline (MAO-A inhibitor) (Philips and Boulton, 1979) indicating that MAO-B is predominantly involved in the oxidation of tryptamine in rat in vivo. In contradistinction to this, the MAO-A inhibitor CGP 11305, caused increased urinary excretion of tryptamine but not β-phenylethylamine in humans, while deprenyl enhanced β-phenylethylamine excretion but not tryptamine excretion (Bieck et al., 1984). In the present study the affinity of MAO-A and MAO-B to tryptamine, however, was found only slightly different (Table 1). From the biphasic clorgyline inhibition response curve (Fig. 2), tryptamine appears to be oxidized in vitro by both MAO-A and MAO-B, and preferably by MAO-A at relatively higher tryptamine concentrations (Fig. 3). It is, therefore, difficult to predict how MAO-A and MAO-B are involved in the in vivo oxidation of tryptamine on the basis of the in vitro inhibitor titration observations. Tranylcypromine does not exhibit a selective preference with regard to tryptamine oxidation, but does have a profound in vivo effect in rat and human (Philips and Boulton, 1979; Bieck et al., 1984). This suggests that there are other factors involved in the accumulation and excretion of tryptamine caused by tranylcypromine.

ACKNOWLEDGEMENTS

Continuing financial support by Saskatchewan Health, and the Canadian Medical Research Council is gratefully acknowledged. I also thank Mrs. D. Schneider and D. Young for their technical assistance.

REFERENCES

Bieck P.R., Nilsson E., Schick C., Waldmeier P.C. and Lauber J. (1984). Urinary excretion of tryptamine in comparison to normetanephrine and β-phenylethylamine in human volunteers after subchronic treatment with different monoamine oxidase inhibitors, in, Neurobiology of the Trace Amines (eds. A.A. Boulton, G.B. Baker, W.G. Dewhurst and M. Sandler) pp. 525-539, Humana Press, Clifton, New Jersey.

Durden D.A. and Philips S.R. (1980). Kinetic measurements of the turnover rates of phenylethylamine and tryptamine in vivo in the rat brain. J. Neurochem. 34, 1725-1732.

Houslay M.D. and Tipton K.F. (1974). A kinetic evaluation of monoamine oxidase activity in rat liver mitochondrial outer membranes. Biochem. J. 139, 645-652.

Philips S.R., Durden D.A. and Boulton A.A. (1974). Identification and distribution of tryptamine in the rat. Can. J. Biochem. 52, 447-451.

Philips S.R. and Boulton A.A. (1979). The effect of monoamine oxidase inhibitors on some arylakylamines in rat striatum. J. Neurochem. 33, 159-167.

Weissbach H., Lovenberg W., Redfield B.G. and Udenfriend S. (1961). The *in vivo* metabolism of 5-HT and tryptamine: Effects of monoamine oxidase inhibition. J. Pharm. Exp. Ther. 131, 26-32.

White H.L. and Tansik R.L. (1979). Characterization of multiple substrate binding sites of MAO, in, Monoamine Oxidase: Structure, Function and Altered Functions (eds. T.P. Singer, R.W. Von Korff and D.L. Murphy) pp. 129-144, Academic Press, New York.

Wu P.H. and Boulton A.A. (1973). Distribution and metabolism of tryptamine in rat brain. Can. J. Biochem. 51, 1104-1112.

Yu P.H. (1979). Effect of lipid-depletion of type A and type B monoamine oxidase of rat heart and bovine liver mitochondria, in, Monoamine Oxidase: Structure, Function and Altered Functions (eds. T.P. Singer, R.W. Von Korff and D.L. Murphy) pp. 233-244, Academic Press, New York.

POSSIBLE INTERFERENCE OF URIC ACID IN THE MEASUREMENT OF CATECHOLAMINES BY HPLC WITH ELECTROCHEMICAL DETECTION.

Weil-Fugazza, J. and Godefroy, F.

Unité de Neurophysiologie Pharmacologique de l'INSERM

2, rue d'Alésia, 75014 Paris, France.

INTRODUCTION

It has been reported that selective detection of catecholamines and indoleamines by differential pulse voltammetry "in vivo" may be contaminated by the simultaneous detection of uric acid (Crespi et al., 1983 ; O'Neill et al., 1984). In addition uric acid has been identified in striatal and nucleus accumbens perfusates (Ungerstedt, 1984) and brain extracts (Aoki et al., 1984) by mean of HPLC-ED. The present report addresses the possibility that uric acid may interfere in the analysis of catecholamines in brain and spinal cord extracts by mean of HPLC-ED.

METHODS AND RESULTS

A variety of chromatographic conditions proposed for the simultaneous analysis of catecholamines and indolamines were studied.

Brain and spinal cord pieces (equivalent to 10 to 1000 µg of proteins) were homogenized by sonification in 100 µl of 0.1 M $HClO_4$ containing 0.1 mM EDTA and 0.01% cysteine. After centrifugation 25 µl of the supernatant (diluted if necessary) were injected into the chromatographic system. C18 reversed-phase columns (Spherisorb ODS 1 or Hypersil ODS, 5 µm particle size),150 or 250 mm lenght, 4.6 mm I.D., were used.The mobile phase was 0.1M

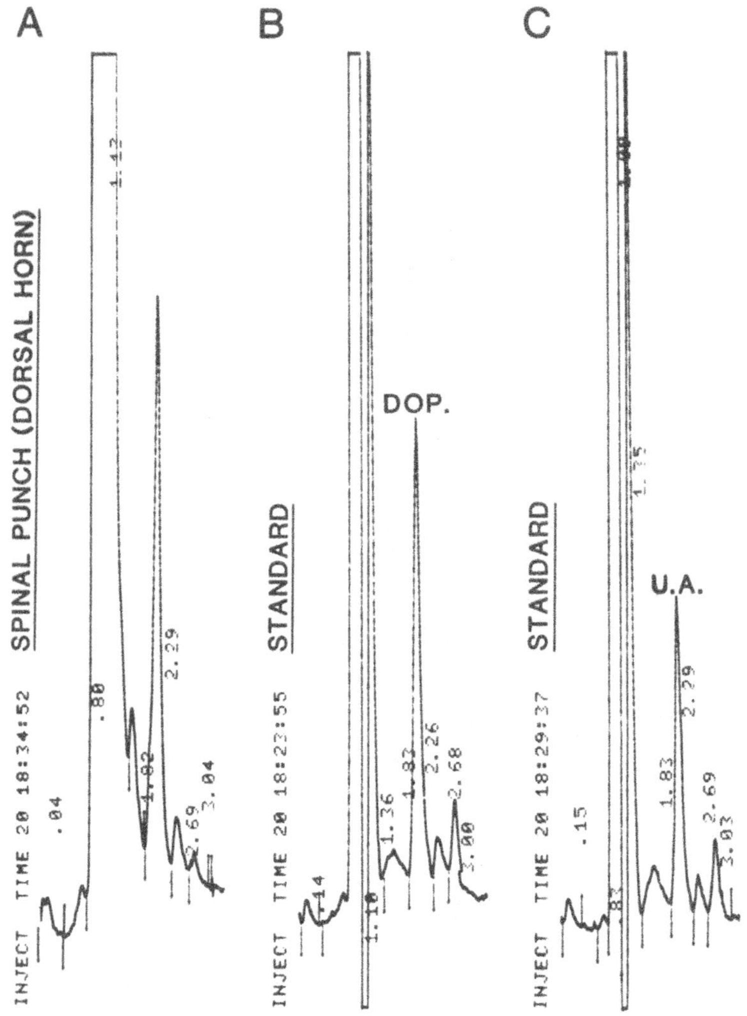

Fig. 1 : Chromatograms of a spinal cord sample (A) and of dopamine (DOP, 100 pg) standard (B) and of uric acid (U.A., 100 pg) standard (C). Mobile phase : 0.1 M phosphate buffer with 5 % methanol ; flow rate : 1.5 ml/min ; oxidation potential : 0.65 V.

phosphate buffer (K_2HPO_4 - citric acid, pH 4.7) containing 0.1 mM EDTA. Various concentrations of methanol and sodium octyl sulfate were added. The glassy carbon working electrode of the detector was set at 0.65 or 0.75 V.

Under different conditions we consistently detected, in brain and spinal cord, a compound the retention time of which corresponded to that of uric acid. Depending of the chromatographic conditions, the peak of uric acid may overlap either dopamine (Fig.1) or norepinephrine peaks (Fig.2). The addition of uricase (0.01 unit) to the standard solution of uric acid or to a tissue extract produced the disappearance of the uric acid peak.

The resolution of the peaks of uric acid and norepinephrine was obtained by using 0.1 M phosphate buffer containing 0.3 mM sodium octyl sulfate and 4% methanol as mobile phase, with a flow rate of 0.8ml/min (Fig.3). In these conditions the retention time of dopamine peak was about 9 min .

This method was applied for studies performed in arthritic rats. A significant increase in the level of uric acid was observed in the spinal cord of these rats whereas the level of norepinephrine did not seem to be significantly modified. Furthermore, the evaluation of the accumulation of 5-HTP after benserazide administration (800 mg/kg, I.P.) confirmed our previous observations (Weil-Fugazza et al., 1984) that the 5-HT synthesis was increased in the spinal cord of arthritic rats. Interestingly, several unidentified peaks were detected on chromatograms of spinal cord extracts of arthritic rats. The levels of these compounds were markedly changed after benserazide administration as shown in Fig. 3B and 3C where the computer reports of chromatograms of spinal samples (equivalent to 25 µg of proteins) are given. Experiments now in progress indicate that the compounds detected after benserazide administration are not metabolites of this inhibitor since they are detected as traces in concentrated extracts (equivalent to 250 µg of proteins) of control arthritic rats. They are also detected in normal rats after benserazide administration but it seems likely that the effect of this inhibitor is less marked than in arthritic rats. This suggests that several metabolic pathways are modified at spinal level in arthritic rats.

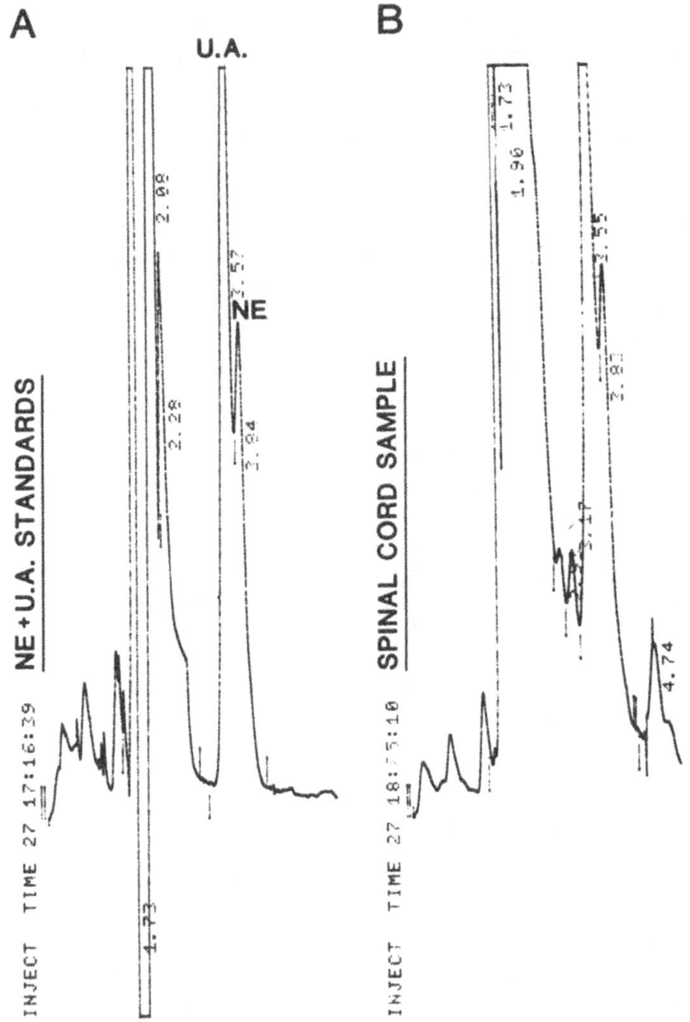

Fig. 2 : Chromatograms of norepinephrine (NE, 100 pg) and uric acid (U.A., 200 pg) standards (A) and of a spinal cord sample (B). Mobile phase : 0.1 M phosphate buffer with 0.5 mM octyl sulfate and 7.5 % methanol ; flow rate : 0.8 ml/min ; oxidation potential : 0.75 V.

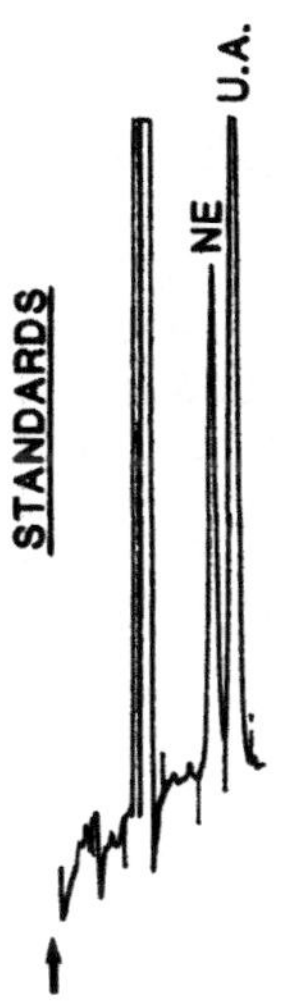

A

	RT	AREA
	1.75	52304
	2.03	95845
	2.84	42991
	3.04	35614
NE	3.53	90925
U.A.	3.95	42240
DOPAC	5.19	25847
5-HTP	7.60	9443
DOP	8.80	6104
5-HIAA	12.95	9410
5-HT	30.56	47334

B

	RT	AREA
	1.96	4491649
	2.69	34624
	3.13	17542
NE	3.53	98039
U.A.	3.91	51958
	-	-
	5.01	4540
	5.56	3028
5-HTP	7.68	6311
DOP	8.78	1203
	-	-
	-	-
5-HIAA	12.97	17239
	-	-
	17.49	2605
	20.35	4767
5-HT	30.16	13776

C

RT	AREA
1.92	4637706
2.69	46062
3.12	21137
3.51	101722
3.91	71368
4.27	49784
4.84	27422
5.56	162527
7.64	20703
-	-
9.05	4290
11.11	6889
12.97	5386
15.16	2086
17.60	923
20.47	3105
30.25	6215

Fig. 3 : Computer reports A) Standard mixture (200 pg), B) Spinal cord sample of arthritic rat, C) Spinal cord sample of arthritic rat pretreated with benserazide for 60 min. Chromatographic conditions described in text.

CONCLUSION

The present findings show that uric acid may interfere with catecholamines detection by HPLC-ED. In addition they indicate that the level of uric acid is increased in the CNS of arthritic rats. It appears consistent to postulate that this increase is related to changes in cellular metabolism due to inflammation. However, several recent reports suggest the role of adenosine as neuromodulator (see ref. in Geiger and Nagy, 1984; Feuerstein et al., 1985) and it has been proposed that variations of uric acid level in the CNS could be a valuable tool in the evaluation of the metabolism of this purine (O'Neill et al., 1984). So one can also hypothesize that the increase in uric acid observed in arthritic rats is related to a change in adenosine release at spinal level in these chronic suffering animals.

REFERENCES

Aoki T., Yoshiura M., Iwamoto, T., and Iriyama, K. (1984) Anal.Biochem. 143,113-118.

Crespi F., Sharp T., Maidment N., and Marsden, C. (1983) Neurosc. Lett., 43, 203-207.

Feuerstein T.J., Hertting G., and Jackisch, R. (1985) Eur. J. Pharmacol.,107,233-242.

Geiger J.D., and Nagy, J.I. (1984) Brain Res. Bull., 13, 657-666.

O'Neill R.D., Fillenz M., Grunewald R.A., Bloomfield M.R., Albery W.J., Jamieson C.M., Williams J.H., and Gray J.A. (1984) Neurosc. Lett., 45, 39-46.

Ungerstedt U. (1984) in Measurement of Neurotransmitter Release in Vivo (Marsden C.A., eds). pp. 81-105. John Wiley & Sons Ltd.

Weil-Fugazza J., Godefroy F., Bineau-Thurotte M., and Besson J.M. (1984) in Progress in Tryptophan and Serotonin Research (Schlossberger H.G., Kochen W., Linzen B., Steinhart H., eds) pp. 405-408. Walter de Gruyter & Co., Berlin - New York.

CLINICAL PHARMACOLOGY

Invited Communications

CHRONIC ADMINISTRATION OF MONOAMINE OXIDASE INHIBITORS: BASIC AND CLINICAL INVESTIGATIONS

Glen B. Baker[1], Ronald T. Coutts[1], Jupita M. Yeung[1], David R. Hampson[1], Gavin J. A. McIntosh[2] and Mary G. McIntosh[2].

Neurochemical Research Unit and PMHAC Research Unit (Psychiatry and Pharmacy), University of Alberta[1], and Royal Alexandra Hospital[2], Edmonton, Alberta, Canada T6G 2G3

INTRODUCTION

Tranylcypromine (Parnate) and phenelzine (Nardil) are frequently prescribed monoamine oxidase (MAO)-inhibiting antidepressants. Although there have been a large number of reports on the effects of these drugs on biogenic amines in laboratory animals, in many cases these studies concentrated on either the classical neurotransmitter amines or the trace amines and did not measure simultaneous changes occurring in both classes of amines. In addition, many of these reports dealt with effects on amines at short time periods after drug administration rather than considering the effects of long-term administration. In our laboratories we have been investigating the effects of long-term administration of these drugs on brain concentrations of several amines in rats and have carried out preliminary studies on the effects of 2-week administration of phenelzine on urine and plasma levels of some amines, their amino acid precursors and their metabolites in patients with dysthymic disorder. Some of these findings will be reported here.

METHODS

Subjects

In the animal studies, male Sprague-Dawley rats (120-140 g initial weight) were used and injected i.p. with either tranylcypromine (TCP), 1 or 10 mg/kg, or with phenelzine (PLZ), 15 mg/kg. The rats were sacrificed by cervical dislocation 1.5 h or 6 h after drug administration and the brains were immediately dissected out and frozen. This tissue was stored at -60°C until the time of analysis.

In the human investigations, patients diagnosed as having dysthymic disorder according to DSM-III criteria were studied before and 2 weeks after treatment with Nardil (15 mg t.i.d.). Urine (24 h) and blood samples (10 ml) were collected after the patients had been on a low amine diet (Klein _et al._, 1980) for at least 48 h before the day of collection. The blood samples were used for preparation of platelets and platelet-rich plasma.

Analytical Techniques

Levels of β-phenylethylamine (PEA) in brain were measured by either of two previously reported electron-capture gas chromatography (GC-ECD) techniques (Coutts _et al._, 1981; Hampson _et al._, 1984) and urinary excretion of PEA was also measured using GC-ECD, but after reaction with pentafluorobenzenesulfonyl chloride (Baker, Coutts, Rao and Kuefler, manuscript in preparation). The procedure of Coutts _et al._ (1981) was utilized for analysis of _m_- and _p_-tyramine (_m_- and _p_-TA), 3-methoxytyramine (3-MTA) and normetanephrine (NME) in urine and brain. Urinary and brain T and 5-HT were quantitated by the procedure of Baker _et al._ (1981), while amino acids in urine and plasma were measured by the procedure of Yeung _et al._ (1984). Dopamine was quantitated in brain using GC-ECD (Baker _et al._, 1983) or a radioenzymatic method (Martin _et al._, 1978), while urinary levels were analyzed using HPLC (with electrochemical detection after diluting the urine sample with 0.1 N perchloric acid).

Monoamine oxidase activity in platelets was measured by a modification of the procedure of Wurtman and Axelrod (1963), using ^{14}C-PEA as substrate.

RESULTS AND DISCUSSIONS

In the initial animal studies, doses of 10 and 15 mg/kg were chosen for TCP and PLZ. Although these are higher, on a mg/kg basis, than the doses used clinically, they are similar to, or lower than, doses of these drugs commonly used in animal studies. Using such doses in long-term studies, MAO-A and MAO-B are inhibited by > 90%

by day 2 of administration. In these investigations, groups of rats were killed 6 h after drug administration on days 1, 2, 8 and 19. At these times intervals, the following amines were analyzed in whole brain: PEA, T, 5-HT, *p*-TA, 3-MTA, NME, DA and NA. The results of these studies have been reported elsewhere (Baker *et al.*, 1982, 1983, 1984b, c), but the most interesting *findings* can be summarized as follows:

1. The levels of DA peaked by day 2 with both drugs and NA exhibited this pattern in the TCP-treated rats. With TCP, there was a decline back towards control levels for both catecholamines on day 19, while in the PLZ-treated rats there was a decline in levels from day 2 to day 8, but a increase again on day 19 compared to day 8. In the case of NA, this level at day 19 represented the maximum concentration attained by this amine in the PLZ-treated rats. The concentrations of the O-methylated metabolites of these substances continued to increase throughout the study (Baker *et al.*, 1983), with levels of both 3-MTA and NME being *significantly* higher in the TCP-treated rats than in those injected with PLZ. This difference in levels between the TCP and PLZ-treated groups may be the result of the stronger actions of TCP on transport of catecholamines at nerve terminal membranes (Hendley and Synder, 1968; Baker *et al.*, 1978, 1980).

2. 5-HT levels continued to increase throughout the study in both the TCP- and PLZ-treated groups (Baker *et al.*, 1984c).

3. Brain concentrations of PEA were much higher in the TCP-treated than in the PLZ-treated rats on days 1, 2 and 8, but by day 19, there was a marked elevation of PEA levels in the PLZ group, with values surpassing the TCP-treated rats (Baker *et al.*, 1984b). It has been postulated that this marked *increase* in PEA at this time may be due in part to PEA formed directly from PLZ as a metabolite (Baker *et al.*, 1984b). Such formation has been demonstrated indirectly (Baker *et al.*, 1982) and directly (using a mass spectrometric *technique*, Dyck *et al.*, 1985).

4. Although on day 2 *levels* of *p*-TA were higher in the TCP-treated rats, by days 8 and 19 the concentrations of *p*-TA in the PLZ-treated groups were significantly higher than in the TCP groups. Since metabolic conversion of PEA to *p*-TA is known to occur (Boulton, 1976), these increases in *p*-TA levels may be related to the substantial increases in PEA levels over the same time frame.

5. Both drugs caused a steady increase in brain levels of T from day 1 to day 19, but the levels were consid-

erably higher at all time intervals in the TCP-treated rats. This seems to be due to some effect other than just inhibition of MAO, since by day 2 both drugs have inhibited MAO-A and MAO-B by > 90%.

The levels attained by T, PEA and *p*-TA in these studies are ones at which these substances could well be having effects on reuptake mechanisms for the catecholamines and 5-HT. Many animal studies utilize much higher doses than we have used here, and under such conditions the concentrations of trace amines may reach even higher levels (Philips and Boulton, 1979). Electrophysiological studies have indicated that these trace amines even at very low concentrations can also alter the sensitivities of receptors for these classical neurotransmitter amines (Jones, 1984).

Some of this research on longterm administration has been repeated with TCP using a dose of 1 mg/kg (Baker *et al.*, 1984a), which is closer, on a mg/kg basis, to the clinical dose of this drug. Groups of rats were killed at 1.5 h after drug administration on days 1, 10 and 21 and brains were assayed for MAO activity and levels of PEA, T, 5-HT and DA. The results for PEA and T are illustrated in Figure 1 and indicate the following:

1. There is a marked increase in brain levels of both amines on day 1 of drug administration.

2. Brain levels of PEA are greater than those of T at all time intervals studied. This is similar to the situation with the higher doses mentioned previously.

3. Levels of T increase steadily from day 10 to day 21, while PEA levels remain relatively steady from days 1 through 21. These effects follow quite similar patterns to those seen at the 10 mg/kg dose of the drug.

The study on DA revealed that DA increased significantly to 1.4 x control levels on day 1. There was a slight increase from day 1 to day 21 (1,030 to 1,207 ng/g). With the 1 mg/kg dose of TCP, a pattern of 5-HT increase similar to that seen with the 10 mg/kg dose was observed although lower absolute values of 5-HT were attained. Values were significantly greater than control levels on day 1, and there was a steady increase from day 1 to 21.

The findings from the study on dysthymic disorder are summarized in Table 1 and Figure 2.

No significant differences were observed between dysthymic disorder patients pretreatment and normal matched controls in the urinary excretion of free PEA, *m*-TA, *p*-TA, DA, T or 5-HT. Urinary levels of the catecholamine metabolites 3-MT and NME were significantly greater in the pa-

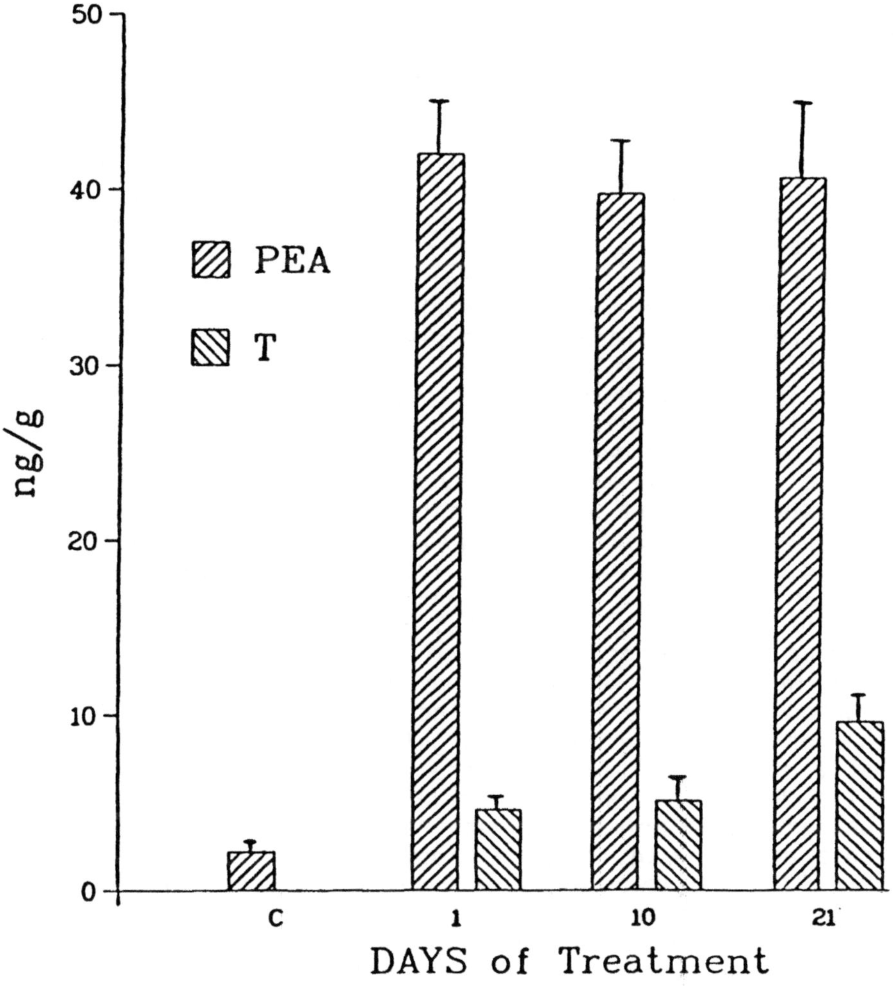

Figure 1. Concentrations of phenylethylamine and tryptamine in rat brain after acute and chronic treatment with tranylcypromine (1 mg/kg i.p.). Control levels of tryptamine were less than 1 ng/g. C=controls. Results represent means ± S.E.M. (N=4-7).

tient group than in the controls. The significance of this increase is unknown at this time since the levels of these two methoxylated amines were further increased after the 2-week treatment with PLZ. Treatment with PLZ for 2 weeks led to a mean inhibition of MAO activity of 73.8% and a significant increase in urinary levels of PEA, *m*-TA, *p*-TA, 3-MT, NME and T, but not 5-HT. As can be seen from Fig. 2, the largest increases in levels were seen with the trace amines PEA and T. The levels of PEA and 5-HT were also measured in platelet-rich plasma--PEA levels doubled after the 2 week drug treatment while 5-HT levels did not change. Analysis of the precursor amino acids phenylalanine, *p*-tyrosine and tryptophan revealed no significant differences in plasma levels pretreatment and after 2 weeks of PLZ administration.

There has been some discussion about the percentage inhibition required to increase amine levels and to cause clinical improvement. Ravaris *et al.* (1976) found that phenelzine at 45 mg/day produced a mean inhibition of 77%, in good agreement with our findings. These investigators further suggested that approximately 85% inhibition of brain MAO is needed before increases in brain levels of monoamines occur. Robinson and coworkers (1978) found that 68% of patients with at least 80% inhibition of MAO had a favourable response to PLZ while 79% of patients achieving 90% inhibition showed improvement. The pilot study reported in the present paper indicates that levels of a number of biogenic amines are significantly increased in urine at a mean inhibition of platelet MAO of < 80%. Murphy *et al.* (1977), in a study in which PLZ was administered to 15 patients with primary depressive disorders for 4 weeks at a dose of 60 mg/kg, found a mean inhibition of platelet MAO of 82% after two weeks of drug administration. Maximum values of urinary excretion of T were generally observed after 7-10 days of treatment and the mean increase was only slightly higher than that observed in our study. These workers also measured NME excretion and found a mean increase of NME levels after PLZ treatment that was approximately twice that observed in our investigation.

In further investigations on dysthymic disorder patients, we are extending the analyses to include acidic and neutral metabolites of the amines and also a number of amino acids which compete with tryptophan for its transport system (DeMyers *et al.*, 1981). Such an approach, combining studies on various types of amines, their precursor amino acids and metabolites may prove to be very valuable in understanding the relationships between the different classes

Table 1 Urinary excretion of free biogenic amines in 13 dysthymic disorder patients (pretreatment) and in 21 age- and sex-matched normal controls.

	PEA	*m*-TA	*p*-TA	3-MTA	NME	DA	T	5-HT
Controls	4.6 ± 3.7	69.1 ± 23.3	349 ± 176	26.5 ± 17.2	25.9 ± 22.8	202.1 ± 113.49	69.3 ± 38.6	112 ± 58.4
Dysthymic Disorder	6.6 ± 7.3	72.5 ± 50.8	269 ± 234	87.6* ± 76.1	44.3† ± 31.9	292.0 ± 151.1	72.0 ± 51.4	82.8 ± 41.0

Values represent µg/g creatinine in a 24-hour urine and are expressed as means ± S.D.
†$p < .05$ and *$p < .001$, Student's *t*-test.

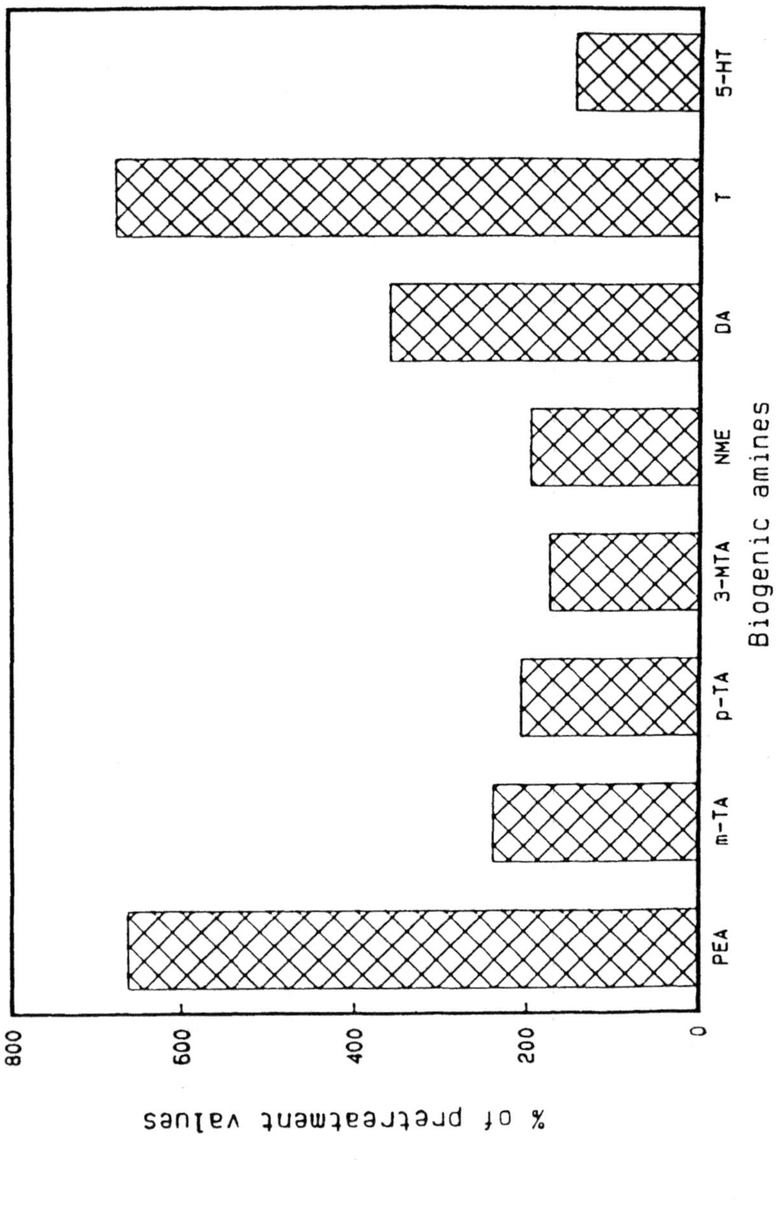

Figure 2. Effects of treatment with phenelzine (15 mg t.i.d.) on urinary excretion of free amines in patients with dysthymic disorder.

of amines, especially if it is combined with multivariate analyses of the data (Hopkinson _et al._, 1983). Such approaches have been applied to the catecholaminergic systems in major depressive disorders (Schildkraut, 1983) and may prove fruitful in future studies on the trace amines and their relationships with other biogenic amines.

In summary, levels of trace amines are altered in remarkable ways after acute and longterm administration of MAO inhibitors. We hope that the results presented here have further reinforced the proposals made by ourselves and others that studies on such drugs should include not only investigations on the classical neurotransmitter amines but also concomitant studies on the trace amines. This seems to be particularly important in the case of chronic administration of these psychotropic drugs.

ACKNOWLEGEMENTS

Funding was provided by the Provincial Mental Health Advisory Council (PMHAC), the Medical Research Council of Canada and the Alberta Heritage Foundation for Medical Research.

REFERENCES

Baker G. B., Boisvert D. P., Coutts R. T, and Hampson D. R. (1983) A sensitive electron-capture gas chromatographic assay for analysis of dopamine, noradrenaline and 5-hydroxytryptamine in regions of rat and monkey brain. _J. Neurochem._ **41**, S80B.

Baker G. B., Coutts R. T., and Martin I. L. (1981) Measurement of amines in the central nervous system using gas chromatography with electron-capture detection. _Progr. Neurobiol._ **17**, 1-21.

Baker G. B., Hampson D. R., and Coutts R. T. (1984a) Acute and chronic effects of the MAO inhibitor tranylcypromine. _J. Pharm. Pharmacol._ **36 Suppl.**, 49W.

Baker G. B., Hiob L. E., and Dewhurst W. G. (1980) Effects of monoamine oxidase inhibitors on release of dopamine and 5-hydroxytryptamine from rat striatum _in vitro_. _Cell. Mol. Biol._ **26**, 182-186.

Baker G. B., LeGatt D. F., and Coutts R. T. (1982) Effects

of acute and chronic administration of phenelzine on 2-phenylethylamine levels in rat brain. Proc. West Pharmacol. Soc. **25**, 417-420.

Baker G. B., LeGatt D. F., and Coutts R. T. (1984b) A comparison of the effects of acute and chronic administration of phenelzine and tranylcypromine on brain concentrations of 2-phenylethylamine, *p*-tyramine and tryptamine in the rat, in Neurobiology of the Trace Amines (A. A. Boulton, G. B. Baker, W. G. Dewhurst, and M. Sandler, eds.), pp. 277-282. Humana Press, Clifton, N.J.

Baker G. B., LeGatt D. F., Coutts R. T., and Dewhurst W. G. (1984c) Rat brain concentrations of 5-hydroxytryptamine following acute and chronic administration of MAO-inhibiting antidepressants. Progr. Neuropsychopharmacol. & Biol. Psychiat. **8**, 653-656.

Baker G. B., McKim H. R., Calverley D. G., and Dewhurst W. G. (1978) Effects of the monoamine oxidase inhibitors tranylcypromine, phenelzine and pheniprazine on the uptake of catecholamines in slices from rat brain regions. Proc. Eur. Soc. Neurochem. **1**, 536.

Baker J. M., Baker G. B., Coutts R. T., and LeGatt D. F. (1983) A comparison of the effects of chronic administration of tranylcypromine and phenelzine on rat brain levels of the catecholamines and their O-methylated metabolites. Proc. West Pharmacol. Soc. **26**, 361-364.

Boulton A. A. (1976) Cerebral arylalkylaminergic mechanisms, in Trace Amines and the Brain (E. Usdin and M. Sandler, eds.), pp. 21-40. Marcel Dekker, New York.

Coutts R. T., Baker G. B., LeGatt D. F., McIntosh G. J., Hopkinson G., and Dewhurst W. G. (1981) Screening for amines of psychiatric interest in urine using gas chromatography with electron-capture detection. Progr. Neuro-Psychopharmacol. & Biol. Psychiat. **6**, 373-376.

DeMeyer M. K., Shea P. A., Hendrie H. C., and Yoshimura N. N. (1981) Plasma tryptophan and five other amino acids and normal subjects. Arch. Gen. Psychiatry **38**, 642-646.

Dyck L. E., Durden D. A., and Boulton A. A. (1985) Formation of β-phenylethylamine from the antidepressant β-phenylethylhydrazine. Biochem.Pharmacol., 1925-1929.

Hampson D. R., Baker G. B., and Coutts R. T. (1984) A rapid and sensitive gas chromatographic method for quantitation of 2-phenylethylamine in brain tissue and urine. Res. Commun. Chem. Path. Pharmacol. **43**, 169-172.

Hendley, E. D., and Snyder S. H. (1968) Relationship between the action of monoamine oxidase inhibitors on the nonadrenaline uptake system and their antidepressant efficacy. Nature (London) **220**, 1330-1331.

Hopkinson G., Baker G. B., Douglass A. B., McKim H. R., and Dewhurst, W. G. (1982) Analysis of urinary excretion patterns of bioactive amines in normal control subjects. Progr. Neuro-Psychopharmacol. & Biol. Psychiat. **6**, 495-498.

Jones R. S. G. (1984) Electrophysiological studies of the possible role of trace amines in synaptic function, in Neurobiology of the Trace Amines (A. A. Boulton, G. B. Baker, W. G. Dewhurst, and M. Sandler, eds.), pp. 205-221. Humana Press, Clifton, N.J.

Klein D. F., Gittelman R., Quitkin F., and Rifkin K. (1980) Diagnosis and Drug Treatment of Psychiatric Disorders: Adults and Children. Second Edition, p. 484. Williams & Wilkins, Baltimore.

Martin I. L., Baker G. B. and Fleetwood-Walker S. M. (1978) Modification of the radioenzymatic assay for the catecholamines. Biochem. Pharmacol. **27**, 1519-1520.

Murphy D. T., Brand E., Goldman T., Baker M., Wright C., van Kammen D. and Gordon E. (1977) Platelet and plasma amine oxidase inhibition and urinary amine excretion changes during phenelzine treatment. J. Nerv. Ment. Dis. **164**, 129-134.

Philips S. R. and Boulton A. A. (1979) The effect of monoamine oxidase inhibitors on some arylalkylamines in rat striatum. J. Neurochem. **33**, 259-267.

Ravaris C. L., Nies A., Robinson D. L., Ives J. O., Lamborn K. R., and Korson L. (1976) A multiple-dose controlled study of phenelzine in depression-anxiety states. Arch. Gen. Psychiatry, **33**, 347-350.

Robinson D. L., Nies A., Ravaris C. L., Ives J. O., and Barlett D. (1978) Clinical pharmacology of phenelzine: MAO activity and clinical response, in Psychopharmacology, A Generation of Progress (M. Lipton, A. Di Mascio, and K. F. Killam, eds.). Raven Press, New York.

Schildkraut J. J., Schatzberg A. F., Mooney J. J., and Orsulak P. J. (1983) Depressive disorders and the emerging field of psychiatric chemistry, in Psychiatry Update (L. Grinspoon, ed.), Vol. II, pp. 457-471. American Psychiatric Press, Inc., Washington, D.C.

Wurtman R. J., and Axelrod J. (1963) A sensitive and specific assay for the estimation of monoamine oxidase. Biochem. Pharmacol. **12**, 1439-1440.

Yeung J. M., Baker G. B., and Coutts, R. T. (1984) Gas-liquid chromatography of isobutyl ester, N(O,S)-pentafluorobenzoate derivatives of free amino acids for quantitative analysis in biological samples. Soc. Neurosci. Abstr. **10**, 232.

FURTHER STUDIES ON PLATELET MONOAMINE OXIDASE AND PHENOLSULFOTRANSFERASE AND PLASMA LEVELS OF ACID METABOLITES IN AGGRESSIVE AND NON-AGGRESSIVE PRISONERS AND CHILDREN WITH ATTENTION DEFICIT DISORDERS

P.H. Yu[1,2], B.A. Davis[1,2], D.A. Durden[1,2] A. Gordon[3], C. Greene[3], D.M. Reid[3], D. Quinn[2], P. Matthews[2], C. D'Arcy[1,2] and A.A. Boulton[1,2]
Psychiatric Research Division, Saskatchewan Health[1], and Department of Psychiatry, University of Saskatchewan[2], Saskatoon, Saskatchewan S7N OWO and Regional Psychiatric Centre[3], Saskatoon, Sasktchewan S7K 3X5

INTRODUCTION

Trace amines are possibly involved in some psychiatric disorders (Boulton, 1980). β-Phenylethylamine, for example, is chemically similar to amphetamine, and can induce behavioural effects in laboratory animals resembling those of amphetamine (Mantegazza and Riva, 1963). Amphetamine abuse in large doses is often associated with homicidal violence (Ellinwood, 19071; Allen et al., 1975). Paradoxically amphetamine can suppress aggressive behaviour (Miczek, 1974) and hyperactivity in children (Arnold et al., 1978). It has been argued that the hyperactive child syndrome and adult psychopathy are related in that a common pathophysiological condition of low CNS arousal states may underlie both conditions (Satterfield, 1978). β-Phenylethylamine has, therefore, become a center of interest in research on mental disorders. Trace amines are metabolized rapidly (Wu and Boulton 1975; Philips and Boulton, 1979; Durden and Philips, 1980). Enzymes involved in the catabolism of the trace amines, monoamine oxidase (MAO) and phenolsulfotransferase (PST), as well as the trace acid metabolites themselves have been analyzed. Sandler et al. (1978) has claimed that the plasma phenylacetic acid (PAA) is significantly elevated in aggressive psychopaths. A subsequent study of ours, however, indicated a reduction in conjugated PAA levels in the plasma of violent offenders (Boulton et al., 1983). During the course of this study it became clear that the classification of violence merely on the basis of offence and court record may not be satisfactory. We have, therefore, in this study included an assessment of aggressiveness in the institution in an attempt to select truly violent subjects. In addition to PAA we have also investigated other acid metabolites arising from the trace amines, catecholamines and indoleamines. We have also assessed a group of children with attention deficit and hyperactive disorders.

MATERIALS AND METHODS

Assay of platelet MAO and PST activities

Platelets and plasma were prepared by differential centrifugation and stored at -60°C as previously described (Yu et al., 1982). MAO activity was determined radioenzymatically using ^{14}C-labelled p-tyramine ($1x10^{-4}$ M), tryptamine ($1x10^{-4}$ M) and β-phenylethylamine ($1x10^{-5}$ M) as substrates (Yu et al., 1982).

Platelet PST activity was also determined radioenzymatically using ^{35}S-labelled phosphoadenosine-phosphosulfate as sulfate group donor as previously described (Yu et al, 1985).

Analytical procedure for plasma trace acids

To plasma (0.5 mL) we added the deuterium-labelled internal standards, anti-oxidants, carrier acids and 0.5 mL of 10% sulfosalicylic acid. After centrifugation the supernatant was saturated with 0.5 g sodium chloride and extracted with ethyl acetate (3x2 mL). The extract was then divided into two portions, each portion was treated with triethylamine (5 µL) and the solvents evaporated in a stream of nitrogen. One portion was derivatized with methanolic HCl followed by pentafluoropropionic anhydride (PFPA), the second portion with trifluoroethanol and PFPA as described by Davis et al. After the derivatization was complete the two portions were combined and concentrated and then dissolved in hexane and washed with phosphate buffer. The hexane layer was concentrated and 0.5 mL injected into a capillary column of a gas chromatograph connected directly to the ion source of a high resolution mass spectrometer (VG-70-70F), which was programmed to detect each compound and its deuterium isotopomer by high resolution selected ion monitoring. PAA, m-HPA, p-HPA and HVA were measured as their TFE-PFP derivatives, and MA, p-HMA and VMA as their methyl-PFP derivatives. DOPAC, m-HMA, MHPG and 5-HIAA can also be measured by this procedure (as their methyl-PFP derivatives).

For the determination of total acid levels, the plasma samples were hydrolyzed in 5% sulfosalicylic acid at 110° C for 18 hours. The conjugated acids were then estimated by subtraction of the unconjugated acid levels from those of the total acid levels.

Research Subjects

One hundred and twenty-two inmates from a federal penitentiary (Regional Psychiatric Centre, Saskatoon) volunteered to participate in this study. Subjects were recruited during their medical examination upon admission to the penitentiary in the period August, 1983 to September, 1984. They were assigned to violent (V, n=82), semi-violent (SV, n=19) and non-violent (NV, n=21) categories as previously described (Boulton et al., 1983). Among these inmate volunteers, 88 of them have been assessed for in-prison aggression, based on the frequency of general verbal

Table 1

Classification of inmates according to the court record and their in-prison behavior

	Current Offences			
In Prison Rating	Violent (V)	Semi-Violent (SV)	Non-Violent (NV)	Totals
Aggressive (A)	13	6	8	27
Non-aggressive (NA)	44	11	6	61
Totals	57	17	14	88

and physical aggressiveness. Table 1 lists the numbers of inmates in these various categories. Unfortunately, the numbers turned out to be unevenly distributed among these different categories. Most of the admitted subjects had committed violent offences, but most were then non-aggressive in the prison. It is interesting that 57% of the non-violent subjects became aggressive in prison, but that only 23% of the violent offenders were recorded as aggressive in this study.

Eleven children (age 9.2 ± 2.2 years, 9 boys and 2 girls) with attention deficit disorder and hyperactivity, who were involved in a methylphenidate treatment program, were also investigated. The platelet MAO and plasma acid metabolite levels before and after methylphenidate (average 20 mg/day) treatment have been assessed.

RESULTS AND DISCUSSION

Violence and aggression studies

Human platelet MAO is thought to be at least in part under genetic control (Wyatt et al., 1973). We have previously demonstrated that platelet MAO activity is reasonably constant over a period of 6 weeks in healthy subjects (Yu et al., 1982). In the present study a group of inmates, who were subjected to a psychosocial type of treatment program, were investigated. Their platelet MAO activity

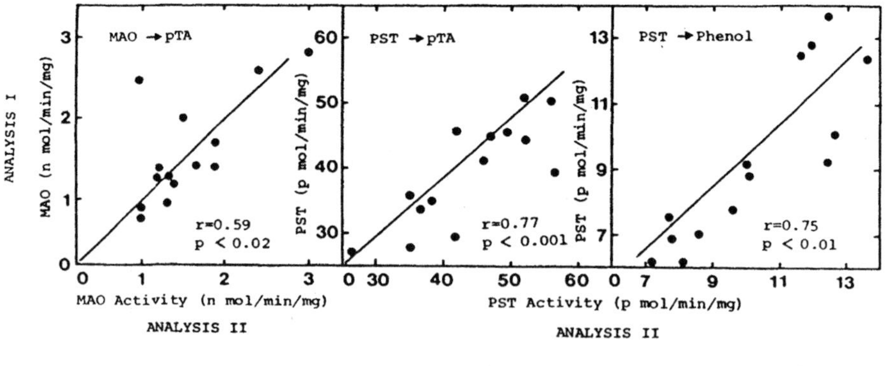

Fig. 1. Platelet MAO and PST activities in a group of inmates before (Analysis I) and 3 months after (Analysis II) a psychosocial type of treatment.

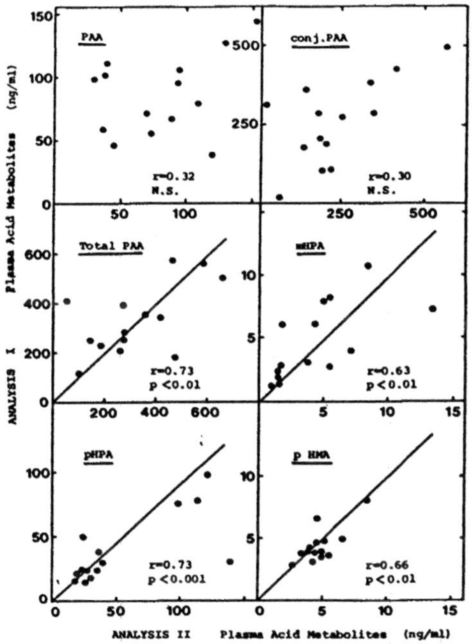

Fig. 2. Correlation of plasma levels of trace acid metabolites in a group of prisoners before (Analysis I) and after (Analysis II) a psychosocial type of treatment. The acids are unconjugated phenylacetic acid (PAA), conjugated phenylacetic acid (conj. PAA), and unconjugated m-hydroxyphenylacetic acid (m-HPA), p-hydroxyphenylacetic acid (p-HPA) and p-hydroxymendelic acid (p-HMA).

before and 3 months after such a treatment was analyzed. As can be seen in Fig. 1, MAO activity towards p-tyramine correlates significantly between the two assays, confirming that platelet MAO is a constant biological factor. There is no indication that platelet MAO activity changed within the 3 months despite the changing psychosocial environment.

Sulfate conjugation may significantly contribute to the catabolism of biogenic amines (Roth and Riveti, 1982) and is possibly involved in the aetiology of mental disorders (Sandler et al., 1983). Platelets possess quite high PST activity and have thus been used in research in psychiatric illness. Both PST-M (using p-tyramine as substrate) and PST-P (using phenol as substrate) in the same group of inmates have been analyzed. As can be seen from Fig. 1, PST activities are also quite constant. This is also consistent with the notion that human platelet PST is under a high degree of genetic control (Reveley et al., 1983).

The acid metabolite levels in plasma also appear to be reasonably consistant between two analyses separated by a 3 month interval (Fig. 2 and Fig. 3) although a considerable individual variation exists. Although unconjugated and conjugated PAA did not correlate in the two analyses, total PAA correlated significantly.

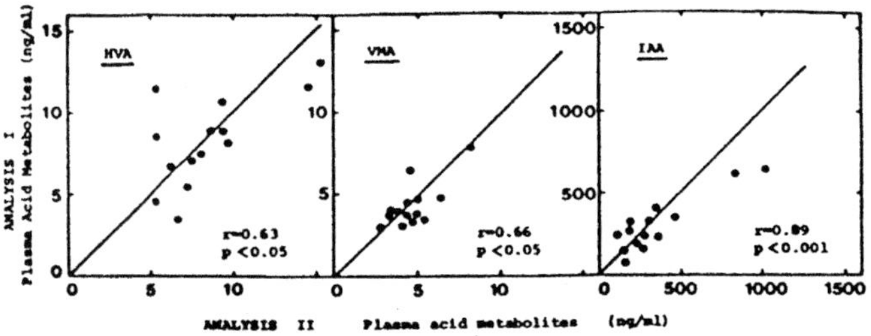

Fig. 3. Correlation of plasma levels of homovanillic acid (HVA), vanilmandelic acid (VMA) and indoleacetic acid (IAA) in a group of prisoners before and after a form of psychosocial treatment.

When age, sentence term and biochemical factors in the 122 inmates categorised according to the court record into three groups, violent (V), semi-violent (SM) and non-violent (NV) were analyzed by ANOVA, none turned out to be significantly different except the longer term of sentence in the violent group which isn't so surprising (see Table 2).

When the comparison is based on the in-prison aggression rating, PAA (unconjugated, conjugated and total) levels were all found to be significantly lower in the aggressive group, while all other biochemical factors remained unchanged (Table 3).

Table 2

A comparison of platelet MAO and PST activities and plasma acid metabolite levels in inmates categorised as violent, semi-violent and non-violent on the basis of their court record.

	NV (n=21)	SV (n=19)	V (n=82)
Age (year)	29.0 ± 1.2	30.4 ± 1.7	28.8 ± 0.9
Sentence (month)	44.7 ± 7.3	76.6 ±15.9*	138.9 ±12.1*
MAO			
pTA	1.84± 0.87	1.79± 0.12	1.85± 0.09
PE	10.21± 0.90	10.22± 0.83	11.00± 1.19
T	0.58± 0.09	0.55± 0.09	0.53± 0.02
PST			
pTA	37.06± 1.71	39.42± 2.27	37.99± 1.53
Phenol	9.82± 0.50	11.51± 0.74	10.16± 0.67
PAA	94.9 ± 9.7	73.7 ± 9.5	102.2 ±10.0
conj. PAA	263.5 ±26.3	246.6 ±26.5	246.4 ±17.5
Total PAA	363.5 ±35.7	320.0 ±34.4	343.7 ±24.8
pHPA	43.7 ± 4.9	47.8 ± 5.8	59.8 ± 6.7
conj. pHPA	20.0 ± 3.4	14.7 ± 7.6	5.5 ± 1.6
mHPA	7.4 ± 1.2	5.3 ± 1.3	5.5 ± 0.5
conj. mHPA	0.9 ± 0.5	1.7 ± 0.8	0.8 ± 0.3
HVA	7.9 ± 0.6	7.3 ± 0.6	8.9 ± 0.3
IAA	201.4 ±16.5	295.6 ±73.9	332.1 ±44.6
pHMA	8.7 ± 0.8	10.0 ± 1.6	8.8 ± 0.4
VMA	4.9 ± 0.3	4.6 ± 0.4	5.4 ± 0.5

mean ± S.E.

Table 3

A comparison of platelet MAO and PST activities and plasma acid metabolite levels in inmates categorised as aggressive or non-aggressive on the basis of their behaviour in prison

	na	a	
	(n=61)	(n=27)	t
Age (year)	29.4 ± 1.0	29.5 ± 1.6	0.02 NS
Sentence (month)	113.2 ±13.9	146.5 ±22.7	1.29 NS
MAO			
pTA	1.85± 0.10	1.86± 0.15	0.04 NS
PE	10.05± 0.56	10.09± 0.76	0.04 NS
T	0.55± 0.05	0.62± 0.09	0.70 NS
PST			
pTA	37.91± 1.98	38.60± 1.58	0.22 NS
Phenol	10.34± 0.86	11.29± 0.62	0.70 NS
PAA	116.9 ±13.1	70.6 ± 5.6	-2.31 $p<0.05$
conj. PAA	275.5 ±21.5	212.6 ±19.6	-1.80 $p<0.05$
Total PAA	391.2 ±31.4	283.1 ±23.7	-2.17 $p<0.05$
pHPA	65.6 ± 8.7	44.5 ± 4.1	-1.58 NS
conj. pHPA	8.9 ± 3.3	5.8 ± 1.8	-0.63 NS
mHPA	6.1 ± 0.6	6.2 ± 1.1	0.14 NS
conj. mHPA	0.9 ± 0.4	1.0 ± 0.6	0.22 NS
HVA	8.5 ± 0.4	8.7 ± 0.5	0.40 NS
IAA	368.9 ±62.2	220.1 ±14.2	-1.58 NS
pHMA	8.7 ± 0.4	9.5 ± 1.2	0.85 NS
VMA	5.7 ± 0.6	4.9 ± 0.2	-0.74 NS

mean ± S.E.

Table 4

A comparison of platelet MAO and PST activities and plasma acid metabolite levels in inmates categorised as non-violent/non-aggressive or violent/aggressive

	NV-na (n=6)	V-a (n=13)	t	
Age (year)	28.5 ± 3.2	30.3 ± 2.7	0.4	NS
Sentence (month)	26.6 ± 5.6	190.1 ±31.9	3.42	p<0.001
MAO				
pTA	1.53± 0.38	1.71± 0.24	0.42	NS
PE	9.06± 2.24	8.98± 1.13	0.03	NS
T	0.58± 0.23	0.55± 0.09	0.03	NS
PST				
pTA	34.51± 4.1	37.8 ± 2.3	0.77	NS
Phenol	10.2 ± 0.9	11.3 ± 1.1	0.68	NS
PAA	126.5 ±21.1	77.9 ± 9.6	-2.43	p<0.05
conj. PAA	325.5 ±35.0	214.3 ±29.0	-2.27	p<0.05
Total PAA	452.0 ±55.1	282.2 ±33.7	-2.77	p<0.01
pHPA	44.9 ±11.0	47.0 ± 7.3	0.16	NS
conj. pHPA	20.5 ±10.5	6.1 ± 3.3	-1.70	NS
mHPA	6.2 ± 1.2	4.6 ± 0.8	-1.17	NS
conj. mHPA	0.3 ± 0.2	1.3 ± 1.0	0.65	NS
HVA	5.3 ± 0.6	9.0 ± 0.8	3.08	p<0.01
IAA	213.3 ±45.2	229.6 ±24.4	0.35	NS
pHMA	7.0 ± 1.7	8.3 ± 0.8	0.82	NS
VMA	4.9 ± 0.6	4.8 ± 0.3	-0.16	NS

mean ± S.E.

If the inmates are grouped differently, i.e., the non-violent-non-aggressive vs violent-aggressive (see Table 1) are compared, the reduction in PAA (Table 4) remained consistant with the previous analysis (Table 3). In addition the HVA levels became significantly higher in the aggressive group.

The finding that PAA was low in aggressive inmates is partly in accord with our observation that PAA and pHPA in plasma is reduced in aggressive mice (Dourish et al., 1982). It is, however, difficult to conclude, at this time, how specific and what mechanism of PAA production underlie this finding or whether in fact it is even related to aggression, especially since plasma PAA levels have been found to be reduced in depressive patients (Yu et al., 1983; Sabelli et al., 1983). It is quite possible that the aggressiveness might be related to depression. The increase of HVA along with the decrease of PAA in the V-a group suggests that phenylalanine hydroxylation might be induced; if this were so the transamination and decarboxylation of phenylalanine would also be reduced and thus the PAA level would decrease.

Attention deficit and hyperactive studies

Methylphenidate has been widely used in the treatment of attention deficit disorder associated with hyperactivity (Barkley, 1977). After administration of methylphenidate the urinary excretion of some aromatic amines was found to be changed (Zametkin et al., 1985). The biochemical effects of methylphenidate are certainly different from those of another stimulant, dextroamphetamine, which is also used in the treatment of hyperkinesis in children. Dextroamphetamine increases the urinary excretion of β-phenylethylamine but methylphenidate does not (Zametkin et al., 1985).

In the pilot study described here, platelet MAO as well as plasma acid metabolite levels in children with attention deficit disorders, who were treated with methylphenidate, have been analyzed. As can be seen from Fig. 4, platelet MAO activity was reduced in 9 out of 11 of the children. Interestingly, dextroamphetamine has been reported to increase platelet MAO activity in children with

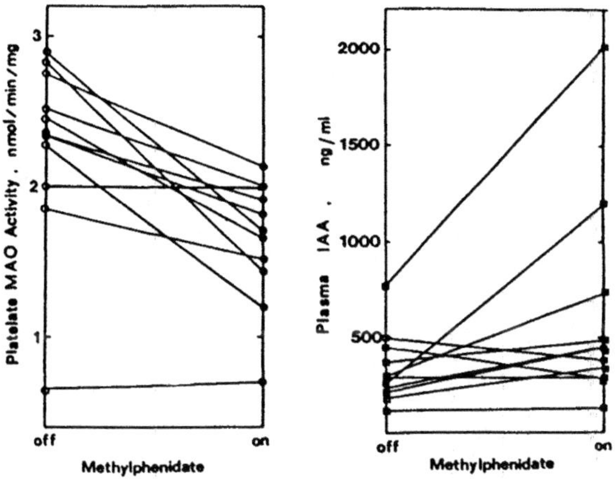

Fig. 4. Platelet MAO activity toward p-tyramine and plasma levels of IAA in children with attention deficit disorder before and during (at least two weeks medication) treatment of methylphenidate (average 20 mg/day).

these same disorders (Shekim et al., 1982). The reduction of platelet MAO activity is not due to direct inhibition of methylphenidate, since methylphenidate does not affect MAO activity, up to 10^{-4} M, *in vitro* (result not shown).

The plasma levels of the acid metabolites were not changed significantly after methylphenidate treatment with the exception of IAA. Plasma IAA levels increased in 7 children, was unchanged in 2 and decreased in 2 of the 11 subjects medicated with methylphenidate (see Fig. 4). IAA concentration in plasma is quite high. It is the major metabolite of tryptamine, but this is not necessarily its sole metabolic source. Further studies on tryptophan and tryptamine metabolism in children with attention deficit and hyperactivity syndromes seem indicated.

ACKNOWLEDGEMENTS

We thank Saskatchewan Health, the Medical Research Council, and the Correctional Service of Canada for financial support.

REFERENCES

1. Allen R.P., Safer D. and Covi L. (1975). Effects of psycho stimulants on aggression. J. Nerv. Ment. Disord. 160, 138-145.
2. Arnold L.E., Christopher J., Huestis R. and Smeltzer D.J. (1978). Methylphenidate vs. dextroamphetamine vs. caffeine in minimal brain dysfunction: controlled comparison by placebo without design with Bayes analysis. Arch. Gen. Psychiat. 35, 463-473.
3. Barkley R. (1977). A review of stimulant drug research with hyperactive children. J. Child Psychol. Psychiat. 18, 137-165.
4. Boulton A.A. (1980). Trace amines and mental disorders. Can. J. Neurol. Sci. 7, 261-263.
5. Boulton A.A., Davis B.A., Yu P.H., Wormith J.S. and Addington D. (1983). Trace acid levels in the plasma and MAO activity in the platelets of violent offenders. Psychiat. Res. 8, 19-23.
6. Davis B.A., Durden D.A. and Boulton A.A. (in press). Simultaneous analysis of twelve biogenic amine metabolites in plasma, cerebral spinal fluid and urine by capillary column gas chromatography-high resolution mass spectrometry with selected ion monitoring. J. Chromat. Biomed. Appl.
7. Dourish C.T., Davis B.A., Dyck L.E., Jones R.S.G. and Boulton A.A. (1982). Alteration in trace amine and trace acid concentrations in isolated aggressive mice. Pharmacol. Biochem. Behav. 17, 1291-1294.
8. Durden D.A. and Philips S.R. (1980). Kinetic measurement of the turnover rates of phenylethylamine and tryptamine in vivo in the rat brain. J. Neurochem. 34, 1725-1732.
9. Ellinwood E.H. (1971). Assault and homocide associated with amphetamine abuse. Am. J. Psychiat. 127, 1170-1175.
10. Mantegazza R. and Riva M. (1963). Amphetamine-like activity of beta-phenylethylamine after monoamine oxidase inhibitor in vivo. J. Pharm. Pharmacol. 15,472-478.
11. Miczek K.A. (1974). Inter species aggression in rats: effects of d-amphetamine and chlordiazepoxide. Psychopharmacol. 39, 279-301.
12. Philips S.R. and Boulton A.A. (1979). The effect of monoamine oxidase inhibitors on some arylakylamines in rat striatum. J. Neurochem. 33, 159-167.

13. Reveley A.M. Bonham-Carter S.M., Reveley M.A. and Sandler M. (1984). A genetic study of platelet phenolsulphotransferase activity in normal and schizophrenic twins. J. Psychiat. Res. 17, 303-307.
14. Roth J.A. and Riveti J. (1982). Does sulfate conjugation contribute the metabolic inactivation of catecholamines in human? Biochem. Pharmacol. 31, 3017-3021.
15. Sabelli H.C., Fawcett J., Gusovsky F., Javaid J. and Edwards J. (1983). Urinary phenyl acetate: A diagnostic test for depressioin. Science 220, 1187-1188.
16. Sandler M., Bonham-Carter S.M. and Walker P.L. (1983). Tyramine and depressive illness, in Neurobiology of the Trace Amines (eds. A.A. Boulton, G.B. Baker, W.G. Dewhurst and M. Sandler) pp. 487-498, Humana Press, Clifton, New Jersey.
17. Sandler M., Ruthven C.R.J., Goodwin B.L., Field H. and Matthews R. (1978). Phenylethylamine overproduction in aggressive psychopaths. The Lancet 2, 1269-1270.
18. Satterfield J.H. (1978). The hyperactive child syndrome: a precursor of adult psychopathy? In, Psychopathic Behavior: Approaches to Research (eds. R.D. Hare and D. Schalling) pp. 329-346, John Wiley, New York.
19. Shekim W.O., Davis L.G., Bylund D.B., Brunngraber E., Fikes L. and Lanham J. (1982). Platelet MAO in children with attention deficit disorder and hyperactivity: a pilot study. Am. J. Psychiat. 139, 936-938.
20. Wu P.H. and Boulton A.A. (1975). Metabolism, distribution and disappearance of injected β-phenylethylamine in the rat. Can. J. Biochem. 53, 42-50.
21. Wyatt R.J., Murphy D.L., Belmaker R., Cohen S., Donnelly C.M. and Pollin W. (1973). Reduced monoamine oxidase activity in platelets: a possible genetic marker for vulnerability to schizophrenia. Science 179, 916-917.
22. Yu P.H., Bowen R., Carlson K., O'Sullivan K. and Boulton A.A. (1982a). A comparison of some biochemical properties of platelet monoamine oxidase in some mentally disordered and healthy individuals. Psychiat. Res. 6, 107-121.
23. Yu P.H., Bowen R., Davis B.A. and Boulton A.A. (1983). Platelet monoamine oxidase activity and trace acid levels in plasma of agoraphobic patients. Acta Psychiat. Sand. 67, 188-194.

24. Yu P.H., Rozdilsky B. and Boulton A.A. (1985). (In press). The sulfate conjugation of monoamines in human brain: purification and some properties of an arylamine sulfotransferase from cerebral cortex. J. Neurochem.

25. Zametkin A.J., Karoum F., Linnoila M., Rapaport J.L., Brown G.L., Chuang L.W. and Wyatt R.J. (1985). Stimulants, urinary catecholamines, and indoleamines in hyperactivity: a comparison of methylphenidate and dextroamphetamine. Arch. Gen. Psychiat. 42, 251-255.

TRACE INDOLEAMINES IN THE CENTRAL NERVOUS SYSTEM

D.L. Murphy, L. Tamarkin, N.A. Garrick,

P. L. Taylor, and S.P. Markey

Laboratory of Clinical Science
National Institute of Mental Health
Bethesda, Maryland 20205 U.S.A.

INTRODUCTION

Tryptamine has been widely accepted as a trace amine in brain, and several reviews of its metabolism, regional localization in brain, and possible neuroregulatory functions have been published (Jones, 1982; Juorio and Durden, 1984). Other endogenously synthesized indoleamines have received only minimal attention in several review monographs and symposium volumes devoted to the trace amines (Baldessarini and Fischer, 1978; Mosnaim and Wolf, 1978, 1980; Boulton et al., 1984). Among these indoleamines are melatonin, N-acetyl serotonin, 5-methoxytryptamine and 6-methoxytryptoline. While these latter substances have been regarded as principally pineal hormones with peripheral actions, recent evidence has suggested that they may also act directly in the brain. Other indoleamines which also have been infrequently discussed in the trace amine literature include the hallucinogenic tryptamines such as N, N dimethyltryptamine(DMT), 5-methoxy DMT and N-methyl DMT.

TRYPTAMINE

Tryptamine is directly formed in brain from the decarboxylation of L-tryptophan. Tryptamine produced or administered peripherally also readily crosses the blood-brain barrier (Lemberger, et al., 1971; Oldendorf and Braun, 1976). In this regard it differs from the major indoleamine in brain, serotonin, which is also formed in the brain from L-tryptophan via tryptophan hydroxylase. While rat brain tryptamine concentrations [approximately 1 ng/g (Young, 1983)] are minute compared to those of serotonin, tryptamine's turnover rate in rodents is much more similar to that of serotonin (Durden and Phillips, 1980), and tryptamine may be a more important product of L-tryptophan in humans than in rodents (Young, 1983). Tryptamine is not stored in vesicles in neurons, but is rapidly metabolized by monoamine oxidase (MAO), forming indoleacetic acid.

As part of a series of studies exploring the role of serotonin, other indoleamines, and drugs which affect serotonin in neuroendocrine function, we recently examined the extent to which tryptamine might affect prolactin and cortisol release. In the last several years, L-tryptophan as well as the serotonin releasing agent, fenfluramine, and the serotonin receptor agonist, m-chlorophenylpiperazine, have been shown to elevate plasma prolactin and cortisol in rodents, non-human primates and humans (Fuller et al., 1981; Mueller, et al., in press; Charney et al., 1982; Siever et al., 1984; Aloi et al., 1984).

L-tryptophan must be given intravenously in large doses (80-100mg/kg) to produce prolactin elevations in humans (MacIndoe and Turkington, 1973; Charney et al., 1982). L-tryptophan's effects in these studies have been primarily interpreted in terms of the well-known action of serotonin to elevate prolactin via an intra-hypothalamic action. However, there is good evidence that L-tryptophan administration leads to dose-proportionate elevations in brain tryptamine (Warsh et al., 1979). After L-tryptophan treatment, the highest concentrations of tryptamine are found in the hypothalamus.

Tryptamine's neuromodulatory properties have been recently reviewed (Jones, 1982): Tryptamine can be released, like serotonin, from brain slices by electrical stimulation and by potassium chloride. Its release, like that of serotonin, is enhanced by serotonin-selective uptake blockers such as fluoxetine. One of the physiologic consequences of tryptamine administration, hyperthermia, can be blocked by the serotonin receptor blocking drug, metergoline, and enhanced by MAO-inhibiting drugs. High affinity binding sites for ^{3}H-tryptamine, which are distinct from those for serotonin, have been described (Kellar and Cascio, 1982).

Table 1. Plasma Prolactin Changes in Rhesus Monkeys Following Intravenous Tryptamine, Serotonin, Fenfluramine, M-Chlorophenylpiperazine and Saline

Drug	Dosage (mg/kg)	Prolactin Change (Δ, ng/ml)
Tryptamine	0.1	1.2 ± 0.7
	1.0	1.5 ± 0.6
	10.0	4.9 ± 1.7
	20.0	13.6 ± 5.4
Serotonin	0.1	2.9 ± 2.6
	1.0	10.2 ± 4.7
Fenfluramine	1.5	10.0 ± 1.8
M-Chlorophenylpiperazine	1.5	9.6 ± 2.2
Saline		-1.5 ± 0.7

*Mean Baseline Plasma Prolactin was 11.28 ± 1.04 ng/ml

Tryptamine's possible actions on neuroendocrine functions such as prolactin or cortisol release do not appear to have been studied in humans or non-human primates, even though tryptamine is known to increase the release of serotonin as well as dopamine, the two major amine neurotransmitter systems which regulate prolactin release.

As indicated in Table 1, tryptamine administered intravenously to rhesus monkeys in our studies produced dose-dependent increases in plasma prolactin measured by radioimmunoassay. Peak prolactin elevations occurred 20 minutes after tryptamine administration, and prolactin levels returned to baseline within 40-50 minutes after injection. The prolactin elevations were comparable to those observed to follow smaller concentrations of serotonin injected intravenously and those produced by two drugs believed to act via serotonergic mechanisms, fenfluramine and m-chlorophenylpiperazine. Baseline prolactin levels were quite stable, as the studies were conducted in monkeys adapted to chair restraint, and the intravenous cannulae

Table 2. Plasma Cortisol Changes in Rhesus Monkeys Following Intravenous Tryptamine, Serotonin, Fenfluramine, M-Chlorophenylpiperazine and Saline

Drug	Dosage (mg/kg)	Cortisol Change (Δ, ng/ml)
Tryptamine	0.1	37 ± 13
	1.0	25 ± 13
	10.0	66 ± 1
	20.0	54 ± 20
Serotonin	0.1	42 ± 11
	1.0	82 ± 13
Fenfluramine	1.5	52 ± 13
M-Chlorophenylpiperazine	1.5	124 ± 13
Saline		0.3 ± 14

*Mean Baseline Plasma Cortisol was 256 ± 20 ng/ml

were inserted the afternoon prior to the day of the study. Control injections of saline given either before or after the active substances elicited no significant prolactin changes (Table 1). These observations taken together suggest that the prolactin changes observed were directly related to the indoleamines and the serotonergic drugs, rather than being non-specific stress responses to the experimental procedure.

Plasma cortisol elevations were also observed to follow intravenous tryptamine, serotonin, fenfluramine and m-chlorophenylpiperazine, but not saline administration to rhesus monkeys (Table 2). Peak cortisol elevations occurred 20 to 30 minutes after tryptamine administration.

These results indicate that tryptamine has dose-related neuroendocrine effects in non-human primates. While the highest tryptamine dose administered (20 mg/kg) was higher than doses administered intravenously previously to humans in non-endocrine studies [277 mg or approximately

4 mg/kg, (Martin and Sloan, 1970)] the fenfluramine and m-chlorophenylpiperazine doses used were also several-fold greater than those required to elicit similar neuroendocrine responses in humans (Siever et al., 1984; Mueller et al., in press). These results demonstrate that tryptamine, if formed in sufficient amounts from L-tryptophan, could contribute to the neuroendocrine responses which follow L-tryptophan administration in humans. However, lower doses of serotonin were needed to elevate prolactin and cortisol significantly, arguing against the likelihood that tryptamine rather than serotonin changes are primarily responsible for L-tryptophan's neuroendocrine effects.

The robust prolactin and cortisol responses to serotonin given intravenously run counter to the generally accepted view (established predominantly from studies in rodents) that serotonin (and now tryptamine) might act on the pituitary to produce prolactin increases. Intravenously administered serotonin in similar doses has, however, been found to elevate prolactin in rhesus monkeys subjected to surgical transsection of the pituitary stalk from the hypothalamus, indicating that at least in this species, a pituitary-level effect of serotonin on prolactin release is possible (Wehrenberg et al., 1980). Whether a similar capacity for indoleamines to affect pituitary prolactin release directly in humans requires further study.

MELATONIN

Melatonin is primarily a product of the pineal gland, although it is also synthesized in the retina. Its most recognized function is the seasonal regulation of reproduction in rodents (Preslock, 1984; Tamarkin et al., 1985). Melatonin, which is released from the pineal gland during darkness, is produced in greater amounts during the longer nights of winter, leading to regression of rodent reproductive organs. Melatonin has a number of other less well-defined actions, including antithyroid and analgesic effects (Lakin et al., 1981; Vaughan et al, 1983). While direct peripheral actions have been demonstrated, implantation of small amounts of melatonin in the anterior hypothalamus and the suprachiasmatic nucleus also produces gonadal suppression (Glass and Lynch, 1981).

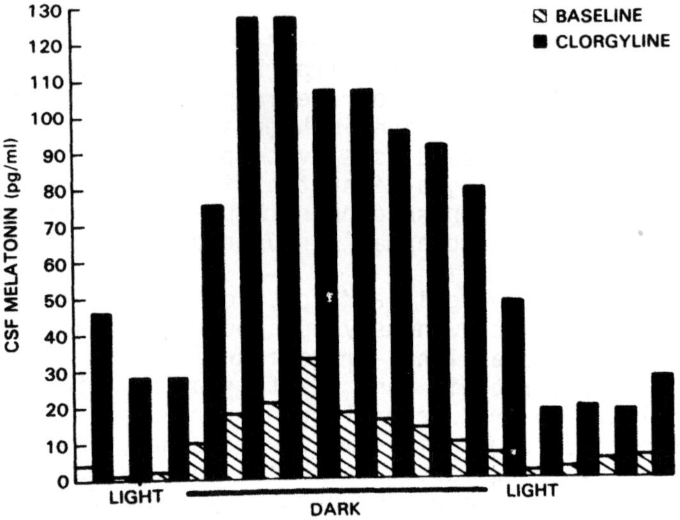

Figure 1. Elevated day time and night time cerebrospinal fluid melatonin concentrations in a rhesus monkey prior to and during treatment for 21 days with clorgyline.

In studies of indoleamines in cerebrospinal fluid CSF), we have become impressed with the possibilities that melatonin and other pineal indoleamines, as well as other substances, may act not only as blood-borne hormones, but also may be distributed in the brain via the cerebroventricular system (Taylor et al., 1982; Garrick et al., 1983; Wood, 1982). While our first reports emphasized the possible importance of CSF serotonin in this regard, (Taylor et al., 1982; Garrick et al., 1983) we have subsequently demonstrated, using a gas chromatographic mass-spectrometric assay, that it is N-acetyl serotonin rather than serotonin which follows melatonin's circadian pattern of release into CSF (Taylor et al., in press).

We have also demonstrated, as indicated in Figure 1, that the selective inhibitor of monoamine oxidase type A, clorgyline, markedly elevates the amount of melatonin released in the CSF of monkeys both during the day and during the night (Garrick et al., submitted). We have similarly found a near doubling of early morning plasma melatonin levels in patients receiving clorgyline and another MAO-inhibitor, tranylcypromine, as antidepressants (Murphy et al., submitted).

Rodent brain melatonin concentrations are quite low [e.g., 1 μM in rat hypothalamus (Koslow and Green, 1973) and 3-80 pg/mg wet tissue in various human brain areas (Kopp et al., 1980)]. These values are 50-100 times lower than those found in the pineal gland. There are several reports of melatonin receptors in bovine and rodent brain (Niles et al., 1979; Cardinali et al., 1979), but the identity and specificity of these receptors for melatonin versus other indoleamines and other substances requires much further work. It is of interest that melatonin has been localized by fluoresence immunohistochemistry in the suprachiasmatic nuclei (SCN), and that the sites in the brain at which melatonin implants were effective in rodents were within 0.1-0.4 mm of the third ventricle in the anterior hypothalamus - SCN area (Glass and Lynch, 1981). There exists in the anterior hypothalamus a melatonin-sensitive dopaminergic system which regulates prolactin release, and in hamsters this mechanism has been implicated in melatonin's gonadal inhibiting effects (Bjorklund et al., 1975; Goldman et al., 1981; Glass and Lynch, 1981).

While the greatest interest currently is in the capacity of melatonin to modulate dopamine function, as reviewed below, it is of note that melatonin has also been demonstrated to alter the uptake and release of serotonin and other neurotransmitter amines in rat brain synaptosomes (Cardinali et al., 1975), to increase the release of somatostatin and gonadotrophin-releasing hormone (Richardson et al., 1981; Kao and Weisz, 1977), to alter brain cyclic AMP and cyclic GMP (Vacas et al., 1981, 1984), to inhibit prostaglandin E_1 release into CSF, (Leach et al., 1982) and increase CSF indoleacetic acid via an apparent increase in the turnover of tryptamine (Young et al., 1984).

Pinealectomy has interesting effects, which have been mostly interpreted as alterations due to the absence of melatonin. For example, reproductive function may be enhanced (Cardinali, 1974), hyopthalamic 5-hydroxyindole-acetic acid (and possible serotonin and serotonin uptake) is reduced (Sugden and Morris, 1979; Cardinali et al., 1975) and an increased susceptibility to seizures develops in several rodent species which may be a counterpart to the sedative and anticonvulsive effects described after melatonin administration in rodents and man (Sugden, 1983; Arendt et al., 1984; Holmes and Sugden, 1982).

There have been several recent reports that nocturnal melatonin release is reduced in some depressed patients, particularly patients with a melancholic symptom cluster and those who have abnormal dexamethasone cortisol suppression tests (Mendlewicz et al., 1979; Claustrat et al., 1984; Brown et al., in press; Beck-Friis et al., in press). Elevated plasma melatonin also follows treatment with the antidepressant, desipramine (Thompson et al., 1983). As melatonin has some subtle effects on mood and performance when administered orally to humans (Arendt et al., 1984; Lieberman et al., 1984), it seems worth considering the neurochemical actions of melatonin to evaluate whether it may act as a physiologically important trace amine in the brain.

The influence of melatonin upon dopamine release and metabolism has recently been studied in some detail. Of special interest are the observations that melatonin's effects have only been demonstrated in certain brain regions and in the retina but not, for example, in the major dopamine systems involved in psychomotor functioning in the corpus striatum and cortex (Zisapel et al., 1983a). In rat hypothalamus slices, calcium-dependent ^{3}H-dopamine release elicited by electrical field stimulation or KCl-induced depolarization was inhibited 50% by less than one nanomolar concentrations of melatonin (Zisapel and Laudon, 1982).

Similar effects of melatonin were also observed upon dopamine release in several other brain areas involved in neuroendocrine modulation, including the ventral (but not dorsal) hippocampus, the preoptic area and the pons-medulla area (Zisapel et al., 1982). Melatonin seems to act by reducing calcium entry into presynaptic dopamine nerve endings (Zisapel and Laudon, 1983). Melatonin's effects on dopamine release are observed at very low concentrations during the estrous and metestrous phases of the rat's menstrual cycle, while at other times higher and equivalent melatonin concentrations were required in both male and female rats (Zisapel et al., 1983b).

In the rabbit retina, which synthesizes melatonin, picomolar concentrations of melatonin have also been observed to inhibit the calcium-dependent release of ^{3}H-dopamine elicited by electrical stimulation (Dubocovich, 1983). Again, similar effects were not observed in the striatium, where apomorphine demonstrated potent antagonism of dopamine release. In these retinal studies, N-acetyl serotonin also had similar ^{3}H-dopamine release-inhibiting properties, but only at significantly higher concentrations than those at which melatonin was effective. Serotonin, in contrast, was ineffective. Tyramine-induced release of ^{3}H-dopamine was also unaffected by melatonin, indicating that only calcium-dependent dopamine release is modulated by melatonin (Dubocovich, 1983).

N-ACETYL SEROTONIN

N-acetyl serotonin (NAS) is formed like melatonin, in the pineal gland, from serotonin. The enzyme responsible for its synthesis, N-acetyl transferase, is modulated by noradrenergic input from the SCN via sympathetic nerves from the superior cervical sympathetic ganglia. Its pineal concentrations are also low relative to serotonin and melatonin. While NAS has some physiological properties, some of which have been noted above, it has generally been regarded as a short-lived precursor of melatonin without major functions.

Nonetheless, because of our interest in the possibility of the cerebroventricular system as an important conduit for neurohormonal regulation of brain functions, we measured NAS concentrations in rhesus monkey CSF over 24-hour periods. We demonstrated that significant amounts of NAS are released into CSF in a generally similar time pattern with melatonin, suggesting that re-evaluation of NAS as a trace indoleamine hormone may be in order (Taylor et al., in press). The concentrations of NAS released in the CSF at night are several-fold higher than those of melatonin (Figure 2).

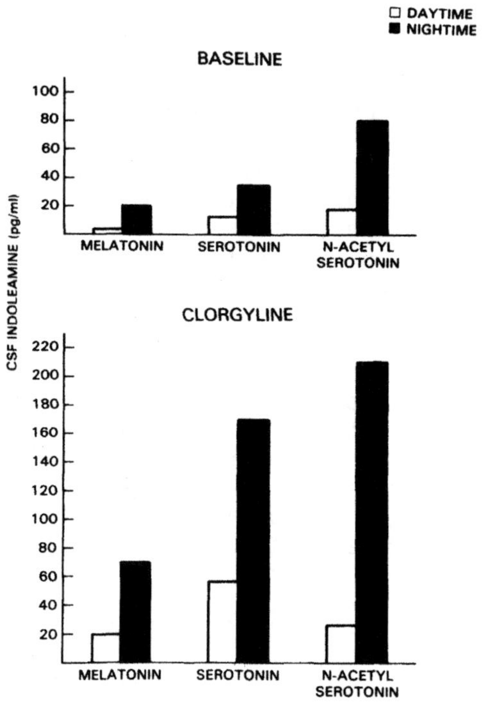

Figure 2. Mean melatonin, serotonin and N-acetylserotonin concentrations in rhesus monkey CSF prior to and during treatment with clorgyline (1 mg/kg/day X 21d).

Limited data currently available indicate that these concentrations, like those of melatonin and serotonin, are markedly increased during treatment with the MAO-inhibitor, clorgyline (Figure 2). While NAS does not possess the broad reproduction-modulating actions of melatonin in rodents, it has been reported to inhibit compensatory ovarian and adrenal hypertrophy in rats, and, like melatonin, to have antithyroid effects (Pang et al., 1981). Other possible functional consequences of its administration have been only minimally examined. One behavioral study demonstrated that the amino acid precursor of NAS, 5-hydroxy-N-acetyltryptophan, was as effective as 5-hydroxytryptophan, the serotonin precursor, in eliciting the behavioral features of the so-called serotonin syndrome in rodents (Burns et al., 1982).

NAS differs from melatonin in that its concentrations in brain are essentially unaltered following pinealectomy (Pang et al., 1982), and a role for NAS as a modulator of brain function has been proposed (Pang et al., 1981). Specific brain receptors have been shown to exist for NAS (Niles et al., 1983). Immunohistochemical studies demonstrated preferential localization of NAS in various nuclei of the brain stem, cerebellum and hippocampus; its distribution in different brain regions is quite different from that of its precursor, serotonin, and very different from that of melatonin (Brown et al., 1981). Further studies are needed to evaluate the extent to which NAS may have an important neuroregulatory role in brain either via its release from the pineal gland or its formation in other brain areas.

5-METHOXYTRYPTAMINE

Although serotonin in the pineal is usually acetylated to NAS before being methylated to yield melatonin, it may also be directly methylated to form 5-methoxytryptamine (5-MT). There appear to be large species differences in the amounts of 5-MT formed in the pineal gland (Pevet, 1983). It has been found in human plasma, CSF and urine, and is higher at night in human plasma (Hooper et al., 1981; Pevet, 1983).

5-MT can produce gonadal regression, but is less effective than melatonin in this regard. Upon administration to rodents, it produces a number of adverse behavioral effects, some of which have been attributed to its conversion to the psychotomimetic, 5-methoxy DMT. Some of its melatonin-like effects have also been attributed to the possibility that it might be acetylated in the pineal or elsewhere to form melatonin. *In vitro*, 5-MT is an effective modulator of serotonin uptake at concentrations of 3-4 M, and there is evidence that it can act at the presynaptic autoreceptor for serotonin to inhibit electrically stimulated serotonin release from hypothalamic slices (De Montigny and Aghajanian, 1977). It also effectively competes for ^{3}H-imipramine binding sites in rodent brain (Segonzac et al., 1985). These characteristics make 5-MT a candidate for being a physiologically important trace amine, but this hypothesis requires additional study.

6-METHOXYTRYPTOLINE

6-methoxytryptoline is present in high concentrations, comparable to those of melatonin, in the human pineal gland (Langer et al., 1984). This tetrahydrobetacarboline (THBC) derivative was the most effective of several other THBC's, and far more effective than melatonin or 5-MT in inhibiting ^{3}H-imipramine binding to human platelets. Langer and coworkers (1984) have recently suggested that it may prove to be an important endogenous ligand for the ^{3}H-imipramine binding site located within the serotonin uptake site in plasma membranes, and thus may play a role in the modulation of serotonin uptake and function.

REFERENCES

ALOI J.A., INSEL T.R., MUELLER E.A., and MURPHY D.L. (1984) Neuroendocrine and behavioral effects of m-chlorophenylpiperazine administration in rhesus monkeys. Life Sci. 34, 1325.

ARENDT J., BORBELY A.A., FRANEY C., AND WRIGHT J. (1984) The effects of chronic, small doses of melatonin given in the late afternoon on fatigue in man: A preliminary study. Neurosci. Ltrs. 45, 317-321.

BALDESSARINI R.J. and FISCHER J.E. (1978) Trace amines and alternative neurotransmitters in the central nervous system. Biochem. Pharmacol. 27, 621-626.

BECK-FRIIS J., LJUNGGREN J-G., THOREN M., VON ROSEN D., KJELLMAN B.F., and WETTERBERG L. Melatonin, cortisol and ACTH in patients with major depressive disorder and healthy humans with special reference to the outcome of the dexamethasone suppression test. Psychoneuroendocrinology, in press.

BJORKLUND A., LINDVAL O., and NOBIN A. (1975) Evidence of an incerto-hypothalamic dopamine neurone system in the rat. Brain Res. 89, 29-42.

BOULTON A.A., BAKER G.B., DEWHURST W.G., and SANDLER M. (eds.) (1984) Neurobiology of the Trace Amines, pp. 449-514. Humana Press, Clifton, NJ.

BROWN G.M., PORIETIS A., and NARASIMHACHARI N. (1981) Identification and quantitation of N-acetylserotonin in rat brain regions by GCMS. Soc. Neurosci. Abstr. 7, 8.

BROWN R., KOCSIS J.H., CAROFF S., AMSTERDAM J., WINOKUR A., STOKES P.E., and FRAZER A. Differences in nocturnal melatonin secretion between melancholic depressed patients and controls. Am. J. Psychiat., in press.

BURNS T.G., SZECHTMAN H., BROWN G.M., and SNIECKUS V. (1982) Behavioral effects of 5-hydroxy-n-acetyltryptophan, a putative synthetic precursor of n-acetylserotonin. Prog. Neuro-Psychopharmacol. and Biol. Psychiat. 6, 359-363.

CARDINALI D.P. (1974) Melatonin and the endocrine role of the pineal organ. Curr. Top. Exp. Endocr. 2, 107-128.

CARDINALI D.P., NAGLE C.A., FREIRE F., and ROSNER J.M. (1975) Effects of melatonin on neurotransmitter uptake and release by synaptosome-rich homogenates of the rat hypothalamus. Neuroendocrinology 18, 72-85.

CARDINALI D.P., VACAS M.I., and BOYER E.E. (1979) Specific binding of melatonin in bovine brain. Endocrinology 105(2), 437-441.

CHARNEY D.S., HENINGER G.R., REINHARD J.F., STERNBERG D.E., and HAFSTEAD K.M. (1982) The effect of intravenous L-tryptophan on prolactin and growth hormone and mood in healthy subjects. Psychopharmacol. 77, 217-222.

CLAUSTRAT B., CHAZOT G., BRUN J., JORDAN D., and SASSOLAS G. (1984) A chronobiological study of melatonin and cortisol secretion in depressed subjects: plasma melatonin, a biochemical marker in major depression. Biol. Psychiatry 19(8), 1215-1228.

DE MONTIGNY C. and AGHAJANIAN G.K. (1977) Preferential action of 5-methoxytryptamine and 5-methyoxydimethyltryptamine on presynaptic serotonin receptors: a comparative iontophoretic study with LSD and serotonin. Neuropharmacol. 16, 811-818.

DUBOCOVICH M.L. (1983) Melatonin is a potent modulator of dopamine release in the retina. Nature 306, 782-784.

DURDEN D.A. and PHILLIPS S.R. (1980) Kinetic measurements of the turnover rates of phenylethylamine and tryptamine in vivo in the rat brain. J. Neurochem. 34(6), 1725-1732.

FULLER R.W., SNODDY H.D., MASON N.R., and OWEN J.E. (1981) Disposition and pharmacological effects of m-chlorophenylpiperazine in rats. Neuropharmacol. 20, 155.

GARRICK N.A., TAMARKIN L., TAYLOR P.L., MARKEY S.P., AND MURPHY D.L. (1983) Light and propranolol suppress the nocturnal elevation of serotonin in the cerebrospinal fluid of rhesus monkeys. Science 221, 474-476.

GLASS J.D. and LYNCH G.R. (1981) Melatonin: Identification of sites of antigonadal action in mouse brain. Science 214, 821-823.

GOLDMAN B.D., MATT K.S., ROYCHOUDHURY P., and STETSON M.H. (1981) Prolactin release in golden hamsters: Photoperiod and gonadal influences. Biol. Reprod. 24, 287-292.

HOLMES S.W. and SUGDEN D. (1982) Effects of melatonin on sleep and neurochemistry in the rat. Br. J. Pharmac. 76, 95-101.

HOOPER R.J.L., SILMAN R.E., LEONE R.M., and YOUNG I.M. (1981) The development of a plasma assay for 5-methoxytryptamine using gas chromatography-mass spectrometry, in Pineal Function (Matthews, C.D. and Seamarks, R.F., eds), pp. 1-6. Elsevier/North-Holland, Amsterdam.

JONES R.S.G. (1982) Tryptamine: A neuromodulator or neurotransmitter in mammalian brainπ Prog. Neurobiol. 19, 117-139.

JUORIO A.V. and DURDEN D.A. (1984) The distribution and turnover of tryptamine in the brain and spinal cord. Neurochem. Res. 9(9), 1283-1293.

KAO L.W.L. and WEISZ J. (1977) Release of gonadotropin-releasing-hormone (Gn-RH) from isolated, perifused medial basal hypothalamus by melatonin. Endocrinology 100, 1723-1726.

KELLAR K.J. and CASCIO C.S. (1982) [^{3}H]Tryptamine: high affinity binding sites in rat brain. Eur. J. Pharmacol. 78, 475-478.

KOPP N., CLAUSTRAT B., and TAPPAZ M. (1980) Evidence for the presence of melatonin in the human brain. Neurosci. Ltrs. 19, 237-242.

KOSLOW S.H. and GREEN A.R. (1973) Analysis of pineal and brain indole alkylamines by gas chromatography-mass spectrometry. Adv. Biochem. Psychopharmacol. 7, 33-43.

LAKIN M.L., MILLER C.H., STOTT M.L., and WINTERS W.D. (1981) Involvement of the pineal gland and melatonin in murine analgesia. Life Sci. 29, 2543-2551.

LANGER S.Z., LEE C.R., SEGONZAC A., TATEISHI T., ESNAUD H., SCHOEMAKER H., and WINBLAD B. (1984) Possible endocrine role of the pineal gland for 6-methoxytetrahydro-β-carboline, a putative endogenous neuromodulator of the [^{3}H]imipramine recognition site. Eur. J. Pharmacol. 102, 379-380.

LEACH C.M., REYNOLDSON J.A., and THORNBURN G.D. (1982) Release of E prostaglandins into the cerebrospinal fluid and its inhibition by melatonin after cervical stimulation in the rabbit. Endocrinology 110, 1320-1324.

LEMBERGER L., AXELROD J., and KOPIN I.J. (1971) The disposition and metabolism of tryptamine and the in vivo formation of 5-hydroxytryptamine in the rabbit. J. Pharmacol. Exp. Ther. 177, 169-176.

LIEBERMAN H.R., WALDHAUSER F., GARFIELD G., LYNCH H.J., and WURTMAN R.J. (1984) Effects of melatonin on human mood and performance. Brain Res. 323, 201-207.

MACINDOE J.H. and TURKINGTON R.W. (1973) Stimulation of human prolactin secretion by intravenous infusion of L-tryptophan. J. Clin. Invest. 52, 1972-1978.

MARTIN W.R. and SLOAN J.W. (1970) Effects of infused tryptamine in man. Psychopharmacologia (Berl.) 18, 231-237.

MENDLEWICZ J., LINKOWSKI P., BRANCHEY L., WEINBERG U., WEITZMAN E.D., and BRANCHEY M. (1979) Abnormal 24 hour pattern of melatonin secretion in depression. Lancet 1362.

MOSNAIM A.D. and WOLF M.E. (1978) Noncatecholic Phenylethylamines - Part 1 - Phenylethylamine: Biological Mecahnsims and Clinical Aspects, Marcel Dekker, Inc., New York.

MOSNAIM A.D. and WOLF M.E. (1980) Noncatecholic Phenylethylamines - Part 2 - Phenylethanolamine, Tyramines, and Octopamine, Marcel Dekker, Inc., New York.

MUELLER E.A., MURPHY D.L., SUNDERLAND, T., and JONES J. A new post-synaptic serotonin receptor agonist suitable for studies in humans. Psychopharmacol. Bull., in press.

MURPHY D.L., CAMPBELL I.C., and COSTA J.L. (1978) The brain serotonin system in the affective disorders. Prog. Neuro-Psychopharmacol. 2, 1.

NILES L.P., WONG Y.W., MISHRA R.K., and BROWN G.M. (1979) Melatonin receptors in brain. Eur. J. Pharmacol. 55, 219-220.

NILES L.P., BROWN G.M., and MISHRA R.K. (1983) Characteristics of high affinity binding of [^{3}H]N-acetylserotonin in rat brain. Neuropharmacol. 22(11), 1311-1314.

OLDENDORF W.H. and BRAUN L.D. (1976) [^{3}H]Tryptamine and ^{3}H-water as diffusible internal standards for measuring brain extraction of radio-labeled substances following carotid injection. Brain Res. 113,

PANG S.F., BROWN G.M., and CAMPBELL S.L. (1981) A radioimmunoassay for N-acetylserotonin in biological tissues. J. Immunoassay 2(3 and 4), 263-276.

PANG S.F., TANG P.L., YU H.S., and YIP M.K. (1982) The level of N-acetylserotonin and melatonin in the brain of male rats: Diurnal variations and effects of pinealectomy. J. Exp. Zool. 219, 271-276.

PEVET P. (1983) Is 5-methoxytryptamine a pineal hormone? Psychoneuroendocrinology 8(1), 61-73.

PRESLOCK J.P. (1984) The pineal gland: basic implications and clinical correlations. Endocrine Reviews 5(2), 282-308.

RICHARDSON S.B., HOLLANDER C.S., PRASAD J.A., and HIROOKA T. (1981) Somatostatin release from rat hypothalamus in vitro: effects of melatonin and serotonin. Endocrinology 109, 602-606.

SEGONZAC A., RAISMAN R., TATEISHI T., SCHOEMAKER H., HICKS P.E., and LANGER S.Z. (1985) Tryptamine, a substrate for the serotonin transporter in human platelets, modifies the dissociation kinetics of [^{3}H]imipramine binding: possible allosteric interaction. J. Neurochem. 44(2), 349-356.

SIEVER L.J., MURPHY D.L., SLATER S., DE LA VEGA E., and LIPPER S. (1984) Plasma prolactin changes following fenfluramine in depressed patients compared to controls: an evaluation of central serotonergic responsivity in depression. Life Sci. 34, 1029.

SUGDEN D. and MORRIS R.D. (1979) Changes in regional brain levels of tryptophan, 5-hydroxytryptamine, 5-hydroxyindoleacetic acid, dopamine and noradrenaline after pinealectomy in the rat. J. Neurochem. 32, 1593-1595.

SUGDEN D. (1983) Psychopharmacological effects of melatonin in mouse and rat. J. Pharmacol. Exp. Therap. 227(3), 587-591.

TAMARKIN L., BAIRD C.J., and ALMEIDA O.F.X. (1985) Melatonin: a coordinating signal for mammalian reproduction? Science 227, 714-720.

TAYLOR P.L., GARRICK N.A., BURNS R.S., TAMARKIN L., MURPHY D.L., and MARKEY S.P. (1982) Diurnal rhythms of serotonin in monkey cerebrospinal fluid. Life Sci. 31, 1993-1999.

TAYLOR P.L., GARRICK N.A., BURNS R.S., TAMARKIN L., MURPHY D.L., and MARKEY S.P. (1982) Diurnal rhythms of serotonin in monkey cerebrospinal fluid. Life Sci. 31, 1993-1999.

TAYLOR P.L., GARRICK N.A., TAMARKIN L., MURPHY D.L., and MARKEY S.P. Diurnal rhythms of N-acetyl serotonin and serotonin in cerebrospinal fluid of monkeys. Science, in press.

THOMPSON C., CHECKLEY S.A., CORN T., FRANEY C., and ARENDT J. (1983) Down-regulation at pineal β-adrenoceptors in depressed patients treated with desipramine? Lancet 1101.

VACAS M.I., KELLER-SARMIENTO M.I., and CARDINALI D.P. (1981) Melatonin increases cGMP and cAMP levels in rat medial basal hypothalamus in vitro. Brain Res. 225, 207-211.

VACAS M.I., BERRIA M.I. CARDINALI D.P., and LASCANO E.F. (1984) Melatonin inhibits β-adrenoceptor-stimulated cyclic AMP accumulation in rat astroglial cell cultures. Neuroendocrinology 38, 176-181.

VAUGHAN M.K., RICHARDSON B.A., JOHNSON L.Y., PETTERBORG L.J., POWANDA M.C., REITER R.J., and SMITH I. (1983) Natural and synthetic analogues of melatonin and related compounds II. effects on plasma thyroid hormones and cholesterol levels in male syrian hamsters. J. Neural. Trans. 56, 279-191.

WARSH J.J., COSCINA D.V., GODSE D.D., and CHAN P.W. (1979) Dependence of brain tryptamine formation on tryptophan availability. J. Neurochem. 32, 1191-1196.

WEHRENBERG W.B., MCNICOL D., FRANTZ A.G., and FERIN M. (1980) The effects of serotonin on prolactin and growth hormone concentrations in normal and pituitary stalk-sectioned monkeys. Endocrinology 107(6), 1747-1750.

WOOD J.H. (1982) Neuroendocrinology of cerebrospinal fluid: peptides, steroids, and other hormones. Neurosurgery 11(2), 293-305.

YOUNG S.N. (1983) The significance of tryptophan, phenylalanine, tyrosine, and their metabolites in the nervous system, in Handbook of Neurochemistry, Vol.3 (Lajtha A., ed), pp. 559-581. Plenum Publishing Corp., New York.

YOUNG S.N., GAUTHIER S., KIELY M.E., LAL S., and BROWN G.M. (1984) Effect of oral melatonin administration on melatonin, 5-hydroxyindoleacetic acid, indoleacetic acid, and cyclic nucleotides in human cerebrospinal fluid. Neuroendocrinology 39, 87-92.

ZISAPEL N. and LAUDON M. (1982) Dopamine release induced by electrical field stimulation of rat hypothalamus *in vitro*: inhibition by melatonin. Biochim. Biophys. Res. Comm. 104(4), 1610-1616.

ZISAPEL N., EGOZI Y., and LAUDON M. (1982) Inhibition of dopamine release by melatonin: regional distribution in the rat brain. Brain Res. 246, 161-163.

ZISAPEL N. and LAUDON M. (1983a) Inhibition by melatonin of dopamine release from rat hypothalamus: regulation of calcium entry. Brain Res. 272, 278-381.

ZISAPEL N., EGOZI Y., and LAUDON M. (1983b) Inhibition by melatonin of dopamine release from rat hypothalamus *in vitro*: variations with sex and the estrous cycle. Neuroendocrinology 37, 41-47.

TRYPTAMINE KINETICS IN HUMAN VOLUNTEERS

Wollmann, H., Nilsson, E., Antonin, K.H. and Bieck P.R.

Human Pharmacology Institute
CIBA-GEIGY GmbH, D-74 Tübingen,
Federal Republic of Germany

Abstract: In 6 healthy, young volunteers the kinetics of L-tryptophan (TRP) following 3 different single oral doses (1.5, 3, 6 g) simultaneously with urinary tryptamine (TA) excretion were examined. Blood and urine samples were collected hourly before and for 14 h after TRP. TRP in plasma and TA in urine were analysed by specific and sensitive HPLC methods using UV and fluorometric detection, respectively. The measurements were repeated in one volunteer after pretreatment with the decarboxylase inhibitor **benserazide** (3 x 125 mg/d for 15 days), and in 2 volunteers after pretreatment with the selective and reversible inhibitor of MAO-A, **brofaremine** (150 mg/d for 10 days).

Results:

	TRP		TA in urine
TRP dose [g]	Cmax [ug/ml]	AUC [$ugxhxml^{-1}$]	[ug/14h]
1.5	67.8	298	205.8
3.0	92.5	444	285.6
6.0	151.2	840	442.6

There was a linear relationship between peak plasma concentrations of TRP, area under the plasma concentration time curves (AUC) and TRP dosage. Urinary TA excretion increased dependent

on the TRP dose but independent on creatinine clearance and urinary pH.
After complete absorption and distribution of TRP there was a linear correlation between TRP plasma concentrations and the urinary excretion rate of TA ($r = 0.95$, $p < 0.01$). During inhibition of decarboxylation by subchronic treatment with **benserazide** basal urinary TA was below the limit of detection (1 ng/ml). After pretreatment with the MAO A inhibitor **brofaremine** the TA excretion was increased by 100%. One volunteer - pretreated with **brofaremine** - developed after 4.5 g TRP neurological symptoms resembling the "serotonin syndrome". In none of the volunteers TA was detectable in plasma.

INTRODUCTION

In 1959 Sjoerdsma et al. observed in some patients treated with a monoamine oxidase (MAO) inhibitor drug and L-tryptophan (TRP) neurological symptoms which they described as "drunkeness", hyperreflexia and clonus. Later on single cases with such neurological changes were reported in patients on combined MAO inhibitor and TRP therapy (Glassman and Platman, 1969; Baloh et al., 1982; Thomas and Rubin, 1984). Because of the similarity with an animal model the changes were labeled "serotonin syndrome" (Gerson and Baldessarini, 1980).

Inhibition of the MAO in the human kidney (Sjoerdsma et al., 1959) or increasing the plasma concentrations of the precursor amino acid TRP (DeQuattro, 1967) results in an increase of the urinary TA excretion. Oates and Sjoerdsma (1966) discussed the possibility, that enhanced concentrations of TA and/or of serotonin (5-HT) in the central nervous system could be responsible for the "serotonin syndrome".

Marsden (1978) applied TRP and TA to rats pretreated with the partially irreversible and nonspecific MAO inhibitor tranylcypromine. They concluded that peripherally formed TA could penetrate into the brain and be partly responsible for the behavioral effects observed. Results discussed by Philips et al (1983) support this suggestion.

In order to test this hypothesis in man the quantitative transformation of TRP to TA after oral loads of the precursor amino acid TRP was investigated. The action of enzymes involved in the formation and degradation of TA was examined by oral pretreatment with a decarboxylase and a MAO inhibitor drug, respectively.

MATERIALS AND METHODS

Subjects

Six healthy male ambulatory volunteers, aged between 18 and 46 years (mean age 31 years) participated. One week prior to the study the volunteers were not allowed to take any other drugs. During benserazide pretreatment no special diet was kept. During brofaremine treatment the volunteers were not allowed to take food or alcohol rich in tyramine. On days with TRP administration the volunteers received a meatless diet.

The design and the procedures of the study were approved by an Ethics Committee. All subjects gave their informed consent.

Dosage

Six subjects took 3 single oral doses of L-TRP (1.5, 3 and 6 g) with 200 ml water on 3 separate days in the morning after an 12 h fasting period. 200 ml mineral water were drunk one hour before drug intake and hourly for 4 hours. Between oral TRP loads there were intervals of at least 7 days.

Pretreatment for 10 days with 150 mg/d **brofaremine** (CGP 11 305 A), a reversible and specific MAO-A inhibitor: two subjects received 3.0 g TRP, one subject additionaly 4.5 g TRP (on day 14).

One volunteer was treated for 10 days with 375 mg/d **benserazide,** a peripherally acting decarboxylase inhibitor drug. On days 7 and 10 he received 3 g TRP and 6 g TRP, respectively.

Sample collection

Blood was collected before and in hourly intervals after administration of L-tryptophan as shown in Fig. 1. 5 ml blood were drawn into EDTA tubes. The samples were centrifuged immediately and stored at - 20°C until assayed. **Urine** samples were collected in hourly intervals for 14 h during the TRP kinetic and stored at - 20°C. 24h urines were collected using sodium disulfite (40 ml of a 10% solution per 24h urine) as preservative. Aliquots were kept frozen until assayed.

Calculations

From plasma concentrations of TRP areas under the plasma level time curves [AUC(0-t)] were calculated using the trapezoidal rule. AUC's were normalized by dividing by the TRP dose given (AUCspec). The plasma half-lives of TRP were calculated by the ln-linear regression of the plasma concentrations during the elimination phase.

Analytical methods

Creatinine in urine and plasma were measured by the Jaffe reaction with Beckman Creatinine Analyzer 2.

Tryptamine in urine was either measured fluorometrically by a modification of the method of Sjoerdsma et al., (1959) or by a more sensitive assay using fluorometric detection (285/360 nm) after HPLC separation (Nilsson, 1985). The samples were made alkaline and extracted with dichlormethane. An u-Bondapak column was used for the chromatographic separation. Both methods used gave comparable results. However, the HPLC assay was more reliable in the lower range.

Tryptamine in plasma was determined by the same HPLC-fluorometric assay with minor changes in order to increase the sensitivity. The limit of detection was 1 ng/ml plasma when 1 ml plasma was used.

Tryptophan in plasma (free + bound) was analyzed by HPLC with UV-detection at 280 nm (Grushka et al., 1977).

RESULTS

Oral Tryptophan Load in Unmedicated Subjects

Plasma concentration of TRP

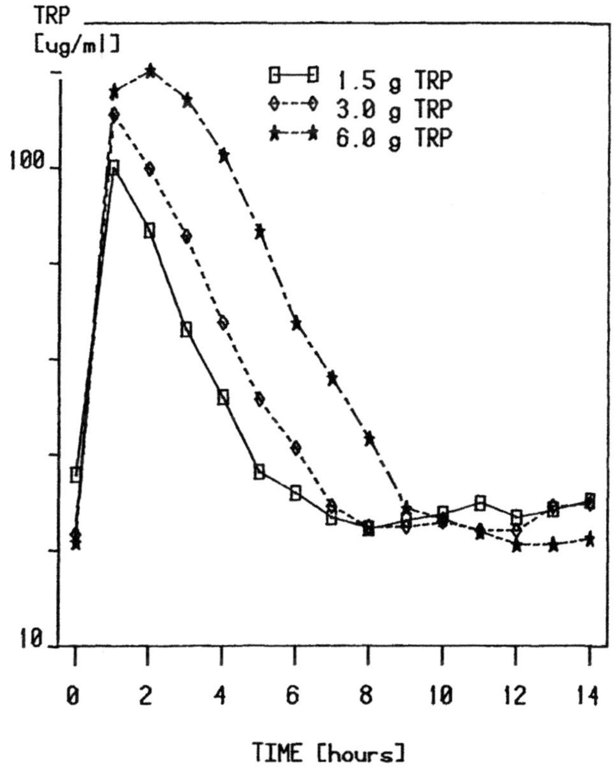

Fig. 1. Plasma concentrations of TRP after oral administration in one subject.

Vol. no.	AUC_{0h-t} [(ug/ml)xh]			AUC_{spec}[1] [((ug/ml)xh)/g]			$T_{1/2}$ [h]		
	1.5	3.0	6.0	1.5	3.0	6.0	1.5	3.0	6.0
	[g TRP]			[g TRP]			[g TRP]		
1	116	157	466	77	52	78	1.8	2.3	2.2
2	133	223	588	89	74	98	2.2	2.2	2.7
3	185	357	795	123	119	133	2.3	2.2	2.6
4	182	292	582	121	97	97	1.9	2.1	2.1
5	58	216	534	39	72	89	3.4	3.0	2.8
6	--	336	787	--	112	131	--	1.7	2.1
mean	135	264	625	90	88	104	2.3	2.3	2.4
± sd	52	78	136	35	26	23	0.6	0.4	0.3

Tab. 1: Pharmacokinetic data of TRP in plasma after an oral load.
See calculations.

Fig. 1 shows an example of the dose-dependent increase of total TRP concentrations in plasma. Peak concentrations for the two smaller doses are reached within 1 hour, for the highest TRP dose within 2 hours. Depending on the dose the elimination phase starts from 1 to 3 hours after administration. The dose-normalized AUC (AUCspec) are not greatly different indicating a **linear elimination process** (Tab. 1). The plasma half-life of TRP is approximately 2.5 hours.

Urinary TA excretion

Fig. 2 examplifies the cumulative urinary TA excretion in one subject. With increasing doses of the precursor TRP a **dose-dependent increase** in TA excretion is observed. The cumulative excretion values during 8 hours were 100 ug (1.5 g TRP), 400 ug (3.0 g TRP) and 550 ug (6.0 g TRP), respectively. The TA excretion is independent of **creatinine clearance** (Fig. 3) and of steady-state **urinary pH** (Fig. 4), respectively.

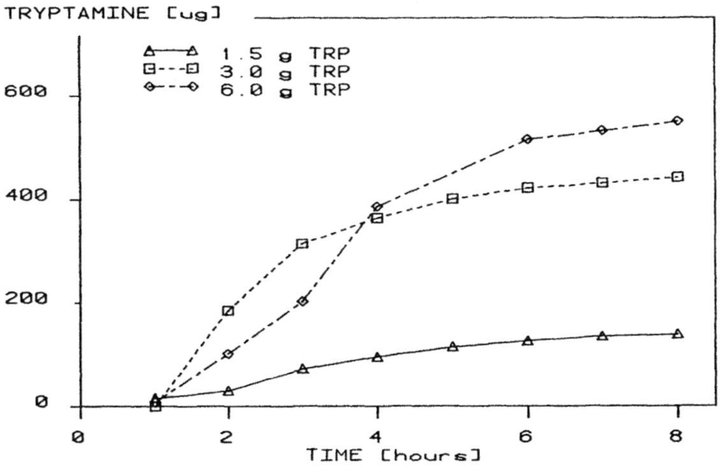

Fig. 2. Cumulative TA excretion in urine after oral TRP in one subject.

TA excretion in relation to TRP in plasma

TRP concentrations in plasma and the corresponding TA excretion values in urine are depicted as

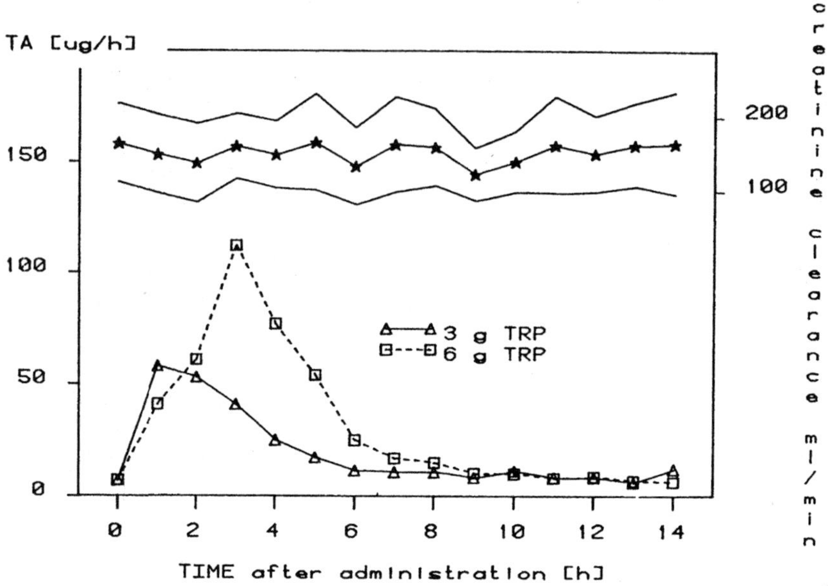

Fig. 3. TA excretion in urine and creatinine clearance after oral TRP in 6 subjects (mean).

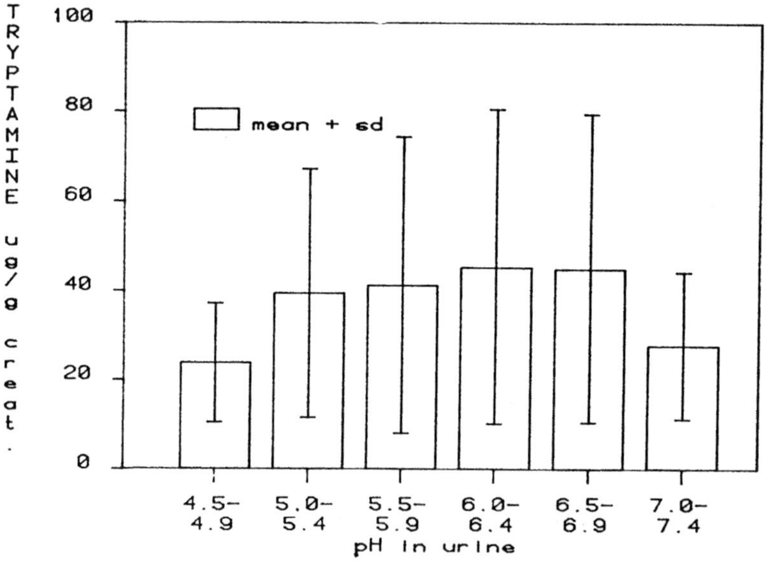

Fig. 4. Basal urinary TA excretion and pH in 9 untreated subjects.

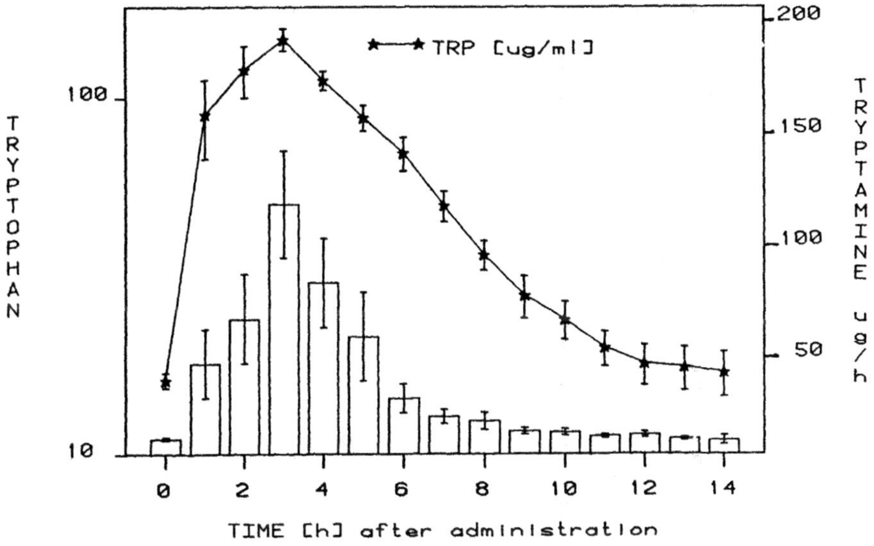

Fig. 5. TRP in plasma and TA excretion in urine after an oral load of 6,0 g TRP. Mean of 5 subjects ± SEM.

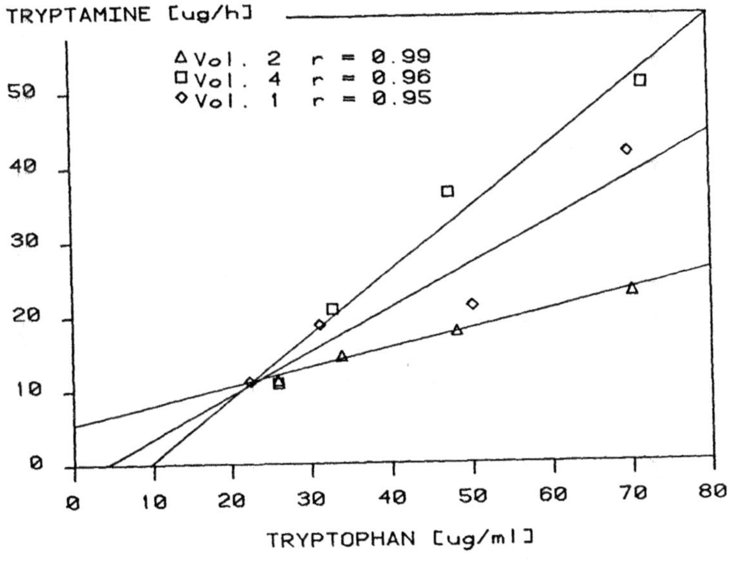

Fig. 6. TA excretion in relation to TRP in plasma at 3, 4, 5 and 6 hours after an oral load of 3.0 g TRP. 3 subjects.

means of 5 subjects in Fig. 5.
After oral administration of 6 g TRP the TRP plasma concentrations show a first-order decline between 3 and 9 hours. Between 3 and 6 hours the urinary excretion rate of TA is linear in the double-linear plot used.
The significant correlations between TRP concentration in plasma and urinary TA excretion hold true also for the smaller TRP dose of 3.0 g. This is shown in Fig. 6.

Inhibition of Decarboxylation

The effect on the TA excretion of **in vivo** decarboxylase inhibition with **benserazide** is shown in Fig. 7. In this subject the TA excretion declined from control levels of about 40 ug/24 hours to 4 ug/24 hours within one day after start of treatment. On the following 5 days no urinary TA could be detected.

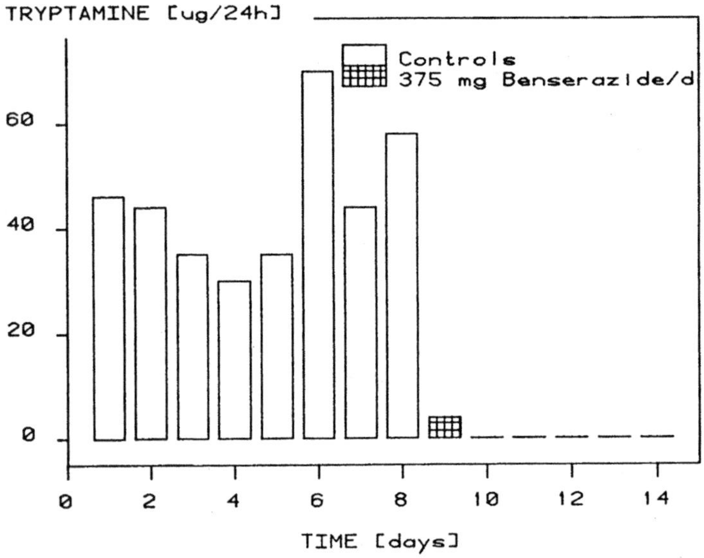

Fig. 7. Urinary excretion of TA after oral decarboxylase inhibition with benserazide in one volunteer.

Inhibition of Oxidative Deamination

The effect of MAO inhibition with **brofaremine** on the urinary TA excretion is shown in Fig. 8. The TA excretion is approximately doubled from control levels of about 100 ug/24 h to more than 200 ug/24 h within two days after start of treatment. This difference is statistically significant ($p < 0{,}002$).

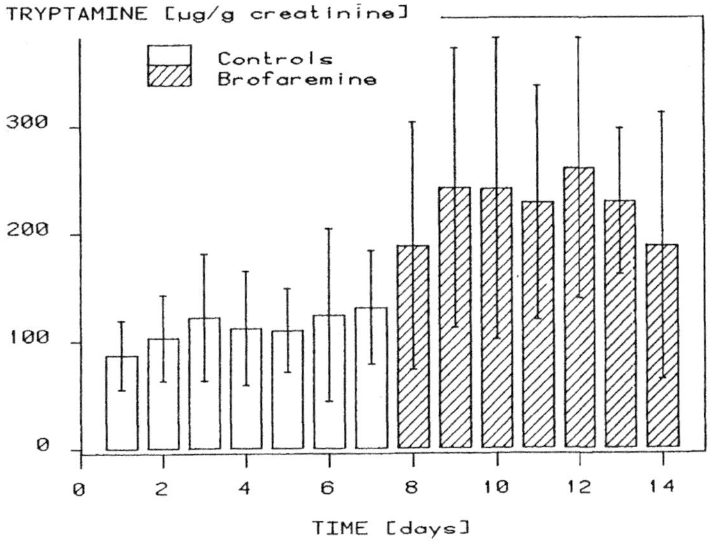

Fig. 8. TA excretion in urine after MAO-inhibition with brofaremine (100 mg/d p.o.; n = 6). Mean ± SEM.

Oral Tryptophan Load in Pretreated Subjects

Plasma concentration of TRP
The plasma concentration-time curves of TRP of a control group and of 3 pretreated volunteers are shown in Fig. 9. All subjects received the same oral load of 3 g TRP. In the control group mean peak plasma concentrations of 90 ug/ml are reached within 2 hours after dosing. The two volunteers pretreated with **brofaremine** showed nearly identical concentration-time curves.

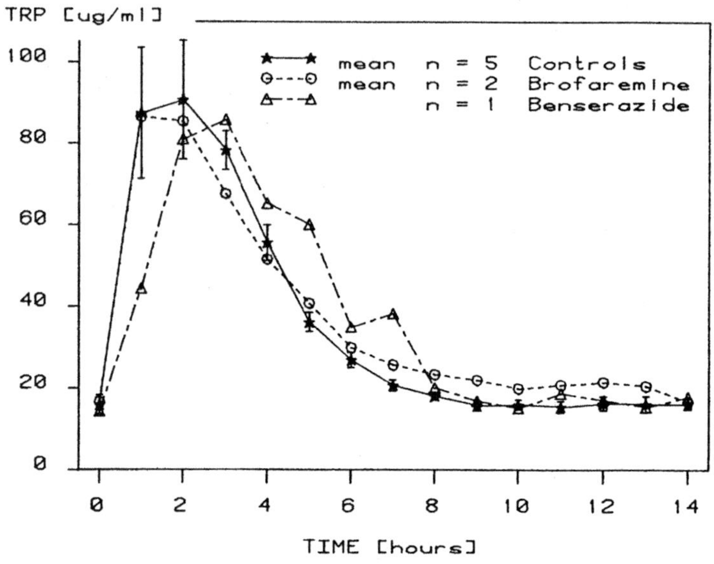

Fig. 9. TRP in plasma after an oral load of 3.0 g TRP. Pretreatment with brofaremine (150 mg/d) or benserazide (375 mg/d). Mean ± SEM.

The volunteer pretreated with **benserazide** had the same peak concentration as the controls but with a delay of one hour. 8 hours after TRP all volunteers had reached their basal plasma concentration values.

Urinary TA excretion

Fig. 10 shows the corresponding urinary TA excretion after administration of 3.0 g TRP. The control group excreted about 250 ug TA within 14 hours. The subject pretreated with **benserazide** did not excrete TA after the 3.0 g TRP load. However, after the higher TRP dose of 6.0 g (not shown in this Fig.) small amounts of approximately 8 ug/14 h TA were found. Pretreatment with **brofaremine** led after the TRP load to a twofold increase of TA excretion with a cumulative amount of 500 ug/14 h. One volunteer excreted 670 ug/24 h after a higher dose of TRP (4.5 g; Fig. 10).

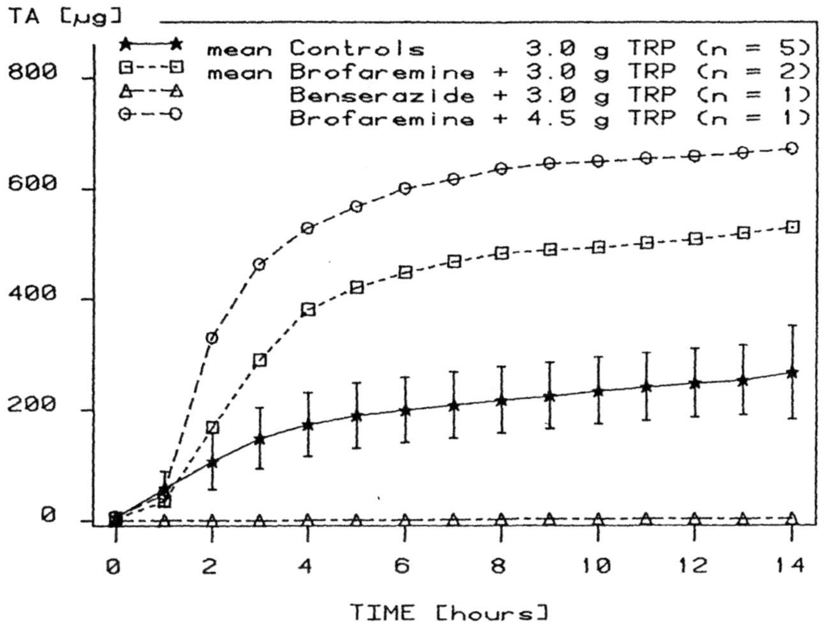

Fig. 10. Cumulative TA excretion in urine after different oral loads of TRP. Pretreatment with brofaremine (150 mg/d) or benserazide (375 mg/d). Mean ± SEM.

TA in plasma

In none of the plasma samples **tryptamine** was detectable, also not after inhibition of monoamine oxidase and administration of 4.5 g TRP.

Neurological changes

After the higest doses of 6.0 g TRP some of the volunteers felt fatigue. The subject pretreated with **benserazide** showed no neurological changes. The volunteer on **brofaremine** and the higher dose of 4.5 g TRP developed jaw clonus, tremor of the upper limbs and unvoluntary finger movements two hours after TRP. These changes disappeared within 2 hours.

DISCUSSION

The dose-normalized plasma AUC values (AUCspec) of TRP after 1.5, 3 and 6 g orally are not significantly different indicating a linear elimination process from the body with a half-life of approximately 2.5 hours (Green and Aronson, 1980).

The part of the metabolic pathway of TRP leading to tryptamine is quantitatively of minor importance; about 99% of tryptophan is metabolized by "other pathways", e.g. via kynurenine to acetyl-coA, or is used for the synthesis of TRP-containing proteins (Fig. 11). Qualitatively however, the formation of TA from TRP and its breakdown is important because this trace amine seems to play an important role in neurotransmission, mood and behaviour (for review see Jones, 1982).

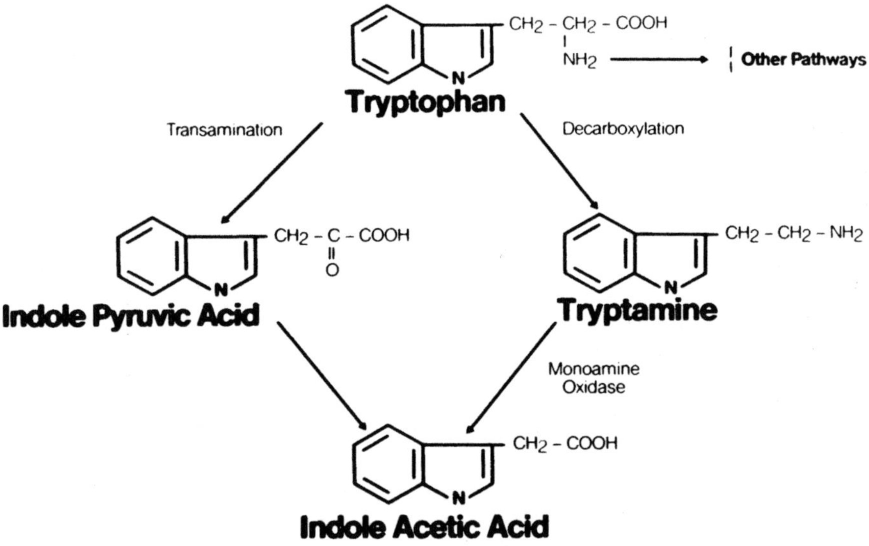

Fig. 11. Tryptamine metabolism.

The results of this study are in agreement with earlier data showing the dependence of urinary TA excretion on the concentrations of the precursor amino acid TRP in plasma (DeQuattro, 1967). After 3 and 6 g TRP, respectively, the plasma TRP concentrations and the TA excretion rates could be shown to be significantly correlated. Apparently, the **enzymatic formation** of TA in the kidney from the precursor amino acid is dependent on the decarboxylase activity.

Benserazide, the peripherally acting inhibitor of this enzyme did not affect absorption and plasma kinetics of TRP but completely abolished the renal TA formation. This inhibition could not be overcome by raising the plasma concentrations of TRP with an oral 3 g load. Small amounts of TA were excreted, however, when producing higher substrate concentrations in the blood by administration of 6 g TRP. This result points towards a competitive enzyme inhibition. Also in favour of an enzymatic process of renal TA formation is the nonlinear kinetics of the urinary TA excretion shown.

Oxidative deamination by MAO is responsible for the rapid turn-over rate and for the short half-life of TA (Durden and Philips, 1980). An enhanced urinary excretion of TA can be taken therefore, as an index of MAO inhibition. In the human kidney TA is mainly a substrate of type A of the MAO (Bieck et al., 1984). During subchronic treatment with **brofaremine,** a specific MAO-A inhibitor, urinary TA excretion was approximately doubled (Fig. 8).

Already in 1960 the question was raised by Oates and Sjoerdsma if elevated TA concentrations in the **central nervous system** could be responsible for symptoms named later "serotonin syndrome". It is occasionally observed in depressive patients after combined MAO-inhibitor treatment and TRP administration (Price et al., 1984). From animal experiments it is known that TA formation and breakdown in the CNS is dependent on precursor concentrations and the activity of the two enzymes involved in its metabolism (Warsh et al., 1979; Marsden and Curzon, 1978). But there are no direct means to measure TA concentrations in the human CNS. If, however, TA formed peripherally, e.g. in the kidney, is transported via the blood into the brain - as suggested by Philips (1984) - one should be able to detect TA with sensitive analytical methods.

In none of the volunteers examined TA was found in plasma - also not in the subject with sustained neurological symptoms caused by combined MAO-A inhibition and TRP load. Therefore, it is not likely that TA formed peripherally is transported by the blood into the brain leading to an enhanced concentration. One could explain the neurological symptoms with a peripheral and central inhibition of TA metabolism occuring simultaneously. Another explanation for the neurological symptoms could be an inhibition of the central 5-HT metabolism by pretreatment with the MAO-A inhibitor brofaremine. Measurements of the major 5-HT metabolite 5-hydroxyindole acetic acid (5-HIAA) in the cerebrospinal fluid of rhesus monkeys before and after treatment with the irreversible MAO-A inhibitor clorgyline, however, showed only minor effects (Garrick et al., 1984). We did not find increases of 5-HTP (5-hydroxytryptophan) or 5-HT concentrations in plasma after oral TRP without or with MAO inhibition. But this does not rule out a central role for 5-HT in causing the "serotonin syndrome".

Acknowledgement: We thank for the assistance of Ms. M. Huegler, B. Lins, C. Rommel and G. Vees.

REFERENCES

BALOH, R. W., DIETZ, J. and SPOONER, J.W. (1982) Myoclonus and ocular oscillations induced by L-tryptophan. **Ann. Neurol.** **2**, 95 - 97.

BIECK, P.R., NILSSON, E., SCHICK, C., WALDMEIER, P.C. and LAUBER, J. (1984) Urinary excretion of tryptamine in comparison to normetanephrine and beta-phenylethylamine in human volunteers after subchronic treatment with different monoamine oxidase inhibitors. In: **Neurobiology of the Trace Amines** (BOULTON, A.A., BAKER, G.B., DEWHURST, W.G. and SANDLER, M. ed.) The HUMANA Press.

DeQUATTRO, V. and SJOERDSMA, A. (1967) Origin of urinary tyramine and tryptamine. **Clin. Chim. Acta** **16**, 227 - 233.

DURDEN, D.A. and PHILIPS, S.R. (1980) Kinetic measurements of the turnover rates of phenylethylamine and tryptamine in vivo in the rat brain. **J. Neurochem. 34**, 1725 - 1732.

GARRICK, N.A., SCHEININ, M., CHANG, W., LINNOILA, M. AND MURPHY, D. (1984) Differential effects of clorgyline on catecholamine and indoleamine metabolites in the cerebrospinal fluid of rhesus monkeys. **Biochem. Pharmacol. 33**, 1423 - 1427.

GERSON, S.C. and BALDESSARINI, R.J. (1980) Motor effects of serotonin in the central nervous system. **Life Sci. 27**, 1435 - 1451.

GLASSMAN, A.H. and PLATMAN, S.R. (1969) Potentiation of a monoamine oxidase inhibitor by tryptophan. **J. Psychiatr. Res. 7**, 83 - 88.

GREEN, A.R., ARONSON, J.K., CURZON, G. and WOODS, H.F. (1980) Metabolism of an oral tryptophan load. I Effects of dose and pretreatment with tryptophan. **Br. J. clin. Pharmacol. 10**, 603 - 610.

GRUSHKA, E. and KITKA, E.J. (1977) Tryptophan and kynurenine determination in untreated urine by reversed-phase high-pressure liquid chromatography. **J. Chromatogr. 143**, 51 - 56.

JONES, R.S.G. (1982) Tryptamine: A neuromodulator or neurotransmitter in mammalian brain ? **Progress in Neurobiology 19**, 117 - 139.

MARSDEN, C.A. and CURZON, G. (1978) The contribution of tryptamine to the behavioral effects of L-trypthophan in tranylcypromine-treated rats. **Psychopharmacology** **57**, 71 - 76

NILSSON, E. (1985) Determination of tryptamine in urine by high-performance liquid chromatography with fluorometric detection. **Internal Report** **T 9**, CIBA-GEIGY, HPI Tübingen.

OATES, J.A. and SJOERDSMA, A. (1966) Neurologic effects of tryptophan in patients receiving a monoamine oxidase inhibitor. **Neurology** **10**, 1076 - 1078.

PHILIPS, S.R. (1984) Analysis of trace amines: Endogenous levels and the effects of various drugs on tissue concentrations in the rat. In: **Neurobiology of the Trace Amines** (BOULTON, A.A., BAKER, G.B., DEWHURST, W.G. and SANDLER, M. ed.) The HUMANA Press.

PRICE, L.H., CHARNEY, D.S. and HENINGER, G. (1984) Effects of i.v. L-tryptophan during tranylcypromine treatment of depression. Abstract from 'American College of Neuropsychopharmacology'. Vanderbilt, 1984.

SJOERDSMA, A., OATES, J.A., ZALTZMANN, P. and UDENFRIEND, S. (1959). Identification and assay of urinary tryptamine: Application as an index of monoamine oxidase inhibition in man. **J. Pharmacol. Exp. Therap.** **126**, 83 - 88.

THOMAS, J.M. and RUBIN, E.H. (1984) Case report of toxic reaction from a combination of tryptophan and phenelzine. **Am. J. Psychiatry** **141**, 2, 281 - 283.

WARSH, J.J., COSCINA, D.V., GODSE, D.D. and CHAN, P.W. (1979) Dependance of brain tryptamine formation on tryptophan availability. **J. Neurochem.** **32**, 1191 -1196.

ELEVATED PHENYLETHYLAMINE EXCRETION IN WOMEN

Lynn E. DeLisi., M.D., Dennis L. Murphy, M.D.,
Jean Hamilton, M.D. Lynn R. Goldin, Ph.D.,
Farouk Karoum, Ph.D., Richard J. Wyatt, M.D.
National Institute of Mental Health Clinical
Center 10-3N220 Bethesda, Maryland 20205

Phenylethylamine (PEA) is an endogenously produced trace amine with chemical similarities to amphetamine. It has been of particular interest to psychiatric researchers as a putative, endogenously produced mood-altering and/or psychotomimetic agent (Fischer and Heller, 1972; Wyatt, 1978).

The urinary excretion of PEA has been quantified in humans using GC-mass spectroscopy (Karoum et al., 1979). Significantly elevated mean phenylethylamine excretion was initially found in a chronic paranoid schizophrenic patient population when compared with controls (Potkin et al., 1979).

Further studies from the same laboratory of affective disorder patients revealed no significant patient group mean differences from controls. However, markedly elevated PEA (several times higher than the elevated values found in the previous study) was present in a subgroup of depressed women (Karoum et al., 1982; DeLisi et al., 1984). In these studies serial (2 to 7) samples from the same individuals revealed, in addition, several fold fluctuations in PEA excretion over time.

In normal individuals, another group of investigators found marked increases in PEA excretion acutely, subsequent to severe stress (Paulos and Tessel, 1982), suggesting that the amount of PEA excreted may be unrelated to specific psychopathology. The most consistent of our recent findings, that marked elevations in PEA excretion may be gender related (found in females only), suggested the possibility that neuroendocrine and neuroregulator fluctuations, such as those occuring during the menstrual cycle, may explain some of the fluctuations in PEA levels. To evaluate these possible menstrual cycle-related changes, the following study of six hospital staff volunteers was completed on urine samples collected serially over a 2 to 2 1/2 month period.

Methods:

Six female volunteers (nurses and laboratory personnel) between 22 and 25 years of age obtained 24-hour urine collections daily for 60-80 days.

All of the subjects were interviewed using the Lifetime Schedule for Affective Disorders and Schizophrenia (SADS-L; Spitzer and Endicott, 1978). Modified Research Diagnostic Criteria (RDC; Spitzer et al., 1977; Mazure and Gershon, 1979) were used to establish any psychiatric diagnoses. Although all of the subjects were originally recruited as "normals", three received diagnoses, and were found to meet RDC criteria for either general anxiety disorder, major depression (unipolar), or depressive personality disorder some time during their lives. One other individual had abused marijuana during adolescence, but not during the 5 years prior to this study. Two of the above subjects had first degree relatives with a history of major depressive disorder, unipolar type (subjects 1 and 3), and one had a parent with a history of bipolar affective disorder (subject 2; Table 1). None of the subjects were receiving any medication, nor did they use oral contraceptives during any time of the study. One subject (#3), however, became pregnant during the last 2 weeks of the sample collections.

Urine was collected into receptacles containing 5 ml of 10% ethylenediamine-tetraacetate(EDTA) and refrigerated during the collection. All samples were then stored at -40°C until assayed. All samples were coded and assay procedures were performed without knowledge of the diagnosis of each subject.

Phenylethylamine (PEA) was quantified using a previously described gas-chromatographic/mass-spectrometric procedure (Karoum et al., 1979). The within-run coefficient of variation (derived from repeated analyses of the same sample) was less than 5%. PEA was expressed as the total amount excreted in 24 hours. Four subjects had PEA quantified from samples from every other day for 2 months; the 2 remaining subjects had samples from every day assayed, as well as serial plasma estradiol and progesterone determinations. The hormone concentrations were determined using radioimmunological assays performed at Hazelton Laboratories, Vienna, Virginia.

Results:

The mean ($\pm$ S.D.) PEA excretion for each subject is listed in Table 1.

Table 1.
Urinary PEA excretion in six volunteers over a 2-2 1/2 month period (means $\pm$ S.D.).

SUBJECTS	PEA EXCRETION (g/24 Hr.)
#1	7.4 $\pm$ 7.6
#2	40.3 $\pm$ 36.8
#3	7.2 $\pm$ 2.8
#4	17.7 $\pm$ 67
#5	9.1 $\pm$ 6.6
#6	2.7 $\pm$ 3.3

Four of the six females had at least one PEA measurement that was greater than the previously published normal sample range (>25 micrograms/24 hours; DeLisi et al., 1984). All of the subjects had peaks in PEA concentrations that ranged from two times to several fold higher than the mean for that individual. Figure 1 presents longitudinal PEA concentrations with time for each individual.

In 4 of the six subjects, the peaks were in the latter half of the menstrual cycle (from days 19-28). The other 2 subjects had their peak values on day 13 or 14, approximately the time of estimated ovulation. Plasma progesterone and estradiol, however, were not correlated with PEA fluctuations (figure 2).

Higher PEA concentrations were not present in the subjects who were found to have lifetime histories of a major psychiatric disorder, although the one subject with a familial history of bipolar affective disorder was the single individual with persisting, high PEA excretion (38 of 63 samples >25 g/24 Hr.).

Discussion:

Elevations in urinary PEA excretion have previously found in patients with major psychiatric disorders, although not specifically associated with any one disorder. It has appeared from recent studies of psychiatric patients that several fold elevations in PEA concentrations have only occurred in female patients, particularly those who have experienced hospitalization for some type of depressive disorder (Karoum et al., 1982; DeLisi et al; 1984). Random samples from normal females have failed to show elevated levels comparable to the high values seen in depressed females.

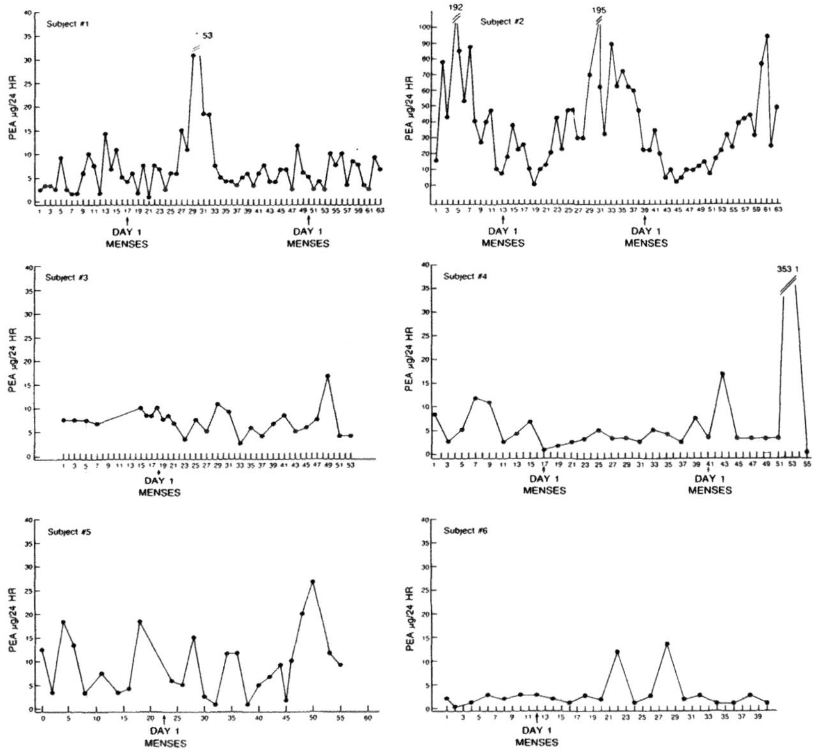

Figure 1.
PEA excretion over time for each individual.

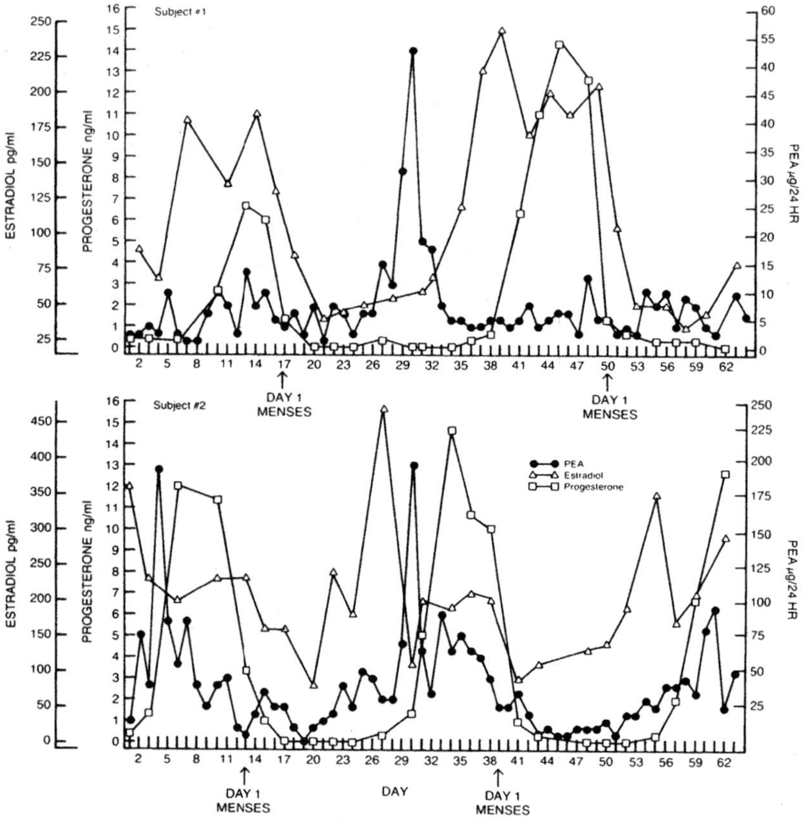

Figure 2
PEA, Estradiol, and Progesterone excretion over time in two female subjects.

It is apparent from the present study that if urinary samples are obtained from the same individuals over a prolonged period (including approximately 2 menstrual cycles per patient), elevated PEA concentrations are more likely to be found. In the present study, elevated values over those obtained during briefer sampling periods in normals were found equally among normal females and those with positive personal or family histories of psychiatric disorders. The one exception was a subject with no personal psychiatric disorder, but a family history of bipolar affective disorder, who had persistent hyperphenylethylamineuria.

Although we do not find these elevations to be closely linked with the menstrual cycle related fluctuations in estradiol and progesterone concentrations, they may be related to other hormonal secretions during the menstrual cycle, as well as the neurotransmitters that regulate release of the hypothalamic regulatory factors. If similar collections are obtained on depressed females such as those from previous studies (Karoum et al., 1982; DeLisi et al., 1984), more frequent elevations might be seen and correlated with the hormonal dyregulation that is often present in affective disorders.

Acknowledgments:

The authors thank Mrs. Joel Hamovit and Drs. Elliot S. Gershon and John I. Nurnberger, Jr. for their help with the clinical evaluations of the subjects in this study.

References:

DeLisi, L.E., Murphy, D.L., Karoum, F., Mueller, E., Targum, S., Wyatt, R.J. (1984) Phenylethylamine in depression. Psychiatry Research 13, 193-201.

Fischer, B. and Heller, B. (1972) Phenylethylamine as a neurohumoral agent in brain. Behav. Neuropsychiat. 4, 8-11.

Karoum, F., Nasrallah, H.A., Potkin, S.G., Chuangg, L., Moyer-Schwing, J., Phillips, I. and Wyatt, R.J. (1979) Mas fragmentography of phenylethylamine, m- and p-tyramine and related amines in plasma, cerebrospinal fluid, urine and brain. J. Neurochem. 33, 251-212.

Karoum, F., Linnoila, M., Potter, W.Z., Chuang, L.W., Goodwin, F.K., and Wyatt, R.J. (1982) Fluctuating high urinary phenylethylamine excretion rates in some bipolar affective disorder patients. Psychiatry Research 6, 215-222.

Mazure, C. and Gershon, E.S. (1979) Blindness and reliability in lifetime psychiatric diagnosis. Arch. Gen. Psychiatry 36, 521-525.

Paulos, M.A. and Tessel, R.E. (1982) Excretion of B-Phenylethylamine is elevated in humans after profound stress. Science 215, 1127-1129.

Potkin, S.G., Karoum, F., Chuang, L.W., Cannon-Spoor, Phillips, I., and Wyatt, R.J. (1979) Phenylethylamine in paranoid chronic schizophrenia. Science 206, 470-471.

Spitzer, R.L. and Endicott, J. (1978) Schedule for affective disorders and schizophrenia (SADS), New York, New York State Psychiatric Institute.

Spitzer, R.L., Endicott, J., and Robins, E. (1977) Research diagnostic criteria (RDC) for a selected group of functional disorders, ed. 3, New York, New York State Psychiatric Institute.

Wyatt, R.J . (1978) Is there an endogenous amphetamine? A testable hypothesis of schizophrenia, in The Nature of Schizophrenia. (Wynne L., Cromwell R. and Matthysse S., eds.), pp. 116-125, John Wiley and Sons, New York.

PHENYLETHYLAMINE AND PERSONALITY

Moises, H.W., P. Waldmeier, H. Beckmann

Central Institute of Mental Health, Mannheim, FRG; Ciba-Geigy AG, Basel, Switzerland; Psychiatric University Clinic, Würzburg, FRG

Central Institute of Mental Health, P.O.Box 5970, D-6800 Mannheim 1, FRG

The phenylethylamine hypothesis of SABELLI & MOSNAIM (1974) postulates that PEA may play a major role in the modulation of affective behavior. Urinary PEA excretion was determined in 32 drug-free healthy volunteers. The MMPI was used for personality measurement. In the female subgroup a significant positive correlation between PEA and hypomania (r_s = .50; $p < .05$) and a significant negative correlation between PEA and depression (r_s = -.65; $p < .01$) was found supporting this hypothesis. Furthermore, PEA correlated significantly negatively with hypochondriasis (r_s = -.58; $p < .01$), paranoia (r_s = - .49; $p < .05$) and social introversion (r_s = - .60; $p < .05$). These results suggest that PEA excretion may be related to mechanisms responsible for the modulation of affective behavior.

INTRODUCTION

Phenylethylamine (PEA) is a trace amine whose chemical structure, and pharmacological and behavioral effects closely resemble that of amphetamine. It has been suggested that PEA may play a role in the etiology of psychiatric disorders. A decrease of PEA has been made responsible for depression (SABELLI & MOSNAIM, 1974), an increase for schizophrenia (SANDLER & REYNOLDS, 1976), paranoid schizophrenia (POTKIN et al., 1979) and aggressive psychopathy (SANDLER et al., 1978). Data in support of these hypotheses have been

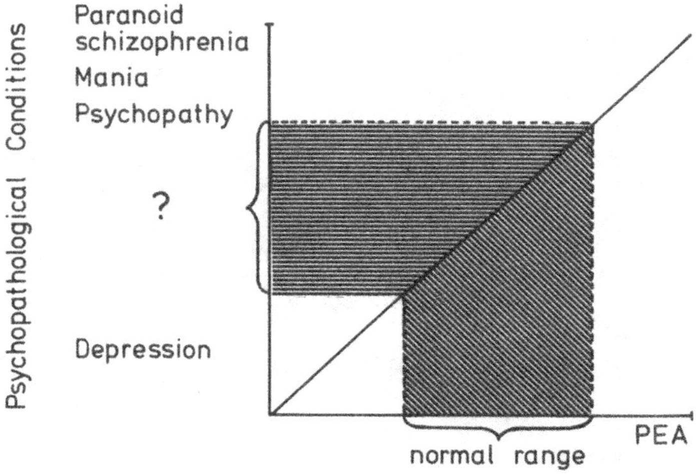

Figure 1: Hypothetical relationship between PEA concentration in the brain und psychopathological conditions.

reported by several investigators, others however failed to confirm these findings (for reviews: WYATT et al., 1977; SAAVEDRA, 1978; WOLF & MOSNAIM, 1983). These hypotheses postulate a relationship between PEA concentrations and psychopathological conditions (figure 1).
Extremely high or low concentrations are thought to cause psychiatric disorders. Nothing is known about the effects of intermediate PEA concentrations. Even in a normal population some signs of depression in individuals with low PEA values and signs of hypomania, aggression and psychopathic and paranoid tendencies in individuals with higher PEA concentrations might be expected. The investigation of normal individuals has the advantage of avoiding the problems of drugs and nonspecific disease effects in psychiatric patients.

The aim of our study was to investigate the relationships between urinary PEA concentrations and personality traits of the MMPI in normal individuals.

MATERIAL AND METHODS

Subjects

To guarantee a wide range of personality variations, 32 healthy volunteers (16 males and 16 females, mean age 26.06. sd 5.43 years) were selected from a heterogeneous sample of 424 individuals, tested in the Psychological Institute of the University Heidelberg by the Freiburger Personality Inventory (FAHRENBERG et al., 1973). Diet, smoking and drinking habits, diseases and family history were evaluated in an interview. All subjects were free of concurrent diseases and drug free according to their own information and to the results of urinary drug screening.

Psychological Trait Measures

For the evaluation of the psychological traits the Minnesota Multiphasic Personality Inventory (MMPI) (HATHAWAY & MCKINLEY, 1951) in its German version by BLASER & GEHRING (1972) was used.

Procedure

24-hour urine samples were collected on three different days within a period of 10 days. Creatinine clearance was determined for each sample to check for the correct collection of the urine. Aliquots were kept frozen at - 70 °C until analyzed.

Biochemical Measures

The concentration of PEA in urine was determined by a highly specific and sensitive method using capillary gas chromatography / chemical ionisation mass spectrometry as described recently (LAUBER & WALDMEIER, 1984). The data obtained were corrected for creatinine values.

Statistical Analysis

For PEA, the median of the values from the three different days was used for further analysis. Spearman rank correlation coefficients (rs) were calculated for the psychological and biochemical measurements. The variable of sex was controlled by examining the correlations for males and females separately as well as for the total group. The significance of sex differences in urinary PEA concentrations was tested by the Wilcoxon rank sum test (one-tailed $p < .10$).
A two-tailed test of significance ($p < 0.05$) was used for the correlations.

RESULTS

Table 1 presents the Spearman rank correlation coefficients between urinary PEA and the MMPI scales. In the total group, PEA correlated significantly negative with depression, paranoia and the masculinity/femininity scale. The latter already indicates some sex differences in the results. This is confirmed by separate calculations for males and females respectively. Most of the significant correlations of the total group are due to the female subgroup. In the male subgroup, none of the correlations reached significance. Moreover, males showed a reduced urinary excretion of PEA (males mean 2.88 sd 2.10, females 4.04. sd. 3.56 µg PEA/g creatinine). This difference however was statistically not significant. In the female group, PEA correlated significantly positive with hypomania and negative with depression (figure 2). Furthermore, a significant negative correlation was obtained with hypochondriasis, paranoia and social introversion.

DISCUSSION

Of the 30 correlations between PEA and psychological variables, 8 were significant on the .05 level. As approximately 2 correlations might have been expected by chance, the results suggest that something more than chance is operating. Furthermore, the patterning of correlations - positive correlation with hypomania and negative with depression - suggets systematic relationships between related variables rather than random ones.If we accept this, two main compo-

Tabelle 1: Correlations (rs) between urinary PEA and MMPI scales

MMPI scales	Total group (N=32)	Females (N=26)	Males (N=16)
Hypomania (a)	.12	.50*	-.29
Depression	-.39*	-.58**	-.13
Paranoia	-.32*	-.49*	-.04
Hypochondriasis (a)	.25	-.65**	.05
Social Introverison	-.24	-.60**	.14
Masculinity/Feminin.	-.34*	-.10	-.40
Conversion Hysteria	-.23	.41	-.18
Psychopathic Deviate	-.20	.05	-.22
Psychasthenia (a)	-.18	-.24	.03
Schizophrenia (a)	-.20	-.14	-.28

(a) = K-corrected; *p < .05; ** p < .01

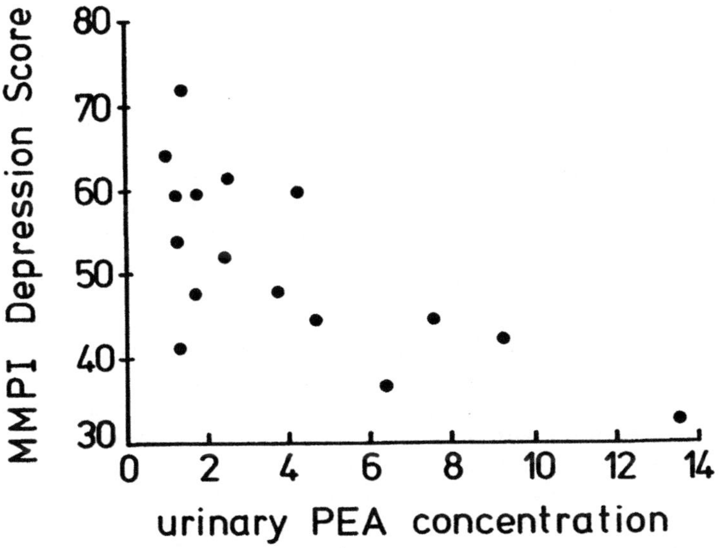

Figure 2: Scatter plot for the correlation between urinary PEA excretion (µg/g creatinine) and depression in females (rs = -.58; p < .01).

nents of our results remain to be explained: the sex differences and the correlations between PEA and personality traits.

With regard to the sex differences, the finding of reduced urinary PEA levels in males is in aggreement with the results of PHILIPS et al. (1978). However, in our sample this difference was not statistically significant. In contrast to the female subgroup no significant correlations were obtained in our male subgroup. This could be due to the lower PEA values in this group or to a male specific confounding variable, like e.g. testosterone which influences mood and possibly PEA metabolism. The significant inverse correlation between PEA and the masculinity-femininity scale, which in turn probably relates somehow to the action of testosterone, could be seen as indirect evidence for the latter assumption.

In females, hypomania correlates significantly positive, whereas depression, hypochondriasis, paranoia and social introversion correlate significantly negative with urinary PEA concentrations. To our knowledge no comparable data have been published concerning PEA and personality traits in normals. However, there have been studies concerned with PEA and related variables in psychiatric patients. Our finding of an inverse correlation between the MMPI depression scale and the 24 hr urinary PEA excretion is well in line with studies indicating a significant reduction of the 24 hr urinary PEA exretion in depressed patients (FISCHER et al., 1972, 1973; MOSNAIM et al., 1973; SABELLI & MOSNAIM, 1974). On the other hand, no positive correlation between PEA and the paranoia or psychopathic deviate scale of the MMPI was found. The results do not agree with those of POTKIN et al. (1979) and JESTE et al. (1981) reporting higher PEA excretion in chronic paranoid schizophrenics as well as those of SANDLER et al. (1978) suggesting PEA overproduction in aggressive psychopaths. It should be noted however that in our study a normal population was investigated while they were concerned with a patient population.

In conclusion, it might be said that our results confirm the predictions made at the beginning of the study, at least in females. As predicted, low PEA concentrations are associated with traits of depression and higher PEA levels with signs of hypomania even in a normal population. These findings could be accounted for SABELLI & MOSNAIM's PEA hypothesis of affective behavior. However, an alternative explanation might be that PEA itself may not modulate affective behavior directly but that PEA metabolism is related to mechanisms involved in the regulation of affective behavior.

Finally, MAO inhibitors are known to increase PEA concentrations and to be especially effective in 'atypical depression' (for reviews: TYRER, 1976; QUITKIN et al., 1979). Further studies are needed to clarify whether the depressive syndrome obtained by negative correlations with PEA - depression, hypochondriasis, paranoia, social introversion - is identical with the 'atypical depression' characterized by 'somatization of symptoms', 'sensitivity to interpersonal rejection' and 'social phobias'.

REFERENCES

Blaser P., and Gehring A. (1972) Ein programmierter Kurs zur deutschsprachigen Ausgabe des Minnesota Multiphasic Personality Inventory von S.R. Hathaway und J.C. McKinley. Hans Huber, Bern

Fahrenberg J., Selg H., and Hample R. (1973) Das Freiburger Persönlichkeitsinventar. Hogrefe, Göttingen

Fischer E., Spatz H., Heller B., and Reggiani H. (1972) Phenylethylamine content of human urine and rat brain. Its alterations in pathogenic conditions and after drug administration. Experientia 28, 307-308

Fischer E., Spatz H., and Fernandex Labriola R. (1973) Quantitative gas chromatographic determination and infrared spectrographic determination of urinary phenylethylamine. Biol.Psychiatry. 7, 161-165

Hathaway S.R., and McKinley J.C. (1951) The Minnesota Multiphasic Personality Inventory Manual. The Psychological Corporation, New York

Jeste D.V., Doonagaji D.R., Panjwani D., Datta M., Potkin S.G., Karoum F., Thatte S., Shet A.S., Apte J.S., and Wyatt R.J. (1981) Cross-cultural study of biochemical abnormality in paranoid schizophrenia. Psychiat.Res. 3, 341-352

Lauber J., and Waldmeier P.C. (1984) Determination of 2-Phenylethylamine in Rat Brain After MAO Inhibitors, and in Human CSF and Urine Capillary GC and Chemical Ionization MS. J. Neural Transmission 60, 247-264

Mosnaim A.D., Inwang E.E., Sugerman J.H., de Martini W., and Sabelli H.C. (1973) Ultraviolet spectrometric determination of 2-phenylethylamine in biological samples and its possible correlation with depression. Biol.Psychiat. 6, 235-257

Philips S.R. (1978) ß-Phenylethylamines: a metabolically and pharmacologically active amine, in Noncatecholic Phenylethylamines, Part I.Phenylethylamines: Biological Mechanisms and Clinical Aspects (Mosnaim A.D., and Wolf M.E., eds.), pp.113-138. M.Dekker, New York

Potkin S.G., Karoum F., Chwang L.W., Cannon-Spoor H.E., Philips I., and Wyatt R.J. (1979) Phenylethylamine in paranoid chronic schizophrenia. Science 206, 470-471

Quitkin F., Rifkin A., and Klein D.F. (1979) Monoamine oxidase inhibitors. A review of antidepressant effectiveness. Arch.Gen.Psychiat. 36, 749-760

Saavedra J.M. (1978) ß-Phenylethylamine: Is this biogenic amine related to neuropsychiatric diseases?, in Noncatecholic Phenylethylamines, Part I.Phenylethylamines: Biological Mechanisms and Clinical Aspects (Mosnaim A.D., and Wolf M.E., eds.), pp.139-157. M.Dekker, New York

Sabelli H.C., and Mosnaim A.D. (1974) The Phenylethylamine Hypothesis of Affective Behavior. Am.J.Psychiatry 131, 695-699

Sandler M., and Reynolds G.P. (1976) Does phenylethylamine cause schizophrenia? Lancet 1, 70-71

Sandler M., Ruthven C.R.J., Goodwin B.L., Field H., and Matthews R. (1978) Phenylethylamine overproduction in aggressive psychopaths. Lancet 2, 1269-1270

Tyrer P. (1976) Towards rational therapy with monoamine oxidase inhibitors. Br.J.Psychiat.128, 354-360

Wolf M.E., and Mosnaim A.D. (1983) Phenylethylamine in Neuropsychiatric Disorders. Gen.Pharmac.14, 385-390

Wyatt R.J., Gillin J.C., Stoff D.M., Majo E.A., and Tinklenberg J.R. (1977) ß-Phenylethylamine and the neuropsychiatric disturbances, in Neuroregulators and Psychiatric Disorders. (Usdin E., Hamburg D.A., and Barchas J.D., eds.), pp.31-45. Oxford University Press, New York

TRACE AMINES/METABOLITE RESPONSES IN STRESS

J. Harris[1], B. A. Davis[2], G. S. Krahenbuhl[3], and A. A. Boulton[2]
[1]Chem. Dept., [3]Health, P.E. Dept., A.S.U., Tempe, AZ 85287 USA; [2]Psych. Res. Div., U. of Saskatchewan, Saskatoon, CN S7N OWO

INTRODUCTION

Few trace amine human stress response data are available compared to the wealth of information on the role of biogenic amines and their relation to stress. While trace amines share in common with biogenic amines, several routes of catabolism, little information is available from experiments in human subjects. In the current experiment the response to stress of several trace amines and their metabolites was measured.

Pilots are among those that carry out demanding skilled and responsible tasks with the lives of others, as well as their own, dependent upon their level of psychomotor performance. The latter involves mental activities, such as reaction time, learning, memory, pattern recognition, etc., and motor responses from pushing simple buttons to that of delicate manipulations.

The state of development in analytical techniques for trace amines and their metabolites is such that these measurements now can be added to previous biochemical, behavioral and physiological indices. The present paper is concerned with demonstrating the range of usefulness of urinary analyses of trace amines and their metabolites as possible complimentary neuroendocrine markers of stress. We report, herein, the results of short term rates of urinary excretion of trace amines and their metabolites in pilots undergoing advanced training and their responses to in-flight psychological stress.

The trace amines studied include the conjugated and unconjugated phenylethylamine (PE), meta- and para-tyramine (m-TA and p-TA) and unconjugated tryptamine (TRYP). The corresponding principal unconjugated trace amine acid metabolites studied include phenylacetic acid (PAA), meta- and para-hydroxyphenylacetic acid (m-HPAA and p-HPAA), indoleacetic acid (IAA), and para-hydroxymandelic acid (p-HMA).

The desire for a non-invasive approach coupled with the practical limitations of analyzing amines and their metabolites in tissue of human subjects directed this study toward measuring rates of urinary excretion.

METHODS

Human Subjects. The subjects were ten male advanced trainee pilots, ages 21-23 years, from whom informed consent was obtained; the research was conducted in conformance with the principles embodied in the Declaration of Helsinki. Each subject, after exposure to syllabus lesson units, was tested in two aircraft sorties which involved 1) their first in-flight check ride (CHECK) and 2) an instrument ride conducted with a darkened canopy, a test performed without environmental visual cues (BLIND). The urinary excretion rates of the trace amines and their metabolites after these two flights were compared with excretion rates obtained in a basal condition, in which the subject is in a non-test and non-active state (BASAL). Collection times for the three trials were matched in time of day so as to control for diurnal variation. Timed urine samples were collected from each subject. Immediately prior to each collection period, subjects emptied their bladders, drank at least 250 ml of water to provide for adequate amounts of urine. At the close of the timed period, normally 30 minute post-experimental conditions, urine samples were collected, the exact length of time and total volume were noted. A 100 ml aliquot of each sample was stabilized with 1 ml of 10% EDTA-4% thioglycolic acid mixture and stored at -90°C until analyzed.

Analyses. To analyze all nine components of the trace amines and their acid metabolites, adaptations of previously reported analytical procedures were used based upon high resolution integrated mass spectrometric analysis of the derivatized amines and acid metabolites and incorporating appropriate internal standards as originally described by Durden et al. (1973), and Davis et al. (1981, 1982). The concentration of each sample was adjusted for time and total volume. Analysis of variance (ANOVA) was applied to the data for each dependent variable of the modalities BASAL, CHECK and BLIND.

Analysis of Amines. Levels of unconjugated amines were determined in an aliquot of urine (1.0 ml) containing a mixture of deuterated internal standards and anti-oxidants. After addition of Na_2CO_3 and acetone, the dansyl derivatives were prepared by reaction with dansyl chloride overnight at room temperature. The dansyl derivatives were extracted with toluene: ethyl acetate (9:1) and dried under a stream of N_2. The residue was dissolved in 300 µl of toluene:acetone (1:1) and spotted on silica gel thin-layer plate along side dansyl amine standards. Chromatographic separation was accomplished with toluene: ethyl acetate (5:2). Further purification resulted by removing the appropriate zones into test tubes, extracting twice with MeOH-acetone (1:1), drying the combined extracts, taking up the residue in toluene:acetone (1:1), spotting on a second plate and rechromatographing in toluene:triethylamine: MeOH(50:5:2). The appropriate zones were transferred into a constricted capillary tube with a glass-fibre filter paper plug at one end (Philips et al. 1979). A second plug was inserted above the column of silica gel to hold the powder in place. About 25 µl of ethyl acetate was drawn into the capillary to elute the dansyl amine. The ends of the capillary tube were sealed with sealing clay and refrigerated until mass spectrometric analysis. High resolution integrated ion current mass spectrometric analysis of the dansyl amines have previously been described in detail by Durden et al. (1973) and Philips et al. (1974,1975).

The very small quantities of phenylethylamine (PE) present in proportion to the relatively large amounts of the other amines made it preferable to analyzed for PE separately. The procedure for PE is identical to that described above except the second TLC developing solvent was toluene:triethylamine (5:1).

Levels of conjugated plus unconjugated (total) amines were determined as above, after hydrolysis of 1.0 ml of urine with 1.0 ml HCl in sealed vials at 100°C for 30 minutes. Tryptamine was destroyed during the hydrolysis. The conjugated amine values were calculated by subtracting the unconjugated value from that of the total (hydrolyzed urine) value.

Unconjugated Acid Metabolites. Levels of the unconjugated acid metabolites were determined in an aliquot of urine (0.2 ml), containing a mixture of deuterated internal standards, carrier acids and anti-oxidants. After deproteinization with sulfosalicylic acid and precipitation of sulfate as well as phosphated ions with $BaCl_2$ solution, the supernatant was saturated with NaCl and extracted (3 times) with ethyl acetate. The extract, divided into two portions converted to the triethylamine salts of the acids was evaporated (under N_2) and traces of water removed by addition of C_6H_6 and evaporating to dryness.

The residue from one portion was derivatized with methanolic HCl, and after evaporation of the solvent, treated with pentafluoropropionic anhydride (PFPA) at 80°C for 60 minutes, to form the pentafluoropropionic-methyl derivatives of *m*- and *p*- HMA and IAA. For PAA and *m*- and *p*HPA, a trifluoroethanol (TFE) derivative was prepared using the residue from the second portion and heating with TFE at 80°C for 60 minutes. After evaporation of TFE (under N_2) PFPA was added and the solution heated to 80°C for 60 minutes. Thw two derivatized solutions were combined, concentrated, dissolved in hexane, washed with 1 M phosphate buffer pH 6.0, then concentrated to 10 µl.

An aliquot was injected onto a bonded phase silica capillary column of a gas chromatograph, the outlet of which was inserted directly into the ion source of a VG 70-70F double focussing mass spectrometer. The acids were quantitated by selected ion monitoring at a resolution of 5000. The molecular ions of the PFP-TFE derivatives of PAA, *m*-and *p*-HPA and the PFP-Me derivatives of IAA, *m*- and *p*-HMA were measured.

RESULTS

The effects of two modes of short-term in-flight training stress on urinary excretion rates (ng ml; $\bar{x}$ + S.E.M.) of trace amines conjugated and unconjugated are shown in Table 1. When examined by themselves, the excretion rates of the amines, by virtue of the wide range of values did not reveal statistically significant ($p < 0.05$) changes, albeit certain trends or changes could be noted. These were shown to be significant upon application of square root transformation ($\sqrt{x}$), which was deemed appropriate because the variances were roughly proportional to the means. Trends and significant changes also could be noted in a few instances when the excretion rate of one substance was coupled in ratio form with other substances; these are shown in Table 3.

In Table 1, it can be seen that the rate of excretion of *m*-TA is always lower than *p*-TA in both conjugated and un-conjugated forms. Under conditions of stress, unconjugated *m*- and *p*-TA showed a trend toward decreased rates of excretion over BASAL rates. Conversely, the conjugated forms of *p*-TA showed a trend toward an increase in excretion rate. Conjugated *m*-TA increased significantly under conditions of the BLIND test while remaining unchanged in the CHECK test. The ratios of both conjugated and unconjugated of *p*-TA/*m*-TA nevertheless remained essentially unchanged in all of the tests (Table 3).

The pattern obtained in the rate of excretion of PE showed a significant decrease in value of the unconjugated form in the CHECK state and a non-significant increase in the BLIND test. The pattern in the BLIND test differed in that the conjugated value was double that of the unconjugated, with the latter value equal to that of the BASAL state.

Unconjugated TRYP was found to decrease significantly in the CHECK test. In Table 2, it can be noted that IAA, the catabolic metabolite of TRYP increased only in the BLIND test and was essentially unchanged in the CHECK test. The ratio of IAA/TRYP was significantly increased in the BLIND test but non-significantly in the CHECK test.

PAA, the major deaminated metabolite of PE, had a significant decrease rate of excretion in the CHECK test and essential an unchanged value in the BLIND test.

Both isomers, *m*- and *p*-HPA, also had significant decreased rates in the CHECK test with no change in rate over BASAL in the BLIND test. No significant change was obtained with unconjugated p-HMA.

DISCUSSION

The training lessons monitored in this investigation are known to be stressors for student pilots (Mefford *et al*. 1971; Harris *et al*. 1984). The results obtained with *m*- and *p*-TA indicated differences between the CHECK and BLIND tests as stress modalities. In all instances the values for *m*-TA were lower than that of *p*-TA, a result which has a similar direction as the relative concentration levels of the two isomers found in human tissue, with the possible exception of the caudate nucleus (Philips *et al*. 1978).

The data with PE also show differences between the two stress tests. In the CHECK condition there was a significant reduction in the rate of excretion of PE in the unconjugated form. This result is in marked contrast to the increased rate of excretion obtained in an inexperienced subject performing an initial parachute jump (Paulos *et al*. 1982). In a more recent report, Paulos *et al*. (1984) obtained no difference in the rate of PE excretion in a subject who was an experienced parachutist. Our data on unconjugated PE in the BLIND test was essentially unchanged, while the data on the conjugated PE in the BLIND indicated an increase, the latter result in keeping with that of Paulos *et al*. (1984).

PAA is the major deaminated metabolite of PE, the later known to have a very high turnover and consequently one would anticipate reciprocal changes in the rates of excretion, i.e., as the value for PAA decreased, there could be an increased value for PE. This seemed not to have occurred. Since the conjugated PAA was not determined, it is not possible to clarify the matter.

The data on urinary excretion of unconjugated catechol biogenic amines and metabolites for these subjects have been reported earlier (Harris *et al.* 1984). The pattern of change obtained with unconjugated amines and their metabolites in this study with those of the unconjugated catechol biogenic amines and metabolites in the previous study differed in several respects. The excretion rates of the adrenergic, noradrenergic, and dopaminergic varied inversely with their respective non-catechol counterparts, PE, *m*- and *p*-TA.

While the adrenergic and noradrenergic excretion rates increased, the non-catechol amine counterpart PE, *m*- and *p*-TA decreased in the two modes of flying-training stress. The trace acid metabolites *m*- and p-HPA, PAA, *p*-HMA had different patterns of change in excretion rate from that of the catecholamine acid metabolites, DOPAC, HVA and HMA. Similarities, however, in the change in excretion rate were found in the serotonergic system in which parallel results were observed between 5HT, 5-HIAA, and their non-catechol counterparts (HT and IAA).

In this study, an attempt was made to determine the usefulness of urinary analyses of trace amines and their metabolites as possible complimentary neuroendocrine markers of stress. The results obtained in this study indicate that urinary excretion of the trace amines and their acid metabolites may reflect various forms of stress. Attention has been directed by others to brain trace amines and their metabolites, with implications of their involvement in mental and neurological disorders.

One cannot relate directly the central nervous system events from peripheral ones, nor are the changes in urinary excretion patterns a reflecting of changes in the brain. With human subjects, however, the advantage of a non-invasive approach to the study of stress as well as mental and neurological disorders may well be of value as a research procedure.

ACKNOWLEDGEMENTS

We thank Mr. R. D. Malchow, Dr. J. R. Stern, Mrs. G. Wheatley, for excellent technical assistance, Mr. W. Pedskalny, Mr. M. Mizuno and Dr. D. A. Durden for the mass spectrometric analyses. We also acknowledge with appreciation the financial support from several sources, namely, Arizona State University Research Fund, the U.S. Air Force (Grant AFSC F33615-80-K-0022), the Saskatchewan Department of Health and the Medical Research Council of Canada.

REFERENCES

Davis, B. A. and Boulton, A. A. (1981) Longitudinal urinary excretion of urine "trace" acids in a human male. J. Chromatograp., 222, 161-169.

Davis, B. A. Durden, D. A. and Boulton, A. A. (1982) Plasma concentrations of *p*- and *m*-hydroxyphenylacetic acid and phenylacetic acid in humans: gas chromatographic-high resolution mass spectrometries analysis. J. Chromatograph. & Biomed. Applic., 230, 219-230.

Durden, D. A., Philips, S. R. and Boulton, A. A. (1973) Identification and distribution of β-phenylethyl amine in the rat. Can. J. Biochem., 51, 995-1002.

Harris, J., Krahenbuhl, G. S., Malchow, R. D. and Stern, J. R. (1984) in "Stress: The Role of Catecholamines and Other Neurotransmitters", E. Usdin, R. Kvetnansky and J. Axelrod (eds.) Vol 2, 991-1001, Gorden and Breach, NY.

Mefford, R. B. Jr., Hale, H. B., Shannon, I. L., Pregmore, J. R. and Ellis, J. P. Jr. (1971) Stress response as a criteria for personnel selection. Aerospace Med., 42, 42-51.

Paulos, M. and Tessel, R. E. (1982) Excretion of phenylethylamine in humans after profound stress, Science, 215, 1127-1129.

Paulos, M., Mosnaim, A., Wolf, M. and Tessel, R. (1984) Phenylethylamine, phenylacetic acid and methionine-entrephalin levels in humans following profound acute stress, in Neurobiology of the Trace Amines (Boulton, A. A., Baker, G. B., Dewhurst, W. G. and Sandler, M., eds.) pp. 563-569, Humana Press, Clifton, N.J.

Philips, S. R., Durden, D. A. and Boulton, A. A. (1974) Identification and Distribution of tryptamine in the rat. Can. J. Biochem., 52, 447-451.

Philips, S. R., Davis, B. A., Durden, D. A. and Boulton, A. A. (1975) Identification and distribution of m-tyramine in the rat. Can. J. Biochem., 53, 65-69.

Philips, S. R., Rozdilsky, B. and Boulton, A. A. (1978) Evidence for the presence of m-tyramine, p-tyramine, tryptamine, and phenylethylamine in rat brain and in several areas of the human brain, Biol. Psychiatry, 13, 51-57.

Philips, S. R. and Boulton, A. A. (1979) Effect of monoamine oxidase inhibitors on some arylalkylamines in rat striatum, J. Neurochem., 33, 159-167.

TABLE 1

Urinary excretion rates (ng/min) of trace amines, conjugated (c) and unconjugated (u) (X ± S.E.M.)

Amines	BASAL	BLIND	CHECK	Significance ($\sqrt{\bar{X}}$)
m-TA (u)	84.8 ± 12.8	77.6 ± 21.2	68.3 ± 16.3	
m-TA (c)	40.2 ± 9.0	95.0 ± 37.6	45.1 ± 9.7	$p < 0.10$ BLIND vs BASAL $p < 0.10$ CHECK vs BLIND
p-TA (u)	413.2 ± 51.7	381.3 ± 86.4	334.9 ± 105.5	
p-TA (c)	238.2 ± 66.6	392.7 ± 117.9	320.3 ± 148.2	
PE (u)	9.0 ± 2.3	11.0 ± 2.8	4.7 ± 1.2	$p < 0.05$ CHECK vs BLIND
PE (c)	11.3 ± 3.8	27.8 ± 15.6	5.3 ± 1.5	
TRYP (u)	61.7 ± 12.7	35.6 ± 12.8	37.2 ± 18.7	$p < 0.10$ CHECK vs BASAL

TABLE 2

Urinary excretion rates (µg/min) of trace amine acid metabolites (unconjugated)
(X + S.E.M.)

Acids	BASAL	BLIND	CHECK	Significance
PAA	2.3 ± 0.3	2.0 ± 0.3	1.2 ± 0.2	$p < 0.05$ CHECK vs BASAL $p < 0.10$ CHECK vs BLIND
m-HPA	7.1 ± 1.1	9.8 ± 3.8	4.1 ± 1.4	$p < 0.10$ CHECK vs BLIND $p < 0.10$ CHECK vs BASAL
p-HPA	17.5 ± 1.9	18.2 ± 1.5	13.8 ± 2.3	$p < 0.10$ CHECK vs BLIND
IAA	4.1 ± 1.1	6.1 ± 2.1	3.6 ± 1.1	
p-HMA	3.6 ± 0.9	5.2 ± 1.3	4.2 ± 0.9	

TABLE 3

Ratio of mean values of urinary excretion rates
X ± S.E.M.

Ratio	BASAL	BLIND	CHECK	SIGNIFICANCE ($\sqrt{\bar{X}}$)
TRYP/PE (u)	7.37 ± 1.49	4.78 ± 1.53	5.30 ± 2.38	
p-TA/*m*-TA (u)	5.15 ± 0.49	5.97 ± 0.75	4.95 ± 0.58	
p-TA/*m*-TA (c)	5.0 ± 1.3	5.1 ± 1.4	5.4 ± 1.7	
PE/*m*-TA (u)	0.11 ± 0.002	0.25 ± 0.10	0.08 ± 0.02	$p < 0.10$ CHECK vs BLIND
PE/*p*-TA (u)	0.02 ± 0.004	0.034 ± 0.009	0.022 ± 0.009	$p < 0.10$ CHECK vs BLIND
PE/*m*-TA (c)	0.32 ± 0.15	0.41 ± 0.24	0.12 ± 0.05	
PE/*p*-TA (c)	0.07 ± 0.02	0.40 ± 0.33	0.05 ± 0.03	
IAA/TRYP (u)	16.7 ± 2.0	21.8 ± 2.0	19.7 ± 2.2	p < BLIND vs BASAL

THE ADRENALINE HYPOTHESIS OF ESSENTIAL HYPERTENSION: AN INVESTIGATION IN PATIENTS AND CONTROLS OF THE MRC HYPERTENSION TRIAL

M.J. Brown, R.C. Causon, P. Brennan

Department of Clinical Pharmacology

Royal Postgraduate Medical School, London

INTRODUCTION

In the adrenaline hypothesis it is proposed that the uptake of circulating adrenaline into sympathetic nerve endings and its resulting conversion from hormone to neurotransmitter may explain how intermittent surges of adrenaline release from the adrenal medulla might result in a sustained increase in sympathetic vasoconstrictor tone (Brown and Macquin, 1981). It was suggested that some support for a possible neurotransmitter action of adrenaline in man was apparent in the persistent tachycardia after adrenaline infusions after the plasma adrenaline concentration had returned to baseline (Brown and Macquin, 1981; Brown and Dollery, 1984). Investigation of the hypothesis required (i) a demonstration that facilitation of noradrenaline release by adrenaline occurred under physiological conditions in man, and (ii) a demonstration that adrenaline release from the adrenal medulla is increased during the development of essential hypertension. The first of these propositions has received qualified support from investigations reported elsewhere (Brown and Dollery, 1984; Brown, Struthers, Burrin et al, 1985). This paper is concerned with a recent study in which we availed ourselves of the opportunity provided by the current MRC Mild Blood Pressure Trial in the UK to study "unlimited" numbers of untreated hypertensives and perfectly matched naive controls, in order to obtain evidence whether adrenaline secretion is increased in patients with hypertension. The adrenaline hypothesis proposes that the concentration of adrenaline in the sympathetic nerve endings is increased as a consequence of an integrated increase in adrenaline secretion from the adrenal medulla over time, and in the present study we therefore used 24 hour urine measurements of adrenaline excretion.

METHODS

The study was organised by the co-ordinating centre of the MRC Trial (MRC Working Party, 1977) and was based in 21 general practices. Each practice recruited 6 male and 6 female hypertensives from the placebo treated group of the main hypertension trial and the same number of age and sex matched controls who were subjects screened at the outset of the main hypertension trial and found to be normotensive. All patients were taken from the two youngest age groups of the trial, namely ages 35-45 and 45-55 at the time of entry to the trial (up to 5 years earlier). At admission to the present study patients and controls had their blood pressure and heart rate recorded and were asked to collect 4 consecutive 24 hour urine samples from the following Sunday to Wednesday. Following this, subjects returned to the clinic for a final blood pressure and heart rate recording. An aliquot of each urine sample was removed and stored at -70°C until subsequent assay for catecholamine and creatinine concentrations. Catecholamines were analysed by the double isotope enzymatic technique (Brown and Jenner, 1981) and creatinine was measured colorometrically. These assays were performed at the Royal Postgraduate Medical School whereas the patients' blood pressure data was sent only to the MRC co-ordinating centre who subsequently received also results of the biochemical analyses. All data was then analysed by an independent statistician who used, as appropriate, t tests and Chi squared tests on multiple regression coefficients. The analysis was performed on the mean of the urine data from days 3 and 4 of the collections.

RESULTS

Blood pressure (mean and standard deviation) was 144/89 ± 16/9 in 268 hypertensive patients and 125/74 ±

15/9 in 268 controls; the mean age in each group was 50. There was no significant difference between excretion of either catecholamine between hypertensives and controls. On days 3 and 4 noradrenaline excretion in the hypertensives was 73.5 ± 88.7 and 68.4 ± 50.9 µg/g creatinine respectively compared with in the controls values of 67.3 ± 46.7 and 68.9 ± 58.7 µg/g creatinine. For adrenaline the excretion on days 3 and 4 in the hypertensives was 20.5 ± 24.9 and 19.1 ± 16.7 µg/g creatinine respectively compared to 19.4 ± 16.0 and 19.6 ± 19.3 µg/g in the controls. There was no significant difference between the groups for either noradrenaline or adrenaline. However there was a highly significant correlation between the excretion of both catecholamines and systolic blood pressure due entirely to a positive correlation in the controls: for adrenaline the Chi squared and p values calculated from a multiple regression analysis were 15.76 and 0.0001 respectively; for noradrenaline the corresponding Chi squared and p values were 11.75 and 0.0006.

DISCUSSION

Numerous previous studies have investigated differences in plasma catecholamine concentrations between hypertensives and normals (Goldstein, 1983). However, measurement of plasma concentrations would not assist in the elucidation of the adrenaline hypothesis in which it is an elevated integrated output of adrenaline over time which is proposed. In the event the study was largely negative despite the large numbers of subjects in each group. The range of adrenaline excretion was higher than anticipated from the initial pilot study so that it is still conceivable that the numbers were insufficient to reveal a difference between the two groups; however, the mean values are two close to leave this is a viable possibility. The explanation of the correlations between blood pressure and plasma catecholamines in the controls is uncertain. Since neither group was re-screened before admission to the urine study, it was anticipated (and indeed happened) that there was considerable overlap between the two groups. Of most interest is a number of patients in the control group who had clearly developed mild hypertension since original screening for the MRC trial, usually within the previous 5 years. The correlation found therefore in the control group may be due to a role of elevated sympathoadrenal activity in the early stages of hypertension, whereas established hypertension is maintained by other factors.

REFERENCES

Brown, M.J., and Macquin, I. (1981) Is adrenaline the cause of essential hypertension? Lancet 2, 1079-1082.

Brown, M.J., and Dollery, C.T. (1984) Adrenaline and hypertension. Clin. Exp. Hyper. 86, 539-549.

Brown, M.J., Struthers, A.D., Burrin, J. et al. (1985) The physiological and pharmacological role of presynaptic adrenoceptors in man. Brit. J. Clin. Pharm. in press.

Brown, M.J., and Jenner, D.A. (1981) A novel double isotope technique for the enzymatic assay of plasma catecholamines permitting high precision sensitivity and sample capacity. Clin. Sci. 61, 591-598.

Goldstein, D.S. (1983) Plasma catecholamines and essential hypertension. An analytical review. Hypertension 5, 86-99.

Medical Research Council Working Party on Mild to Moderate Hypertension (1977) Randomised controlled trial of treatment for mild hypertension: Design and pilot trial. Brit. Med. J. 1, 1437-1440.

TYRAMINE KINETICS BEFORE AND AFTER MAO INHIBITION WITH THE REVERSIBLE MAO INHIBITOR BROFAREMINE (CGP 11 305 A).

Bieck, P.R., Schick, Ch., Antonin, K.H., Reimann, I., and Moerike, K.

Human Pharmacology Institute CIBA-GEIGY GmbH, D-7400 Tuebingen, Federal Republic of Germany

Abstract: The so-called "cheese-reaction" - a hypertensive crisis during treatment with monoamine oxidase (MAO) inhibitors - is generally considered to result from ingestion of food containing high amounts of tyramine (TYR). In order to evaluate this effect in a controlled setting as close as possible to the clinical situation TYR has to be given orally. Additional information on the extent of intestinal MAO inhibition can be obtained during this test by measuring TYR kinetics. After oral doses of TYR causing systolic blood pressure increases of 30 mm Hg plasma and urine concentrations of both TYR and para-hydroxyphenylacetic acid (HPAA) were measured in 9 healthy subjects. The areas under the plasma level-time curves (AUC) as a measure of the amount of bioavailable substance in the body were calculated for TYR and HPAA considering the different TYR doses given (AUCspec). AUCspec (TYR) was 2.5fold higher during MAO inhibition, but AUCspec (HPAA) remained essentially unchanged. This means - contrary to an earlier assumption - that HPAA itself or the ratio HPAA/TYR is not a good index of the inhibition of intestinal TYR metabolism. From plasma concentrations the total clearance (Cltot/f) of TYR and from urinary excretion the TYR fraction eliminated in unchanged form (Fel) were calculated.

Cltot/f after oral TYR load was significantly lowered by 48% ($p < 0.0002$) and Fel was enhanced by 200% ($p < 0.0016$) during MAO inhibition with brofaremine (100 mg/d x 15 d). Both kinetic parameters from plasma and urine were significantly correlated ($n = 9$, $r = 0.806$, $p < 0.01$). The mean relative bioavailability of TYR increased up to 300% during brofaremine treatment. It is concluded that the non-invasive measurement of urinary TYR excretion in addition to the oral TYR pressor test gives valuable information on the extent of gastrointestinal MAO inhibition. Plasma concentration data and analysis of HPAA are not needed.

INTRODUCTION

Hypertensive crises have been well documented following ingestion of foodstuffs with high tyramine (TYR) content in patients treated with monoamine oxidase inhibitors (MAOI) (for review see Marley and Blackwell, 1970). Because of the severity of some of these incidents one consequence has been a limitation of the use of MAOI drugs in the treatment of depression. The possible risk of blood pressure (BP) increase is usually assessed by measuring the sensitivity to intravenous TYR (for review see Ghose, 1980). Since the invariable route of TYR in the clinical setting is oral we started to use systematically the oral tyramine pressor testing (Bieck et al., 1983). First results with tranylcypromine, the partially irreversible MAO-A+B inhibitor showed a 4fold larger increase of oral vs i.v. TYR sensitivity (Bieck et al., 1984).

Inactivation of TYR in the body has been attributed to first-pass intestinal and hepatic metabolism. The first direct evidence of its metabolism in the intestine came from work by Ilett et al., (1980) using **in situ** jejunal loop preparations in dogs. We combined the oral TYR pressor testing with measurements of TYR kinetics and assessed the changes caused by brofaremine (CGP 11 305 A), a selective and reversible inhibitor of MAO-A (Waldmeier et al., 1983) in order to gain additional information to help increasing the safety of MAOI treatment.

MATERIAL AND METHODS

No objections against the design and procedures of these studies were raised by the Ethical Committee of the Human Pharmacology Institute CIBA-GEIGY, Tuebingen

Subjects

The data of nine healthy volunteers, one female and 8 males, aged between 24 and 30 years who participated in two studies were evaluated. All were shown to be healthy on the basis of clinical and laboratory examination. None of the volunteers took any other medication for one week prior to and during the studies. During the experiments no foodstuffs rich in tyramine or alcohol were permitted. All volunteers were ambulatory and carried out their usual activities.

Oral tyramine pressor test

Control period		MAO I period	
No drug		Brofaremin 100 mg/d (n=6) 150 mg/d (n=3)	
d1	d2	d14	d15
PD30 finding	TYR-Kinetics (TYR + HPAA*)	PD30 finding	TYR-Kinetics (TYR + HPAA*)

* in plasma and urine
TYR ≙ tyramine HPAA ≙ p-hydroxyphenylacetic acid

Fig. 1. Treatment schedule. PD30 = mg TYR required to increase systolic BP by at least 30 mm Hg.

Dosage

In the two studies there were initial **control periods** followed by a **MAOI treatment period** for 15 days with brofaremine. Six subjects received daily oral doses of 100 mg (50mg b.i.d.) and 3 subjects doses of 150 mg (single a.m. doses), respectively (Fig. 1).

Oral Tyramine Pressor Test (Fig. 1)

Control period: On day 1 the untreated volunteers took tyramine capsules together with 100 ml water in hourly intervals. BP was recorded at 3 min intervals by the indirect auscultation method using an automatic apparatus (Boso-Digimat P). The lowest TYR-dose given was 200 mg p.o. The amount of amine (mg TYR) required to increase the systolic BP by at least 30 mm Hg (= PD30) was the dose aimed for (**dose-finding**). On day 2 the volunteers took this amount of TYR as single dose to measure the TYR kinetics.

MAOI period: Volunteers treated with brofaremine received oral TYR starting from 12 mg two h after the last medication to estimate the PD30 on day 14. On day 15 this amount of TYR was given as single dose.

Tyramine Kinetics

Blood samples were obtained before and at frequent intervals (for details see Fig. 3) after oral TYR administration. The blood was centrifuged immediately, plasma was separated and kept frozen at -20°C until analysis. Urine was collected for 6 - 10 h in bottles which contained 5 ml HCl (32%). Aliquots of the urine were kept frozen at -20°C until analysis.

Analytical Methods

Unchanged TYR and p-hydroxyphenylacetic acid (p-HPAA) were measured by modifications of the methods of Oates (1961) and Ilett et al., (1980). After extraction into ethyl acetate of 1 ml acid-/alkaline plasma and urine aliquots, respectively, analyses were performed using HPLC with electrochemical detection. The potential was set to 0.87 V with 100 ng TYR and p-HPAA corresponding to 4 nA and 2 nA, respectively. To compensate for the loss data were calculated using internal standards (3-methoxy-tyramine and 0-hydroxyphenylacetic acid) carried through the entire procedure. Typical chromatograms are seen on Fig. 2. TYR and p-HPAA show retention times of 6.6 and 7.0 min, respectively.

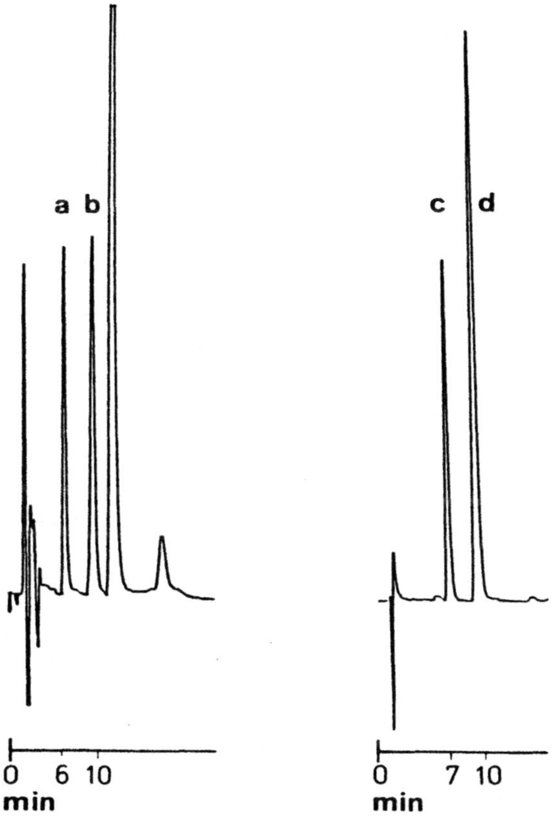

Fig. 2. Chromatograms of TYR and p-HPAA. One ml plasma spiked with 100 ng TYR (a), 200 ng 3-methoxy-TYR (b), 500 ng p-HPAA (c) and 5000 ng o-HPAA (d).

Calculations

From plasma concentrations of TYR and HPAA areas under the plasma level-time curves (AUC(o-t)) were calculated using the trapezoidal rule. The AUC were normalized by dividing by the TYR dose given (AUCspec). The total clearance of TYR after oral administration is given by the equation: Cltot/f = dose/AUC. f = the fraction of the dose reaching the systemic circulation. From the amount of urinary TYR excretion (Ae) and the dose given the TYR fraction eliminated in unchanged form (Fel) is calculated: Ae(o-t)/dose x f. The relative bioavailability (BAVrel) was obtained from plasma and urine by calculating:

$$\textbf{BAVrel (plasma):}\quad \frac{\text{Cltot / f (control)}}{\text{Cltot / f (brofaremine)}}$$

$$\textbf{BAVrel (urine):}\quad \frac{\text{Fel (brofaremine)}}{\text{Fel (control)}}$$

For all kinetic calculations see Ritschel (1980). The paired t-test of Student was applied to test for statistical differences. Correlations were calculated using linear regression analysis.

RESULTS

Plasma Concentrations of TYR and HPAA

The mean plasma concentration-time curves of TYR after administration of an oral TYR load to six subjects before and during subchronic MAO-A inhibition with brofaremine (100 mg/d) are shown in Fig. 3. It can be seen that the concentrations rise within 15 - 30 min to maxima of 94 ng/ml and 59 ng/ml, respectively. The decline is very fast with an approximate mean half-life of 0.5 h. The AUC amount to 101 and 22 ng/ml x h^{-1} and the dose-normalized AUC (AUCspec) to 0.19 and 0.47 ng/ml x h^{-1}, respectively. This is a 2.5fold increase of systemic availability of TYR caused by MAO inhibition.

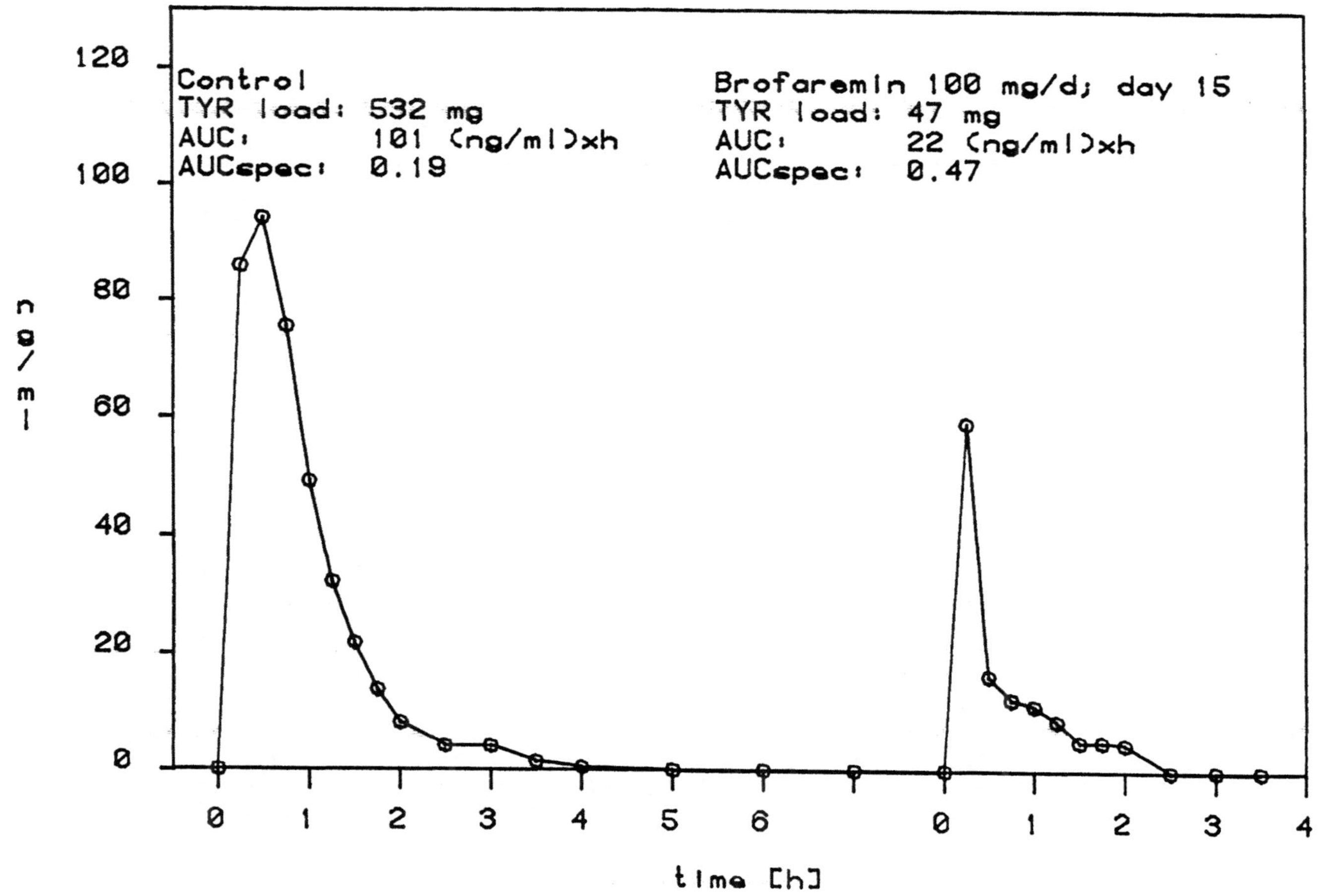

Fig. 3. Plasma concentrations of TYR after oral TYR before and during subchronic MAO-A inhibition. Mean of 6 subjects.

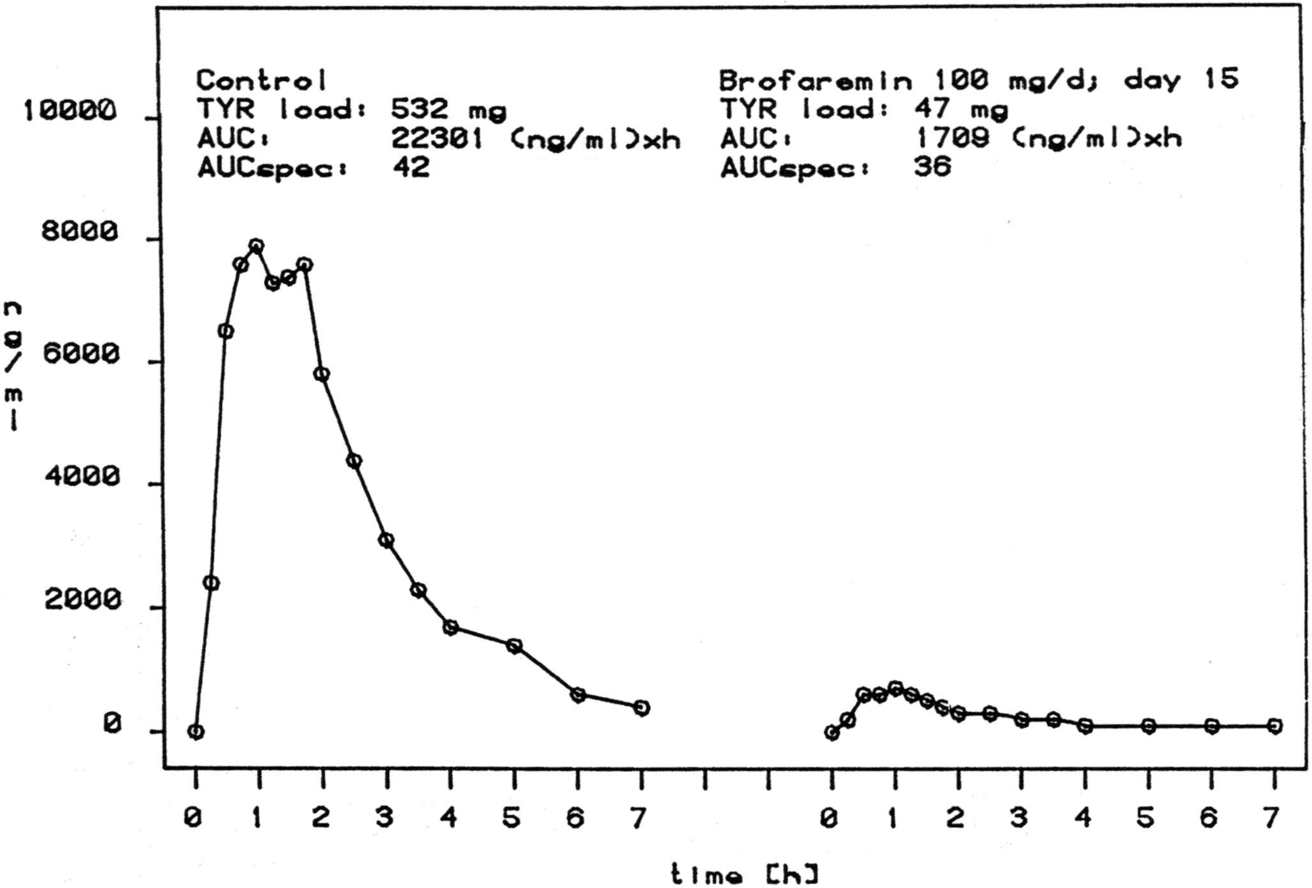

Fig. 4. Plasmaconcentrations of HPAA after oral TYR before and during subchronic MAO-A inhibition. Mean of 6 subjects.

Fig. 4 shows the corresponding mean HPAA plasma concentrations. As to be expected the peaks of the metabolite appear 15 - 30 min later than those of the parent substance. During the control period (left part) Cmax values of 7900 ng/ml and during the MAOI treatment period Cmax concentrations of 700 ng/ml were measured. The AUC normalized for the dose (AUCspec) were 42 ng/ml x h^{-1} (control) and 36 ng/ml x h^{-1} (MAOI), respectively. The difference is not significant.

Urinary TY Excretion

Tab. 1 summarizes the fractions of TYR eliminated in urine in unchanged form (Fel) after an oral load given to 9 subjects. The mean oral TYR doses given were 553 ± 165 mg (control) and 57 ± 22 mg (brofaremine).

Vol. No.	control (C) [x 10^{-3}]	brofaremine (B) [x 10^{-3}]	ratio C/B
1	2.38	7.91	0.30
2	1.15	3.01	0.38
3	2.15	9.68	0.22
4	1.06	1.61	0.66
5	3.54	11.68	0.30
6	2.60	8.98	0.29
7	1.49	2.90	0.51
8	0.77	3.72	0.21
9	2.18	6.57	0.33
mean	1.92	6.23	0.36
± sd	0.88	3.56	0.14
paired t-test		$P \leq 0.0016$	

Tab. 1. Fraction of TYR (Fel) in urine eliminated in unchanged form after an oral load. See Calculations.

From the individual data the means of Fel [x 10^{-3}] were calculated: 1.92 (control) and 6.23 (brofaremine), respectively. This is a highly significant increase ($p < 0.002$) of urinary TYR excretion from 2 to 6‰ of the dose caused by MAO inhibition. The change of Fel caused by subchronic treatment with brofaremine is also expressed as ratio (Tab. 1).

Comparison of Individual Plasma and Urinary TYR Data

In order to evaluate an influence of MAOI treatment on the total plasma clearance after oral administration of TYR (Cltot/f) the clearance was calculated for the same 9 subjects. Tab. 2 shows the results.

Vol. No.	control(C)	brofaremine(B)	ratio B/C
	[(ml/h x 10^6]		
1	9.65	5.38	0.56
2	5.60	2.39	0.43
3	5.69	2.58	0.45
4	4.23	3.50	0.83
5	3.42	0.86	0.25
6	7.94	4.17	0.53
7	7.08	4.58	0.65
8	2.38	0.52	0.22
9	2.45	1.35	0.55
mean	5.38	2.81	0.50
± sd	2.51	1.71	0.19
paired t-test		$p \leq 0.0002$	

Tab. 2. Total plasma clearance (Cltot/f) of TYR after an oral load. See Calculations.

Cltot/f decreased significantly by 48% from 5.38 to 2.81 ml/h x 10^6. If the individual changes of Fel and Cltot/f caused by MAO inhibition are compared a significant correlation of $p < 0.01$ ($n = 9$, $r = 0.806$) can be calculated (Fig. 5).

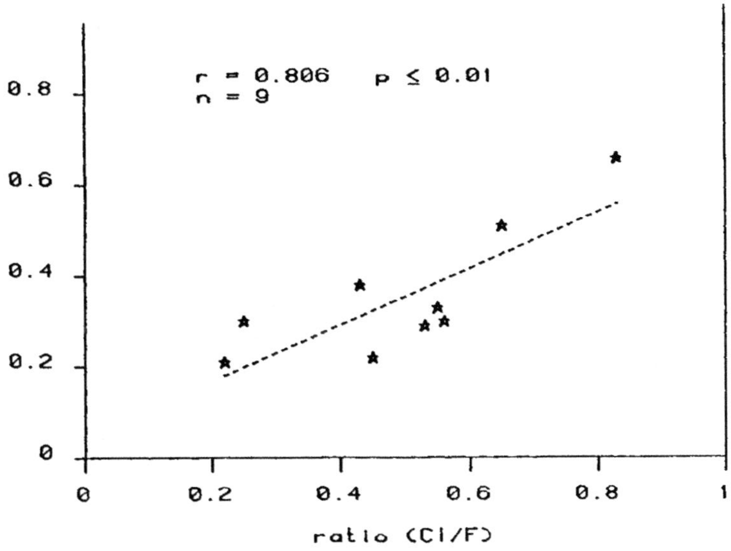

Fig. 5. Correlation of urinary TYR excretion and plasma clearance during subchronic MAO inhibition.

Plasma clearance of TYR

$$\frac{Cl_{tot}}{f} = \frac{dose}{AUC}$$

BAVrel = 238 %

Fraction of unchanged urinary TYR

$$Fe_l = \frac{Ae(0-t)}{dose \times f}$$

BAVrel = 308 %

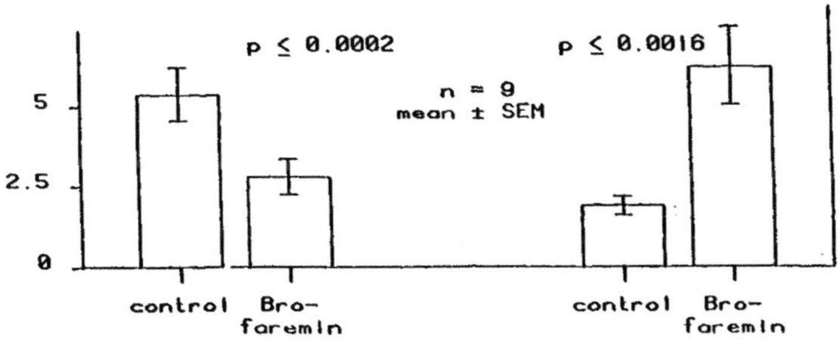

Fig. 6. Kinetic parameters of TYR. see Calculations.

Fig. 6 elucidates the reciprocal relationship between plasma and urinary data for the mean of all volunteers.

DISCUSSION

The MAO of **intact intestine** and liver is metabolizing TYR to the main metabolite p-HPAA when it passes through the intestinal wall and liver tissue (Fig. 7, left part).

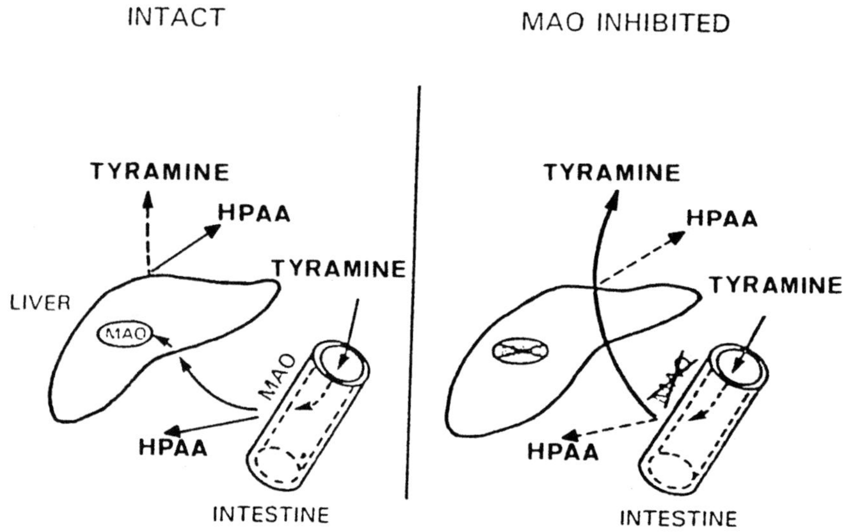

Fig. 7. Schematic graph of intestinal and hepatic TYR metabolism.

Ilett found in **venous plasma** from the isolated jejunal loop of dogs a mean ratio of 5 : 1 for HPAA/TYR. The metabolite HPAA accounted for some 80% of the orally administered TYR, thus identifying the intestine as the major site of metabolism for this compound. MAO inhibition significantly lowered this ratio to 0.2 - 1.1 since more TYR reaches the systemic circulation (Fig. 7, right part).

Using this information from animal experiments it seemed interesting to know if this finding could be reproduced in man. Both substances were measured in plasma after an oral TYR load as evidence for TYR metabolism in the gastrointestinal tract. We assumed that the metabolic ratio of HPAA/TYR would be a valuable tool for the non-invasive assessment of MAO inhibition (Schick and Bieck, 1984). Subsequently, in using the HPAA/TYR ratio as an index of MAO inhibition it was found that the metabolite HPAA was not always lowered after treatment with MAO inhibitors (Bieck and Schick, 1984). Therefore, TYR kinetics were evaluated more in depth.

The maximal amount of TYR which can be given safely in human studies is limited by its pharmacodynamic effects on the cardiovascular system. The PD30 decreased about tenfold from 550 to 57 mg during MAO inhibition with brofaremine. However, by using equally effective oral pressor doses of TYR one can assume that the protective intestinal MAO enzyme system is saturated not only during the control but also during the MAOI period therewith allowing a valid comparison.

The time-related appearance of first TYR and somewhat later of HPAA after oral administration supports the assumption of HPAA being the major metabolite also in man. However, opposite to the animal data from the isolated loop much higher HPAA/TYR ratios of 221 (control) and 78 (brofaremine) were found. According to the expectation the AUC normalized for the dose **(AUCspec) of TYR increased about 2.5fold.** But the **AUCspec of HPAA did not decrease.** This means that any alterations of the HPAA/TYR ratio depend only on the amount of TYR escaping deamination. Another factor which seriously limits the validity of such a ratio is the large concentration difference of HPAA and TYR (Cmax of 7900 ng/ml vs 94 ng/ml). For these reasons kinetic parameters were calculated only for TYR.

Total plasma clearance characterizes clearing of the hypothetical plasma volume of a substance. This includes all pathways of excretion as well as renal elimination and **metabolism.** After oral administration f, the **fraction** of TYR absorbed after mucosal metabolism and **reaching systemic circulation** in presence of a first-pass effect in the liver has to be considered. Both the equation for Cltot/f and for fel contains f. Therefore, our data do not allow differentiation between metabolism in the intestinal tract and/ or in the liver. However, the **in vitro** finding of an abundance of MAO in the intestine of animals (Squires, 1972) and of man (Levine and Sjoerdsma, 1962) are in favour of a major contribution by mucosal metabolism. During subchronic treatment with brofaremine, a reversible MAO-A inhibitor TYR total plasma clearance is diminished and the amount of TYR eliminated unchanged in urine is elevated. Both parameters show a reciprocal relationship. Smaller intraindividual changes of Cltot/f correspond to smaller changes of Fel. By taking the reciprocals (see calculations) of these plasma and urinary ratios one arrives at the **relative bioavailability** (**BAVrel**) of TYR during brofaremine treatment. Both plasma and urinary data show comparable results with BAVrel values of 238 $\pm$ 114% and 317 $\pm$ 107%, respectively.

We conclude from our results that the noninvasive measurement of **urinary TYR excretion** in addition to the oral TYR pressor test gives valuable information on the extent of gastrointestinal MAO inhibition. Neither measurement of HPAA concentrations nor calculation of HPAA/TYR ratios contribute to a more valid assessment of TYR kinetics.

Acknowledgement: We thank for the assistance of Ms. M. Huegler, C. Rommel und G. Vees.

REFERENCES

BIECK, P.R., CREMER, G., GLEITER, C. and ANTONIN K.H. (1983) Effect of the reversible MAO-inhibitor CGP 11 305 A on blood pressure response after oral tyramine. **II. World Conference on Clinical Pharmacology and Therapeutics** (1983) Washington.

BIECK, P.R., SCHICK, Ch. (1984) Urinary Tyramine (TYR) and p-Hydroxyphenylacetic acid (HPAA) after oral TYR load as index of MAO inhibition. **J. Pharm. Pharmacol.** **36**, Workshop Suppl. 69 W.

BIECK, P.R., ANTONIN K.H. and CREMER, G. (1984) Tyramine pressor effects of CGP 11 305 A in comparison to tranylcypromine after prolonged treatment of human volunteers. In **Monoamine Oxidase and Disease** (TIPTON, K.F., DOSTER, P. and BENEDETTI, M.S. ed.)pp 505 - 513. Academic Press, London.

GHOSE, K. (1980) Biochemical assessment of antidepressive drugs. **Br. J. Clin. Pharmacol.**, **10**, 539 - 550.

ILETT, K.F., GEORGE, C.F. and DAVIES, D.S. (1980) The effect of monoamine oxidase inhibitors on first pass metabolism of tyramine in dog intestine. **Biochem. Pharmacol.** **29**, 2551 - 2556.

LEVINE, R.J. and SOERDSMA, A. (1962) Monoamine Oxidase activity in human tissues and intestinal biopsy specimens. **Proc. Soc. Exper. Biol. Med.** **109**, 225 - 227.

MARLEY, E. and BLACKWELL, B. (1970) Interactions of monoamine oxidase inhibitors, amines and foodstuffs. In **Advances in Pharmacology**, (GARATTINI, S., GOLDIN, A., HAWKIN, F. and KOPIN, I., ed.) pp 185 - 239. Academic Press, New York.

OATES, J.A. (1961) Measurement of urinary tryptamine, tyramine and serotonin. **Year Book Medical Publisher,** Chapter 9, 171 - 172.

RITSCHEL, W.A. (1980) **Handbook of Basic Pharmacokinetics.** Drug Intelligence Publ., Inc., Hamilton.

SCHICK, Ch., BIECK, P.R. (1984) Presystemic tyramine (T) elimination in man assessed by measurement of T and its metabolite p-hydroxyphenylacetic acid (HPAA). **Naunyn-Schmiedeberg's Arch. Pharmacol.,** Suppl. 325, R 87.

SQUIRES, R.F. (1972) Multiple forms of monoamine oxidase in intact mitochondria as characterized by selective inhibitors and thermal stability: a comparison of eight mammalian species. **Adv. Biochem. Psychopharmacology.** 5, 355 - 370.

WALDMEIER, P.C., FELNER, A.E.and TIPTON, K.F. (1983) The monoamine oxidase inhibiting properties of CGP 11 305 A. **Eur. J. Pharmacol.** 94, 73 - 83.

TYRAMINE-CONJUGATION DEFICIT AS A TRAIT-MARKER IN ENDOGENOUS DEPRESSIVE ILLNESS

M. Sandler*, A.S. Hale+, P.L. Walker* and P.K. Bridges+

*Bernhard Baron Memorial Research Laboratories, Queen Charlotte's Hospital, London W6 0XG, UK.

+Geoffrey Knight Psychosurgical Unit, Brook General Hospital, London SE18 4LW, UK.

ABSTRACT

A highly significant decrease in ability to metabolize an oral load of tyramine to its sulphate conjugate compared with controls is characteristic of patients with endogenous unipolar depressive illness. Neurotic depressives do not differ from controls. The biochemical deficit persists after clinical recovery from depression and is also present in about half the first-degree relatives of endogenously depressed probands. Thus, it is likely to be a trait marker for the disease and a predictor of vulnerability to it.

About half of a small group of acute schizophrenics without overt depression also manifested with a tyramine conjugation deficit. A number of methodological questions related to precision of diagnosis, drug status and completeness of urine collection need to be answered before these findings can form the basis of a unitarian hypothesis of functional mental illness.

INTRODUCTION

The early history of the tyramine conjugation deficit as a marker for depressive illness was fully described in the proceedings of the 1st International Trace Amine Symposium (Sandler et al., 1984). A substantial minor pathway for orally-administered tyramine disposition occurs via conjugation with sulphate. In normal subjects, 10-15% of the ingested amine is metabolized by this route. In depressive illness, however, not only is there a highly significant decrease in degree of conjugation but the deficit remains unchanged after clinical recovery (Sandler et al., 1975; Bonham Carter et al., 1978a). Thus, a prima facie case can be established for the finding being a trait marker for the disease, an interpretation supported by a prospective study in which low conjugation ability correlated with lifetime history of depressive illness (Bonham Carter et al., 1980b).

At the time of the first symposium, the mechanism of the biochemical defect was obscure and even now remains so. In affected subjects, no deficit can be identified, for instance, in activity of the platelet sulphate-conjugating enzyme, phenolsulphotransferase (Bonham Carter et al., 1981). Nor is any disturbance of sulphate economy (Bonham Carter et al., 1980a) or gut motility (Bonham Carter et al., 1978a) apparent. Since that time, there have been two important developments, however. Harrison et al. (1984) have confirmed our earlier findings in every particular and, indeed, added to them; we ourselves have carried out further extensive studies and now present evidence to show that the conjugation deficit is a clean trait marker for endogenous unipolar depression.

a. Recent Investigations in Depression

Using a gas chromatographic assay procedure (Walker and Sandler, 1985), we (Hale at al., 1985) have recently performed a tyramine-conjugation test on several different categories of subject (Table 1). Highly significant differences were apparent not only between endogenous (Newcastle scale) unipolar depressives and controls but between endogenous and neurotic depressives (Hamilton depression rating scale). The difference between groups was very clean: all individual tyramine excretion values in the endogenous unipolar depression group fell at or below a level of 4.1mg per 3h. The only two control subjects with a value below this cut-off point were also the only two in the whole group with a family history of depressive illness (see below). When eight endogenous depressives were retested two days after their first tyramine test, none changed category across the cut-off point as a result. These findings are so striking that we are almost tempted to use the test as a

TABLE: Showing mean tyramine excretion values, age and Hamilton depression (HamD) score (± SD) in the different clinical groups.

	Mean tyramine excretion (mg/3h) ± SD	Mean age (yr) ± SD	Mean HamD score ± SD
Endogenous depression	2.75 ± 1.40[ab]	52.9 ± 14.9	27.8 ± 6.9
Neurotic depression	5.57 ± 1.25	37.3 ± 10.8	21.3 ± 6.4
Schizophrenia	3.60 ± 1.43[a]	35.6 ± 12.9	10.1 ± 5.0
First degree relatives	4.20 ± 1.18[a]		
Controls	6.12 ± 1.85	44.0 ± 12.33	-

a. $p < 0.0001$ compared both with controls and neurotic depression.
b. $p < 0.045$ compared with schizophrenia.
(From Hale et al., 1985).

marker for "endogenousness". Age, sex, menopausal status, severity of depression, severe weight loss or gastro-intestinal symptoms did not affect tyramine excretion. Nor was there any relationship between post-dexamethasone (1mg) cortisol and tyramine excretion values.

The specificity of our findings to endogenous depression of 90% was higher than the 70% reported by Harrison et al. (1984). This may partially be due to choosing different cut-off levels but probably stems mostly from the greater homogeneity of our depressive groups due to the availability of longitudinal data, in most cases, for many years prior to the index admission. Preliminary results with and without imipramine administration indicate that values are not affected by tricyclic administration. Drugs of this kind would not account, therefore, for differences between our two groups.

We only had access to five bipolar patients, two of whom were currently depressed. Here, the results are less clear-cut than those for the larger group reported by Harrison et al. (1984) and further data from different phases of the cycle are obviously needed.

b. Family Studies

One of the most interesting findings in our recent study was that the conjugation deficit was present in about half the first degree relatives of endogenously depressed probands. Thus, this group showed a highly significant difference from controls or from neurotic depressives. None of the relatives with values below the cut-off point were currently depressed nor gave any history of depressive illness. If the excretion abnormality does not actually have an intimate aetiological relationship to depression, it seems to be inherited in a similar fashion. We think, on balance, that the finding is likely to reflect vulnerability to depressive illness which can be provoked, presumably, by some unknown triggering factor; thus, it appears to represent a trait-marker, probably the first clear-cut one yet to be identified. It may well prove useful in clinical practice eg. in identifying predisposition to attacks of postnatal depressive illness in a pregnant population, so that appropriate psychiatric and social support can be provided.

c. Schizophrenics

Perhaps the most baffling finding, although potentially of the greatest theoretical importance, was the highly significant difference we identified between acute schizophrenics and controls. About half the schizophrenics showed values below the cut-off point but this finding did not appear to reflect any depressive component in their mental state, at least according to the Hamilton rating scale used for assessment. Only one of the schizophrenics gave a family history of affective disorder.

What we do not know is the possible effect of neuroleptics on tyramine excretion; although the patients we investigated had been drug-free for three weeks, some long-term hangover effect cannot be ruled out. Another important methodological question with a group such as this is of ensuring the collection of a complete 3h urine sample. Schizophrenic thought disorder and negativistic behaviour are particularly likely to lead to failure to comply with instructions for urine collection, despite supervision, although such a problem, together with that of frank incontinence, are by no means specific to this diagnostic group.

If these methodological artefacts can be ruled out, and the precision of diagnosis confirmed, there remains, another, and more intriguing possibility, that the tyramine excretion test is picking up some abnormality common to all the endogenous psychoses. If this were the case, the question must once again be raised as to whether some underlying basic lesion, eg. a membrane transport deficit (see Bonham Carter et al., 1978a) predisposes both to endogenous depression and schizophrenia.

REFERENCES

Bonham Carter S.M., Glover V., Sandler M., Gillman P.K. and Bridges P.K. (1981) Human platelet phenolsulphotransferase: separate control of the two forms and activity range. *Clin.Chim.Acta* 117, 333-344.

Bonham Carter S.M., Goodwin B.L., Sandler M., Gillman P.K. and Bridges P.K. (1980a) Decreased conjugated tyramine output in depression: the effect of oral L-cysteine. *Br.J.Clin.Pharmac.* 10, 305-308.

Bonham Carter S.M., Reveley M.A., Sandler M., Dewhurst J., Little B.C., Hayworth J. and Priest R.G. (1980b) Decreased urinary output of conjugated tyramine is associated with lifetime vulnerability to depressive illness. *Psychiat.Res.* 3, 13-31.

Bonham Carter S.M., Sandler M., Goodwin B.L., Sepping P. and Bridges P.K. (1978a) Decreased urinary output of tyramine and its metabolites in depression. *Br.J. Psychiat.* 132, 125-132.

Bonham Carter S.M., Sandler M., Sepping P. and Bridges P.K. (1978b) Decreased conjugated tyramine outputs in depression: gastrointestinal factors. *Br.J.Clin.Pharmac.* 5, 269-272.

Hale A.S., Walker P.L., Bridges P.K. and Sandler M. (1985) Tyramine conjugation deficit: a trait-marker for endogenous depression. *Submitted for publication.*

Harrison W.M., Cooper T.B., Stewart J.W., Quitkin F.M., McGrath P.J., Leibowitz M.R., Rabkin J.R., Markowitz J.S. and Klein D.F. (1984) The tyramine challenge test as a marker for melancholia. *Arch.Gen.Psychiat.* 41, 681-685.

Sandler M., Bonham Carter S.M., Cuthbert M.F. and Pare C.M.B. (1975) Is there an increase in monoamine oxidase activity in depressive illness? *Lancet* i, 1045-1049.

Sandler M., Bonham Carter S.M., and Walker P.L. (1984) Tyramine and depressive illness. In: *Neurobiology of the Trace Amines* (Boulton A.A., Baker G.B., Dewhurst W.G. and Sandler M., eds.) pp.487-498. Humana Press, Clifton, USA.

Walker P.L., and Sandler M. (1985) The analysis of urinary meta- and para-tyramine using gas chromatography with electron-capture detection. *Submitted for publication.*

THE ORIGIN OF CENTRAL AND PERIPHERAL P-HYDROXY-PHENYLACETIC ACID IN MAN AND RATS

Farouk Karoum, Ph.D.

Neuropsychiatry Branch, National Institute of Mental Health, St. Elizabeths Hosp., Washington, D.C. 20032

ABSTRACT

The effects of a number of monoamine oxidase, two types of dopadecarboxylase inhibitors and neomycin on the production of p-hydroxyphenylacetic acid (PHPA) and catecholamine metabolites were evaluated in an attempt to determine the origin of central and peripheral PHPA in rats. Acute intragastric (I.G.) administration of pargyline as well as chronic I.G. carbidopa, alpha-methyldopa (dopadecarboxylase inhibitors) and neomycin treatments failed to reduce PHPA concentration in the brain and urine, suggesting minor roles of gut flora and endogenously produced p-tyramine in the overall body production of PHPA. Neomycin reduced p-Ty and increased PHPA excretion. Catecholamine metabolites, phenylethylamine and p-tyramine (p-Ty) excretions were altered according to the expected mode of action of the drugs employed. Paradoxically, carbidopa (like alpha-methyldopa) significantly reduced hypothalamic norepinephrine and its metabolism, suggesting a central influence of carbidopa. Chronic administration of three types of monoamine oxidase (MAO) inhibitors, pargyline, clorgyline and deprenyl failed to reduced urine and brain PHPA. These drugs produced changes in phenylethylamine and catecholamine metabolite's excretion and brain content that are consistent with effective inhibition of either or both MAO type A and B. By a process of elimination, it is concluded that while most body p-Ty is derived from p-tyrosine decarboxylation, most central and peripheral PHPA in rats and possibly man originate from p-tyrosine transamination to p-hydroxyphenylpyruvic acid followed by decarboxylation to PHPA. This conclusion was confirmed by demonstrating that the administration of p-hydroxyphenylpyruvic acid significantly elevated the excretion of PHPA as well as p-hydroxyphenyllactic acid and homogentisic acid. Blockade of p-tyrosine decarboxylation by carbidopa was also demonstrated to elevate the excretion of deuterated PHPA derived from administered deuterated p-tyrosine, thus adding support to the above conclusion. The contribution of p-tyramine metabolism towards total body output of PHPA is less than 30%. The role of the gut flora in the production of phenylethylamine and catecholamine metabolites was also concluded to be minor. A number of new observations related to p-Ty, PHPA and PEA were made. p-Ty was found to be almost completely excreted in rat in the conjugated form. In contrast most urine PHPA (around 80%) and PEA (70-90%) are excreted unconjugated. It is concluded that although p-Ty may be an important biogenic amine, its metabolism and turnover rate, unfortunately, cannot be assessed from the assay of PHPA.

INTRODUCTION

A number of recent reports on the central and peripheral concentration of p-hydroxyphenylacetic acid (PHPA) suggest a precursor/metabolite relationship between p-tyramine (p-Ty) and PHPA in both man and the experimental animal (Durden and Boulton, 1981, Yu et al., 1982; Boulton et al., 1983). Furthermore, PHPA excretion appears to be abnormally low in aghoraphobic patients and in violent and sexual offenders (Boulton et al., 1983; Davis et al., 1983). There is also a tendency for PHPA concentration in the cerebrospinal fluid to be low in patients with neurological and psychiatric disorders (Kobayashi et al., 1983), a finding that may be of diagnostic importance if confirmed and proven to be consistent. Against this interesting background, however, are the observations of poor correlations between in vivo monoamine oxidase (MAO) inhibition (as determined by the reduction in catecholamine metabolism and increased p-tyramine excretion) and PHPA excretion in man (Karoum et al., 1982; Linnoila et al., 1982). The possibility may, therefore, exist that most urinary PHPA in man is derived from a source other than p-tyramine deamination.

To specifically determine the metabolic origin of PHPA, a number of experiments were conducted to evaluate the contribution of p-Ty metabolism towards central and peripheral PHPA production. As will be shown, and in accordance with human observation, most urine and brain PHPA in the rat does not originate from p-Ty.

MATERIAL AND METHODS

Urine was collected from rats placed in "Econo Metabolic Units" (American Scientific Products, Springfield, Va.). Capillary columns were purchased from Supelco Inc., Pa. Brain parts were dissected according to Glowinski and Iversen, 1966. All chemicals and reagents used were of the highest grades commerically available.

Brain tissues were homogenized in 0.5 ml of 1N HCl containing 50 ng of each of the following deuterated compounds: p-tyramine ($^2H_4$-p-Ty), 3,4-dihydroxyphenylacetic acid (DOPAC) ($^2H_5$-DOPAC), homovanillic acid (HVA) ($^2H_5$-HVA), $^2H_4$-PHPA and phenylethylamine (PEA) ($^2H_9$-PEA). Ten or five μl of the homogenates were pipetted out for protein determination (Lowry et al., 1951). The rest of the homogenates were centrifuged at 15,000 g (Eppendorf 5412 Centrifuged, American Scientific Products) and the clear supernatant stored at -10°C until analyzed.

Urine creatinine was measured by a modification of the original method of Here, 1950.

Unconjugated dopamine, NE, p-Ty and PEA were measured mass fragmentographically as previously reported (Karoum, 1983). A 30 m fused silica capillary column, 0.32 mm i.d. bond coated with SPB-5 (methylphenyl-vinyl-silicone gum) was used for the assay of catecholamines, their metabolites and PHPA. A 30 m fused silica capillary column bond coated with a mixture of SP-2250 (methyl-phenylpronyl silicone gum) SPB-5 and SP 2401 (fluoropropyl silicone) was used for the assay of PEA, m- and p-tyramine.

Total PEA, p- and m-Ty are measured as follows: 50 μl of urine is mixed with 100 ng $^2H_9$-PEA and $^2H_4$-Ty followed by the addition of 100 μl 10 N HCl. The mixture is heated at 100°C for 60 minutes. After hydrolysis, 100 μl of 10 N NaOH is added, mixed and the amines assayed as for free Ty and PEA (Karoum, 1983).

Dopamine metabolites, DOPAC and HVA, p-hydroxyphenyllactic acid (PHPLA), homogentisic acid and PHPA in urine or brain tissues were assayed as previously published (Karoum, 1983).

A model 4000 Finnigan gas chromatograph-quadrupole mass spectrometer (Finnigan Corp., Sunnyvale, California) was used for the mass-fragmentographic measurement of the various amines and metabolites.

Monoamine oxidase inhibitors, pargyline, clorgyline and deprenyl were dissolved in saline and administered intraperitoneally (I.P.) or intragastrically (I.G.). Alpha-methyldopa and carbidopa were admistered I.G. from a saline suspension. The doses and treatment regiments are described in the appropriate Tables.

RESULTS

Preliminary studies on the presence of total (assayed after hydrolysis of urine in 6N HCl and at 100°C for one hour (Martin et al., 1979) and free PHPA concentration in rat urine revealed about 20% of total PHPA is excreted conjugated. Further, intravenous administration of deuterated PHPA ($^2H_9$-PHPA), 10 mg/kg, failed to appear in both the hypothalamus and caudate nucleus after periods ranging from 0 to 2 hours following its administration. It is therefore assumed that peripherally formed PHPA does not efficiently cross the blood brain barrier and that most rat urine PHPA is excreted unconjugated. Furthermore, while over 80% of total p-Ty in urine is conjugated most urine PEA in rats is free (70-90%).

The Role of the Gut Flora in the Production of PHPA and Other Metabolites of Biogenic Amines

To assess the extent of drug inhibition of amino acid decarboxylase and MAO, PEA, p-Ty, NE and DA as well as their metabolites (except that of PEA) were measured in addition to PHPA. To isolate the effects of gut flora metabolism of p-Ty from that of peripheral tissues, PHPA excretion was measured in rats after two weeks chronic neomycin treatment. To further investigate into the contribution of the gut flora to the overall body production of PHPA via p-Ty metabolism, the effects of intragastric (I.G.) administrations of pargyline (MAO inhibitor), carbidopa (peripheral amino acid decarboxylase inhibitor) and alphamethyldopa (peripheral and central amino acid decarboxylase inhibitor) were evaluated. As shown in Table 1 and 2, neither destruction of the gut flora with neomycin nor inhibition of p-Ty production and metabolism (both intragastrically and within the periphery and the brain) significantly reduced the excretion of PHPA. In contrast to what was expected, neomycin significantly increased PHPA excretion while p-Ty excretion was significantly reduced.

The results in Table 1 also illustrate the effects of the above I.G. treatments on PEA and catecholamine metabolites excretion. As can be deduced from the results MAO and amino acid decarboxylase activities within the body were markedly reduced. Thus, the two amino acid decarboxylase inhibitors employed, significantly reduced PEA excretion. MAO was also markedly inhibited because the excretion of HVA and MHPG were also reduced.

None of the above treatments changed brain PHPA excretion, Table 2.

TABLE I

THE EFFECTS OF INTRAGASTRIC (I.G.) ACUTE PARGYLINE AND CHRONIC CARBIDOPA, ALPHA-METHYLDOPA AND NEOMYCIN TREATMENTS ON URINE CONCENTRATIONS OF BIOGENIC AMINES AND METABOLITES IN RAT

Description	Total PEA	Total p-Ty	Free PHPA	Free HVA	Total MHPG
Urine					
Controls (μg)	0.314 ± 0.030	16.5 ± 3.7	62.6 ± 10.8	6.6 ± 0.3	5.48 ± 0.50
Pargyline (μg)[a]	3.081 ± 0.275**	41.4 ± 12.8**	44.3 ± 10.8	4.8 ± 0.6*	3.42 ± 0.38**
Carbidopa (μg)[b]	0.169 ± 0.025**	5.92 ± 0.93**	46.0 ± 6.2	5.8 ± 0.8	2.92 ± 0.40**
Alpha-Methyl-dopa (μg)[b]	0.066 ± 0.015**	2.87 ± 0.35**	36.3 ± 1.9**	3.7 ± 0.4*	2.02 ± 0.11**
Neomycin (μg)[c]	0.290 ± 0.042	6.26 ± 1.12**	127.6 ± 38.1*	—	—

[a]Pargyline was administered at a dose of 50 mg/kg and the rats placed in metabolic cages for four hours to collect urine. After four hours the rats were decapitated and their brains remove and analyzed.

[b]Carbidopa, 100 mg/kg, and alpha-methyldopa, 400 mg/kg, were administered I.G. once daily for five days. On the fifth day urines were collected for four hours after the last injection and rats decapitated soon after urine collection. The brains were removed dissected out and analyzed.

[c]Neomycin, 125 mg/kg, I.G. was administered twice daily for two weeks and urine collected overnight after the last treatment. Baseline urines were collected on the same rates before neomycin treatment. All statistical evaluation for the neomycin treated rates were carried out and paired t-test. Neomycin baseline values were comparable to those of the controls and therefore not shown. Five rats were included in each group.

Results are expressed in μg/mg creatinine.

*$p < 0.05$, **$p < 0.001$ compared to the appropriate controls or baselines

TABLE 2

THE EFFECTS OF INTRAGASTRIC (I.G.) ACUTE PARGYLINE AND CHRONIC CARBIDOPA, ALPHA-METHYLDOPA AND NEOMYCIN TREATMENTS ON CAUDATE NUCLEUS CONCENTRATIONS OF BIOGENIC AMINES AND METABOLITES IN RAT

Description	Free PHPA	Free DOPAC	Free HVA
Controls (ng)	0.500 ± 0.0058	8.75 ± 0.44	6.50 ± 0.56
Pargyline (ng)	0.515 ± 0.042	1.18 ± 0.35***	1.56 ± 0.83***
Carbidopa (ng)	0.481 ± 0.036	10.45 ± 0.53	9.12 ± 0.65
Alpha-Methyldopa (ng)	0.451 ± 0.029	7.82 ± 0.39**	5.0 ± 0.49

Brain Results are expressed in ng/mg protein

**p < 0.005, ***p < 0.001 compared to the appropriate controls or baselines

See Table 1 for explanation of experiment and dosage.

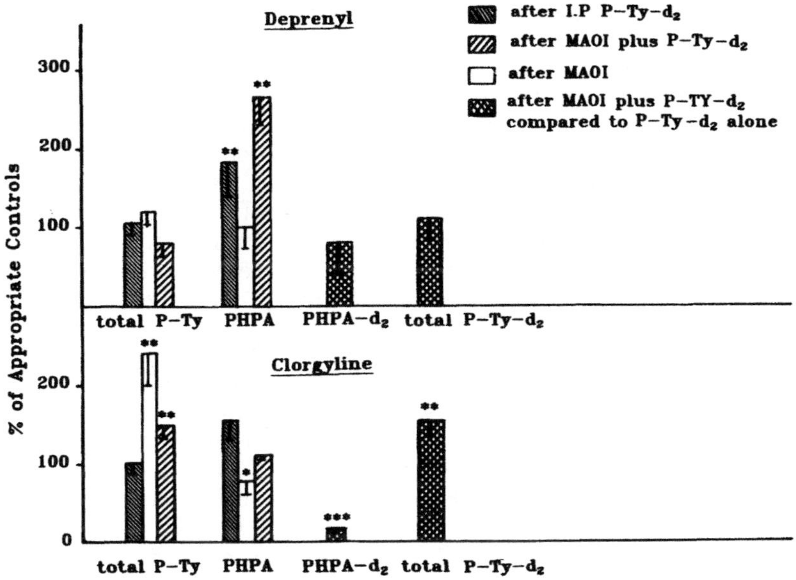

FIGURE I

Changes in total p-tyramine (p-Ty), p-hydroxyphenylacetic acid (PHPA) and their deuterated isomers excretions in rats challenged with 2 mg deuterated p-Ty (p-Ty-d_2) before and after six days treatments with deprenyl 10 kg, and clorgyline 25 mg/kg, I.P. There MAOIs were administered 12 hours and 1 hour before challenging with p-Ty-d_2.

The percent changes in the non-deuterated compounds are as compared to the appropriate baselines. Those of p-Ty-d_2 and PHPA-d_2 (shown) represent the percent changes as compared to the excretions of these compounds after 2 mg p-Ty-d_2 challenge before MAOI.

Evaluation of MAO Inhibitions by Clorgyline and Deprenyl and Their Correlations to PHPA Excretion

The results so far presented (Tables 1 and 2) suggest that *in vivo* inhibition of MAO by pargyline does not significantly reduce PHPA excretion. Since pargyline (at a high dose) is a mixed MAO type A and B inhibitor and since MAO activities within the rat bodies were not measured, the possibility may exist that MAO was not efficiently inhibited, hence the poor reduction in PHPA (Table 1). To assess MAO inhibition with specific reference to p-Ty, the metabolism of I.P. administered $^2H_2$-p-Ty to $^2H_2$-PHPA was evaluated before and after I.P. treatments with clorgyline (a potent MAO type A inhibitor) and deprenyl (an MAO type B inhibitor) (Knoll, 1980). The results are summarized in Figure 1. As shown, clorgyline markedly inhibited $^2H_2$-p-Ty metabolism but reduced PHPA excretion by only about 30%. Deprenyl did not significantly reduce PHPA excretion nor $^2H_2$-p-Ty metabolism. Interestingly, the administration of $^2H_2$-p-Ty significantly increased PHPA excretion, an effect that may be due to stimulation of PHPA production. The results shown in Figure 1 clearly indicate that while p-Ty is efficiently metabolized to PHPA (as represented by $^2H_2$-p-Ty metabolism) most urine PHPA does not apparently originate from p-Ty metabolism. Furthermore, the production of PHPA seems to be enhanced by the availability of free p-Ty in the body. While clorgyline significantly increased both total p-Ty and ($^2H_2$)-p-Ty, deprenyl did not affect total p-Ty nor total ($^2H_2$)-p-Ty excretions. Deprenyl markedly increased total PEA excretion signifying a potent inhibition of MAO type B (results not shown).

The Effects of Chronic MAO Inhibition on PHPA and Other Biogenic Amines Metabolites Excretion and Brain Concentrations

It could be argued that in spite of the relatively high doses of pargyline, clorgyline and deprenyl used, long term inhibition of MAO is needed to affectively reduce PHPA excretion. For this reason, the effects of chronic I.P. administrations of pargyline, clorgyline and deprenyl on the excretion and brain concentrations of a number of amines and/or their metabolites were assessed after two weeks of treatments. The results are summarized in Table 3 and 4. As shown, marked inhibition of NE and DA metabolism by these three MAO inhibitors, was not accompanied by a similar reduction in PHPA in urine or brain. In fact a slight, but significant increase in caudate nucleus PHPA was found after clorgyline. As expected, pargyline and deprenyl but not clorgyline significantly increased free PEA excretion. Based on the observed increase in PEA excretion, the doses of pargyline, clorgyline and deprenyl employed in the present chronic study appear to be within

TABLE 3

THE EFFECTS OF CHRONIC I.P. ADMINISTRATION OF THREE TYPES OF MONOAMINE OXIDASE INHIBITIONS ON URINE EXCRETION OF A NUMBER OF BIOGENIC AMINES AND METABOLITES

Description (unit)	Free PEA	Free p-Ty	Free PHPA	Free DOPAC	Free HVA	Total MHPG
Urine						
Controls (µg)	0.272 ± 0.029	1.64 ± 0.25	59 ± 8	0.658 ± 0.110	4.1 ± 0.4	5.5 ± 0.4
Pargyline (µg)	2.873 ± 108**	2.44 ± 0.88	58 ± 8	0.424 ± 0.044*	1.9 ± 0.2**	1.8 ± 0.2**
Clorgyline (µg)	0.287 ± 76	2.09 ± 0.33	62 ± 8	0.345 ± 0.049**	1.5 ± 0.1**	0.9 ± 0.1**
Deprenyl (µg)	0.838 ± 116**	0.87 ± 0.16	41 ± 8	0.544 ± 0.053	2.1 ± 0.1**	2.0 ± 0.2**

Pargyline, 7 mg/kg, clorgyline, 6 mg/kg and deprenyl, 5 mg/kg were administered I.P. once a day for two weeks and urine collected for four hours after the last injection. The rats were decapitated four hours after the last injection and the brains analyzed as described in the text. Five rats were included in each experiment.

Results are expressed as mean ± SEM per mg creatinine.

*$p < 0.05$; **$p < 0.005$ by unpaired t-test as compared to the appropriate controls.

TABLE 4

THE EFFECTS OF CHRONIC I.P. ADMINISTRATION OF THREE TYPES OF MONOAMINE OXIDASE INHIBITIONS ON BRAIN CONCENTRATION OF A NUMBER OF BIOGENIC AMINES AND METABOLITES

	Caudate Nucleus			Hypothalamus
Description (unit)	Free PHPA	Free DOPAC	Free HVA	Total MHPG
Caudate Nucleus[c]				
Controls (ng)	0.570 ± 0.050	9.9 ± 0.70	8.65 ± 0.76	1.48 ± 0.1
Pargyline (ng)	0.588 ± 0.064	1.7 ± 0.23**	2.84 ± 0.45**	1.11 ± 0.05**
Clorgyline (ng)	0.753 ± 0.048**	0.72 ± 0.04**	0.49 ± 0.2**	0.02**
Deprenyl (ng)	0.676 ± 0.034	3.43 ± 0.34**	4.3 ± 0.32**	1.19 ± 0.02

Pargyline, 7 mg/kg, clorgyline, 6 mg/kg and deprenyl, 5 mg/kg were administered I.P. once a day for two weeks and urine collected for four hours after the last injection. The rats were decapitated four hours after the last injection and the brains analyzed as described in the text. Five rats were included in each experiment.

[c]Results are expressed as mean ± SEM per mg protein.

**$p < 0.005$ by unpaired t-test as compared to the appropriate controls.

the appropriate ranges that induce specific MAO type A or type B inhibitions.

Conversion of P-Hydroxyphenylpyruvic Acid to Homogentisic Acid, P-Hydroxyphenyllactic acid (PHPLA) and PHPA

Theoretically, it most of urine and brain PHPA does not arise from the metabolism ot p-tyramine as concluded from the data shown in Tables 1-4, a plausible precursor of PHPA could be p-hydroxyphenylpyruvic acid; a compound reported to be converted in the rabbit to PHPA (Kirberger and Bucher, 1952). This latter phenolic acid is derived from p-tyrosine transamination. p-hydroxyphenylpyruvic acid is known to be efficiently hydroxylated at the ortho (two ring position) position by phenylpyruvic acid oxidase to homogentisic acid followed by ring cleavage by homogentisic acid oxidase to give 4-maleyl-acetoacetic acid (Medes, 1932). However, p-hydroxyphenylpyruvic acid decarboxylation to PHPA has not been well documented in rats nor has the identity of PHPA derived from administered p-hydroxyphenylpyruvic acid been structurally confirmed.

To investigate into the possible conversion ot p-hydroxyphenylpyruvic acid to PHPA, 5 mg of this acid was administered intravenously to five rats and their urines collected for four hours immediately after treatment. The following compounds were measured: PHPA, p-hydroxyphenyllactic acid (PHPLA), homogentisic acid, and p-tyrosine. The results are summarized in Table 5. As shown, about the same amounts of PHPA and PHPLA were formed from p-hydroxyphenylpyruvic acid but the percent increase in PHPLA is 20 fold higher than that found for PHPA. Although the design ot the experiment was not intended to critically trace the metabolism of p-hydroxyphenylphenylpruvic acid, it has nevertheless been established here to be metabolized to PHPA in rats.

Conversion of P-Tyrosine to PHPA

The results in Table 5 together with those reported by Kirberger and Bucher, 1952 regarding the conversion of p-hydroxyphenylpyruvic acid to PHPA can be criticized on account of the chemically unstable nature of p-hydroxyphenylpyruvic acid. Thus one could argue that the elevations observed in PHPA following p-hydroxyphenylpyruvic acid are artifacts resulting from the chemical decomposition within the body of p-hydroxyphenylpyruvic acid to PHPA. Furthermore, administration ot p-tyrosine alone and evaluating the change in PHPA excretion as performed by Kirberger and Bucher, 1952 cannot differentiate between the contributions arising from p-Ty metabolism and that arising from p-hydroxyphenylpyruvic acid decarboxylation. For these reasons

TABLE 5

METABOLISM OF P-HYDROXYPHENYLPYRUVIC ACID (HPPA) IN RAT[a]

Description	Free PHPA	Free PHPLA	Free Homogentisic Acid	Free p-Tyrosine
Baseline	46.0 ± 5.2	2.58 ± 0.83	None	7.14 ± 1.52
HPPA	70.9 ± 9.6*	27.73 ± 10.0**	111.3 ± 27.4	9.23 ± 0.97
% Change[b]	43 ± 18.4	1204 ± 594	---	48.6 ± 27.8
Absolute change	24.8 ± 11.8	25.20 ± 9.44	111.3 ± 27.4	2.09 ± 1.72

[a]Urine was collected from five rats overnight (baseline). On the following day 5 mg of HPPA was administered I.V. and urine collected immediately for four hours.

[b]The results represent the means ± SEM of % change in each rat.

All results are in μg/mg creatinine (mean ± SEM).

*$p < 0.05$; **$p < 0.005$ by paired t-test as compared to the appropriate baselines.

TABLE 6

THE FORMATION OF $^2H_2$-PHPA FROM $^2H_2$-P-TYROSINE IN RATS BEFORE AND AFTER CARBIDOPA CO-ADMINISTRATION

Description	P-Tyrosine	$^2H_2$-P-Tyrosine	PHPA	$^2H_2$-PHPA
Baseline	129.4 ± 23.3	---	59.3 ± 3.3	---
$^2H_2$-P-Tyrosine[a]	49.1 ± 15.3*	247.0 ± 70.5	38.7 ± 4.5*	5.9 ± 1.0
$^2H_2$-P-Tyrosine[b] plus Carbidopa	86.6 ± 11.7	625.6 ± 58.1*	53.3 ± 5.6	29.7 ± 4.1*

[a]$^2H_2$-p-tyrosine, 5 mg was administered intragestrically twice daily for three days. On the third day and following the last drug administration, urine was collected overnight.

[b]$^2H_2$-p-tyrosine, 5 mg plus carbidopa, 100 mg/kg was administered twice daily for three days. On the third day and following the last treatment, urine was collected overnight.

Five rats were included in each experiment. Results in μg/mg creatinine ± SEM.

*$p < 0.005$ as compared to baseline (non-deuterated compounds) or to $^2H_2$-p-tyrosine (deuterated compounds) by paired t-test.

the conversion of deuterated p-tyrosine to $^2H_2$-PHPA was assessed after three days daily administration of $^2H_2$-p-tyrosine alone and together with 100 mg/kg carbidopa. Co-administration of carbidopa with $^2H_2$-p-tyrosine resulted in about six-fold increase in $^2H_2$-PHPA as compared to when $^2H_2$-p-tyrosine was given alone. These results are summarized in Table 6. Carbidopa administration significantly increased p-tyrosine and $^2H_2$-p-tyrosine excretions. Interestingly, PHPA excretion was significantly reduced after $^2H_2$-p-tyrosine probably as a result of competition with p-tyrosine for the various enzymatic pathways leading to the production of PHPA.

DISCUSSION

Theoretically, there are four feasible sources of PHPA in the body. 1) Dietary PHPA; 2) Deamination of dietary p-Ty by MAO followed by oxidation of the product; 3) Decarboxylation of p-tyrosine (Boulton and Juorio, 1983) to p-tyramine followed by its metabolism as in "2" and finally; 4) Transamination of p-tyrosine to p-hydroxyphenypyruvic acid followed by its decarboxylation to PHPA (Medes, 1932). The 4th source of PHPA may follow the same biochemical sequence as those found in phenylalanine transamination pathway (Curtius et al., 1972; Karoum et al., 1984) and L-DOPA (Calne et al., 1969). The second, third and fourth sources of body PHPA may originate in the gut and/or endogenously within the body. In this context it should be mentioned that decarboxylation of p-tyrosine is not the only source of endogenous p-Ty. Pathways which can lead to the production of p-Ty include p-hydroxylation of phenylethylamine (Boulton et al., 1974; Silkaitis and Mosnaim, 1976) and dehydroxylation of dopa and dopamine (Boulton and Dyck, 1974).

The failure of I.G. administrations of pargyline, neomycin and amino acid decarboxylase inhibitors (Table 1) to alter PHPA urinary excretion strongly suggest a minor role of gut flora p-Ty synthesis or metabolism in the overall body production of PHPA. The marked reductions in PEA, m-Ty and p-Ty excretion observed after carbidopa and alpha-methyldopa also suggest that most of these amines are derived respectively from phenylalanine and p-tyrosine decarboxylation and not from dietary sources (Boulton and Juorio, 1983). The failure of neomycin to reduce PHPA excretion in rats is consistent with several reports in the literature on the effect of gut sterilization on p-tyramine and PHPA excretion (Sandler et al., 1969; Boulton, 1976; DeQuattro and Sjoerdsma, 1976). Furthermore, the significant reduction in total p-Ty and the no change in m-Ty excretion (results not shown) after neomycin are curious but nevertheless in line with a number of recent reports of reciprocal changes in m- and p-Ty productions after various pharmacological manipulations (Juorio, 1982). Whether this latter observation is coincidental or due to the removal of p-Ty inhibitory effects on m-Ty synthesis is open to future exploration.

Inhibition of p-Ty metabolism by clorgyline (a type A MAO inhibitor) produced a modest but a significant reduction in PHPA excretion, Table 2. This result together with the observed efficient metabolism of deuterated p-Ty to deuterated PHPA clearly demonstrated that while at best only about 30% of total urine PHPA originate from p-Ty, p-Ty itself once formed is rapidly and efficiently metabolized to PHPA. For example, about 40% of the deuterated p-Ty administered was excreted in the form of deuterated PHPA within four hours after treatment (compare results in Table 4).

p-Tyramine was not measured in the brain but its metabolite was estimated after both acute and chronic pargyline (Table 2 and 4) as well as after chronic treatments with two amino acid decarboxylase inhibitors (Table 1), clorgyline and deprenyl (Table 4). None of these drug treatments reduce PHPA concentration in the caudate nucleus. These results appear at first sight inconsistent with reports of decreased mouse caudate nucleus m-and p-hydroxyphenylacetic acid following pargyline administration (McQuade et al., 1981). While species difference may be responsible, it should be noticed that the dose of acute (50 mg/kg) and chronic (7 mg/kg) pargyline used here are considerably lower than the 100 mg/kg employed by these workers. Even at this high dose both m- and p-hydroxyphenylacetic acid concentrations were reduced by less than 30%. Therefore the results reported by McQuade et al. (1981) are in general not very contradictory to those reported here. It should also be emphasized that failure of MAO or amino acid decarboxylase inhibitors to significantly reduce PHPA brain concentrations does not in any way imply that p-Ty is not endogenously produced nor that it is devoid of any physiological function. What these data signify is that, the contribution of p-Ty produced in the periphery and the brain towards total body PHPA production is small compared to other metabolic pathways of tyrosine. In fact it could be argued that the availability of an efficient alternative pathway to p-tyrosine metabolism (as will be discussed later) offers a safe mechanism that prevents excessive production of p-Ty from p-tyrosine. p-Tyrosine administration has been shown to increase striatal concentration of p-ty (Juorio and Boulton, 1982). Further, it is possible that a breakdown in this safety mechanism may be associated with the various disorders that are characterized by abnormal p-Ty production.

Although it is difficult to conclusively eliminate the contribution of dietary PHPA towards PHPA urinary excretion in rats, such a possibility is probably not of major consideration in humans to whom chronic MAO inhibitors treatments produced inconsistent reductions in PHPA excretion (Karoum et al., 1982; Linnoila et al., 1982; additional unpublished results). Further, if most urinary PHPA is derived from the diet, this should not influence brain PHPA concentration because peripheral PHPA at relatively low concentrations does not efficiently cross the blood brain barrier (McQuade and Juorio, 1983; see Result section). Thus, making the reasonable assumption that dietary PHPA contribution towards peripheral tissue production and ultimately excretion of PHPA is small, the major source of urinary PHPA can therefore be best attributed to p-tyrosine transamination (source 4). For this reason changes in PHPA excretion in rats (after pharmacological manipulations) and by extrapolation in humans, will most likely be associated with abnormal p-tyrosine transamination than with alteration in p-tyramine metabolism. Formation of PHPA from p-hydroxyphenylpyruvic acid is supported by the results in Tables 5 and 6. As shown, p-hydroxyphenylpyruvic acid administration significantly increased PHPA excretion as well as p-hydroxyphenyllactic acid (PHPLA); a stable metabolite of p-hydroxyphenylpyruvic acid (Karoum et al., 1975). PHPA, PHPLA and p-hydroxyphenylpyruvic acid excretions are elevated in both tyrosinemia (increased blood tyrosine) and tyrosyluria (increased urine excretion of both p-tyrosine and its metabolites) (Mede, 1932; Wong et al., 1967; Partington, 1968; Karoum, 1970; Karoum et al., 1975) of the newborn. These two conditions are associated with a transient reduction in p-hydroxyphenylpyruvic acid oxidase in the newborn period of life especially in premature infants (Bloxam et al., 1960). The results presented in Table 5 reflect upon the metabolism of exogenously administered p-hydroxyphenylpyruvic acid and as such may not faithfully mimic the metabolism of the endogenously produced compound. Considering all the results presented (Tables 1-6), the ability of the body to decarboxylate p-hydroxyphenylpyruvic acid to PHPA can be regarded successfully demonstrated.

The results summarized in Tables 1 on the effect of neomycin on PEA excretion can also be used to argue against a major role of the gut flora in the production of PEA in rats this effect is similar to observations in humans (Karoum et al., 1980).

The data presented here reveal a number of interesting characteristics of pargyline, clorgyline and deprenyl. For example, 10 mg/kg deprenyl had very little effect on p-Ty metabolism as compared to clorgyline (Table 4) but like the other two MAO inhibitors it very affectively reduce NE and DA metabolism both peripherally (as deduced from urine analysis) and centrally (Tables 4). The failure of deprenyl to reduce p-Ty metabolism may be related to the absence of the so-called "chesse effects" following deprenyl treatment (Elsworth et al., 1978). Further, while pargyline markedly elevated PEA excretion it also shares many similarities with clorgyline. Again all three MAO inhibitors tested produced potent central effects on NE metabolism.

In conclusion, although the origin of most urine PHPA could have been predicted from some old literature (Medes, 1932; Kirberger and Bucher, 1952) these reports did not address certain critical areas such as the structural identities of the compounds measured and the effects pharmacological manipulations have on p-Ty production and metabolism. Furthermore, these old reports did not evaluate the origin of central PHPA. Thus as presented in this communication, a number of pharmacological manipulations were conducted to assess the central and peripheral origin of PHPA in rats. The results of these experiments suggest that most central and peripheral PHPA does not originate from p-tyramine metabolism. Transamination of p-tyrosine to p-hydroxyphenylpyruvic acid followed by its decarboxylation is probably the major source of PHPA in rats and possibly humans too.

REFERENCES

Bloxam H.R., Day M.G., Gibbs N.K. and Woolf L.J. (1960) An inborn defect in the metabolism of tyrosine in infants on a normal diet. Biochem. J. 77, 320-326.

Boulton A.A. and Dyck L.E. (1974) Biosynthesis and excretion of meta and para tyramine in the rat. Life Sci. 14, 2497-2506.

Boulton A.A., Dyck L.E. and Durden D.A. (1974) Hydroxylation of β-phenylethylamine in rat. Life Sci. 15, 1673-1683.

Boulton A.A. (1976) The pink spot in parkinsonism. Prog. Neurogenetics 1, 437-441.

Boulton A.A. and Jurio A.V. (1983) Cerebral decarboxylation of meta and para-tyrosine. Experientia 39, 130-134.

Boulton A.A., Davis B.A., Yu P.H., Wormith J.S. and Addington D. (1983) Trace acid levels in the plasma and MAO activity in the platelet of violent offenders. Psychiat. Res. 8, 19-23.

Calne D.B., Karoum F., Ruthven C.R.J. and Sandler M. (1969) The metabolism of orally administered L-dopa in parkinsonism. Br. J. Pharmacol. 37, 57-68.

Curtius H.-CH., Vollmin J.A. and Baerlocher K. (1972) The use of deuterated phenylalanine for the elucidation of the phenylalanine-tyrosine metabolism. Clin. Chim. Acta 37, 277-285.

Davis B.A., Durden D.A. and Boulton A.A. (1982) Plasma concentration of p- and m-hydroxyphenylacetic acid and phenylacetic acid in humans. J. Chromotog. 230, 219-230.

Durden D.A. and Boulton A.A. (1981) Identification and distribution of m- and p-hydroxyphenylacetic acid in the brain of the rat. J. Neurochem. 35, 129-135.

Elsworth J.D., Glover V., Reynolds G.P., Sandler, P., Shaw K.M., Stern G.M. and Kumar P. (1978) Deprenyl administration in man: A selective monoamine oxidase B inhibitor without the "cheese effect". Psychopharmacology 57, 33-38.

Glowinski J. and Iversen L.L. (1966) Regional studies of catecholamines in the rat brain. J. Neurochem. 13, 655-669.

Here, R.S. (1950) Endogenous creatinine in serum and urine. Proc. Soc. Exp. Biol. Med. 4, 148.

Juorio A.V. (1982) A possible role of tyramines in brain function and some mental disorders. Gen. Pharmacol. 13, 181-183.

Juorio A.V. and Boulton A.A. (1982) The effects of some precursor amino acids and enzyme inhibitors on the mouse striatal concentration of tyramines and homovanillic acid. J. Neurochem. 39, 859-863.

Karoum F. (1970) Ph.D. Thesis. London University.

Karoum F., Ruthven C.R.J. and Sandler M. (1975) Urinary phenolic acid and alcohol excretion in the newborn. Arch. Disease Childhood 50, 586-594.

Karoum F., Potkin S.G., Murphy D.L. and Wyatt R.J. (1980) Quantitation and metabolism of phenylethylamine and tyramine's three isomers in humans, in Non-Catecholic Phenylethylamines, Part 2. Phenylethanolamine, Tyramines and Octopamine (Mosnaim A.D. and Wolf M.E. eds.) pp. 177-191. Marcel Dekker, New York.

Karoum F. and Neff N.H. (1982) Quantitative gas chromatography-mass spectrometry (GC-MS) of biogenic amines: Theory and practice, in Modern Methods in Pharmacology (Spector S. and Back N., eds), pp. 39-54. Alan R. Liss Inc., New York.

Karoum F., Chuang L.-W., Eisler T., Calne D.B., Liewbowitz M.R., Quitkin F.M., Klein D.F. and Wyatt R.J. (1982) Metabolism of (-) deprenyl to amphetamine and methamphetamine may be responsible for deprenyl's therapeutic benefit: A biochemical assessment. Neurology 32, 503-509.

Karoum F. (1983) Mass fragmentography in the analysis of biogenic amines: A clinical, physiological and pharmacological evaluation, in Methods in Biogenic Amine Research (Parveg S., Nagatsu T., Nagatsu I. and Parvez H., eds.), pp. 237-255. Elsevier Science Publishers, New York.

Karoum F., Potkin S., Chuang L.-W., Murphy D.L., Liebowitz M.R. and Wyatt R.J. (1984) Phenylacetic acid (PAA) excretion in schizophrenia and depression: The origin of PAA in man. Biol. Psychiatry 19, 546-550.

Kirberger E. and Bucher T. (1952) Separation of p-hydroxyphenylacetic acid after administration of p-hydroxyphenylpyruvic acid in rabbits. Biochem. et Biophys. Acta 8, 294-301.

Knoll J. (1980) Monoamine oxidase inhibitors: Chemistry and pharmacology, in Enzyme Inhibitors as Drugs, (Sandler M., ed.), pp. 151-171. MacMillan Press Ltd., London.

Kobayashi K., Imazu Y. and Shohmori T. (1983) p-Hydroxyphenylacetic acid concentration in cerebrospinal fluid of patients with neurological and psychiatric disorders, in Trace Amines And The Neurosciences p. 6. A Satellite Meeting of the 9th International Society of Neurochemistry, Alberta.

Kretchmer N., Levine S.Z., McNamara H. and Barnett H.L. (1956) Certain aspects of tyrosine metabolism in the young. I. Development of the tyrosine oxidizing system in human liver. J. Clin. Invest. 35, 236-240.

Linnoila M., Karoum F. and Potter W.Z. (1982) Effects of low-dose clorgyline on 24-hour urinary monoamine excretion in patients with rapidly cycling bipolar affective disorder. Arch. Gen. Psychiatry 39, 513-516.

Lowry O.H., Rosenbrough N.J., Farr A.L. and Randall R.J. (1951) Protein measurement with folin phenol reagent. J. Biol. Chem. 193, 265-275.

Martin M.E., Karoum F. and Wyatt R.J. (1979) Phenylacetic acid excretion in man. Anal. Biochem. 99, 283-287.

McQuade P.S., Juorio A.V. and Boulton A.A. (1981) Estimation of the p and m isomers of hydroxyphenylacetic acid in mouse brain by a gas chromatographic procedure: Their regional distribution and the effects of some drugs. J. Neurochem. 37, 735-739.

McQuade P.S. and Juorio A.V. (1983) Some factors affecting the concentrations of para-hydroxyphenylacetic acid and meta-hydroxphenylacetic acid in the mouse caudate nucleus. J. Neurosci. Res. 10, 127-134.

Partington M.W. (1968) Neonatal tyrosinaemia. Biologica Neuronatorum 12, 316-330.

Medes G. (1932) A new error of tyrosine metabolism: tyrosinaris. The intermediary metabolism of tyrosine and phenylalanine. Biochem. J. 26, 917-940.

Silkaitis R.P. and Mosnaim A.D. (1976) Pathways linking L-phenylalanine and 2-phenylethylamine with p-tyramine in rabbit brain. Brain Res. 114, 105-115.

Wong P.W.K., Lambert A.M. and Komrower G.M. (1967) Tyrosinemia and tyrosinuria in infancy. Develop. Med. Child Neurol. 9, 551-562.

Yu P.H., Davis B.A., Bowen R. and Boulton A.A. (1983) The catabolism of trace amines in psychiatric disorders, in Trace Amines And The Neurosciences p. 17. A Satellite Meeting of the 9th International Society of Neurochemistry, Alberta.

Free Communications

URINARY EXCRETION OF TRYPTAMINE, MHPG, HVA, VMA, AND HIAA BY VOLUNTEERS TREATED WITH CGP 4718, A COMBINED INHIBITOR OF MAO-A AND 5-HT UPTAKE.

Antonin*, K.H., Nilsson*, E., Waldmeier**, P.C., and Bieck*, P.R.

*Human Pharmacology Institute CIBA-GEIGY GmbH, D-7400 Tübingen, Federal Republic of Germany

**Research Departement, Pharmaceuticals Divison, CIBA-GEIGY Ltd., CH-4002 Basle, Switzerland

Abstract: The new substance CGP 4718 A (4-(5-chloro-2-benzofuranyl)-1-methylpiperidine hydrochloride) in animal experiments shows both 5-HT-uptake blocking and (preferentially) MAO-A inhibiting properties. CGP 4718 A was studied in man to evaluate its effects on the noradrenergic, dopaminergic and serotonergic systems by measuring urinary metabolites. In a randomized placebo-controlled cross-over study 6 healthy volunteers (3 males, 3 females) were treated for two weeks with CGP 4718 A (2 x 50 mg/d p.o.) or placebo, respectively. In between was an interval of two weeks without drugs. Daily 24 h urine was collected during and for 2 weeks after treatment. Tryptamine was measured by GC with packed columns, 3-methoxy-4-hydroxyphenyl-ethylene glycol (MHPG), homovanillic acid (HVA), vanillylmandelic acid (VMA) and 5-hydroxyindolacetic acid (HIAA) by capillary GC.
The mean tryptamine excretion increased 1.9 fold and the HIAA excretion 1.5 fold. The mean MHPG excretion decreased by 31%. The mean HVA and VMA excretion was unchanged. The increase in urinary tryptamine as an index of renal MAO-A inhibition and the slight decrease of urinary MHPG excretion support the **in vitro** and **in vivo** characteri-

sation of CGP 4718 as an inhibitor of MAO-A. Urinary 5-HIAA was not distinctively changed. Large interindividual variations in urinary HIAA output may primarily reflect intrarenal and not central 5-HT metabolism.

INTRODUCTION

In neurobiochemical, behavioral and psychopharmacological test systems, CGP 4718 A (4-(5-chloro-2-benzofuranyl)-1-methylpiperidine HCl) displays properties indicative of antidepressant-like activity (Waldmeier et al. 1984, 1985 Delini-Stula et al. 1984). The neurobiochemical activity profile of CGP 4718 A is characterized by its inhibitory effects on monoamine oxidase A (MAO-A) and serotonin (5-HT) uptake in the same dose range. In man, CGP 4718 A leads to small increases in tyramine pressor sensitivity which are rapidly reversible (Bieck et al. 1984). The objective of this study was to investigate the effects of subchronic treatment with CGP 4718 A upon the noradrenergic, dopmaninergic and serotonergic system in man by measuring urinary metabolites of catecholamines and 5-HT. The MAO-A inhibitory effect was characterized by measuring urinary tryptamine excretion (Sjoerdsma et al. 1959).

MATERIALS AND METHODS

Subjects

Six healthy volunteers (3 males, 3 females) aged 26 - 28 years and weighing 53 - 90 kg participated in the study. All were shown to be healthy on the basis of clincal examination and hematological and biochemical screening. None of the volunteers took any other medication for two weeks prior to and during the study. No food rich in tyramine or alcohol were permitted. All volunteers were ambulatory and carried out their usual activities. The protocol was approved by the Ethics Committee of the Human Pharmacology Institute CIBA-GEIGY Tübingen.

The volunteers were treated with CGP 4718 A (2 x 50 mg/d p.o.) or placebo, respectively, according to a randomized and cross-over scheme for two weeks. In between there was an interval of two weeks without drug. Daily 24-h urines were collected using sodium disulfite (Merck No.6357; 40 ml of a 10% solution per 24-h urine) as preservative. Aliquots were frozen at -20° C until analysis. Creatinine in urine was determined by the Jaffe reaction (Beckman Creatinine Analyzer 2).

Analytical methods

Tryptamine was measured by GC according to Baker et al. (1979). MHPG, HVA, VMA, and HIAA were determined by capillary GC according to De Jong et al. (1983)

Calculations

The 24-h urine values of amines metabolites were calculated per g creatinine. The values are means ± SD. Statistical differences were tested using Student's t-test.

RESULTS

Tryptamine: During the placebo period the volunteers excreted 91 ± 25 ug tryptamine/g creatinine on the average. During administration of CGP 4718 A urinary tryptamine excretion significantly increased to 166 ± 55 ug/g creatinine (1.9 fold $p < 0.0001$) (Fig. 1).

MHPG: During treatment with CGP 4718 A the mean MHPG excretion decreased by 30% compared to placebo-values (placebo: 3.6 ± 2.6 mg/g creatinine, CGP 4718 A: 2.5 ± 2.1 mg/g creatinine).

HVA and VMA: On the average HVA (placebo: 3.2 ± 1.0 mg/g creatinine, CGP 4718 A: 3.2 ± 1.5 mg/g creatinine) and VMA (placebo: 4.1 ± 2.9 mg/g creatinine, CGP 4718 A: 3.9 ± 2.3 mg/g creatinine) excretion was not affected by treatment with CGP 4718 A.

EFFECTS OF CGP 4718 A ON URINARY EXCRETION OF MHPG, 5-HIAA, HVA, VMA AND TRYPTAMINE

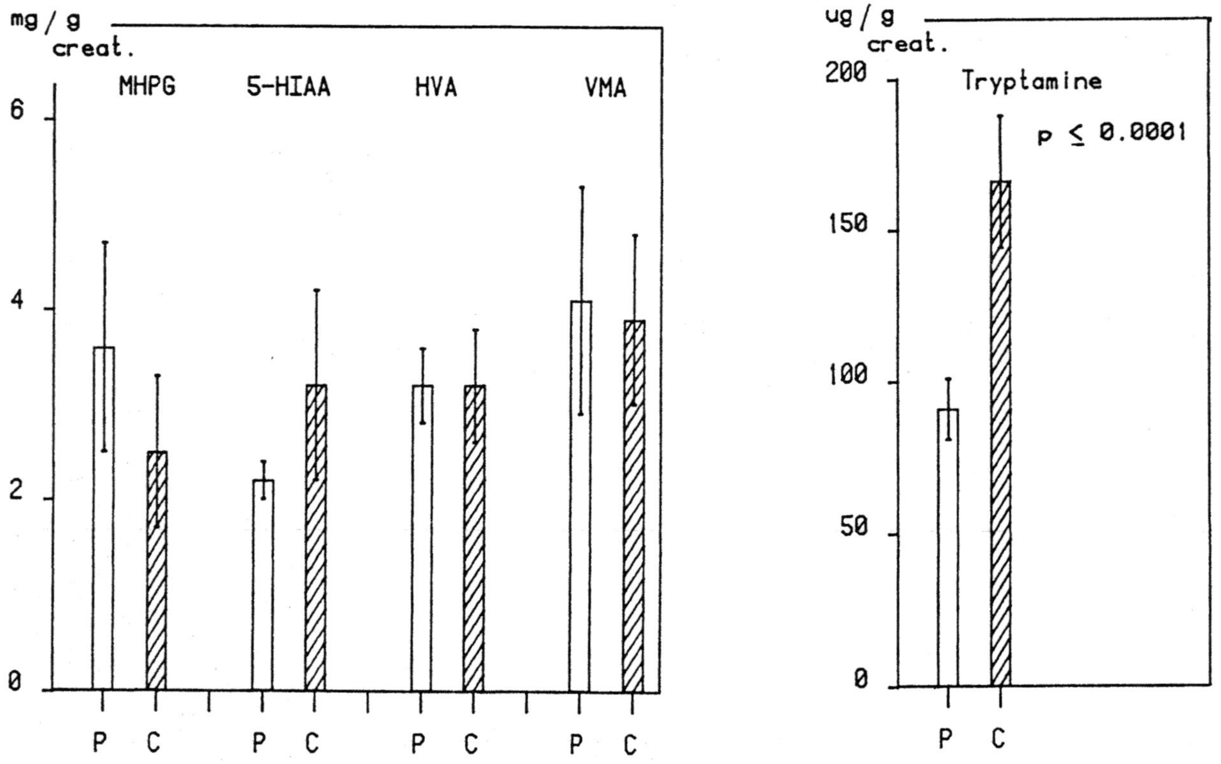

Fig. 1 Effects of subchronic treatment with Placebo (P) and CGP 4718 (C) on urinary excretion of 5 neurotransmitter metabolites. Bars show mean values ± SD in 6 volunteers.

HIAA: Mean HIAA urinary excretion increased during treatment with CGP 4718 A 1.5 fold compared to placebo-values (placebo: 2.2 ± 0.5 mg/g creatinine, CGP 4718 A: 3.2 ± 2.4 mg/g creatinine).

DISCUSSION

The development of new MAO inhibitors has renewed the interest in noninvasive methods for the characterization of their effects in man. An obvious possibility consists in the measurement of the urinary excretion of substrates or products of the enzyme. Already in 1959, the urinary excretion of tryptamine was found to increase in response to treatment with MAO inhibitors (Sjoerdsma et al., 1959). Subsequently, decreases in deaminated catecholamine metabolites (Dollery et al., 1982) and increases in O-methylated catecholamines (Linnoila et al., 1982; Waldmeier et al., 1983) were reported under treatment conditions where MAO A was inhibited specifically or preferentially.

Therefore, CGP 4718 A, being a week, preferenterial inhibitor of MAO-A (Waldmeier et al., 1985), would be expected to cause an increase in tryptamine and a decrease in MHPG excretion, if the pharmacological effects in humans (Delini-Stula, et. al., 1984) were related to MAO A inhibition. The anticipated results were indeed found; the effects of CGP 4718 A were rather weak, which is, however, in line with previous results in man showing only small changes in tyramine sensitivity (Bieck, et al., 1984).

On the other hand, CGP 4718 A did not cause a reduction of the excretion of HVA and VMA, which might have been expected with an MAO inhibitor. Likewise, the 5-HT metabolite 5-HIAA was not reduced. This later result is not surprising, however, since Linnoila et al. (1984) were unable to see a reduction in 5-HIAA after the preferential MAO A inhibitor clorgyline. HVA excretion, however, has been found to be strongly reduced after brofaremine (CGP 11 305 A; Maitre, et al., 1984), a specific inhibitor of MAO A. A possible explanation of the lack of the effect of CGP 4718 A on the urinary excretion of HVA and VMA might be a lower sensitivity of these parameters compared to MHPG. However, this remains speculative in the absence of additional data.

Acknowledgement: We thank for the assistance of Ms. M. Huegler and C. Rommel.

REFERENCES

BAKER, G.B., CALVERLEY, D.G., DEWHURST, W.G., and MARTIN, I.L. (1979) A sensitive gas chromatographic technique for quantification of urinary tryptamine. **Br. J. Pharmacol. 67, 469 P.**

BIECK, P.R., CREMER, G., ANTONIN, K.H., SCHICK, Ch., and NILSSON, E. (1984) Assessment of peripheral effects of the new monoamine oxidase/5-HT uptake inhibitor CGP 4718 A in healthy man. **J. Pharm. Pharmacol. 36,** Workshop Suppl. 69W.

DE JONG, E.B.M., HORSTEN, B.P.M., and GOLDSCHMIDT, H.M.J. (1983) Determination of nine catecholamine metabolites and 5-hydroxyindolacetic acid in urine by capillary gas chromatography. **J. Chromatogr. 279,** 563 - 572.

DELINI-STULA, A., RADEKE, E., VASSOUT, A., JAEKEL, J., GLATT, A., BUECH, O., and BERNASCONI, R. (1984) Pharmacological properties of CGP 4718 A, a new selective MAO-A and 5-HT-uptake inhibitor. 14. **C.I.N.P. congress** P-98.

DOLLERY, C.T., DAVIES, D.S., and STROLIN-BENEDETTI, M. (1982) Clinical pharmacology of MD 780515, a selective and reversible MAO-A inhibitor. In: Kamijo, K., Usdin, E., Nagatsu, T. (eds.) **Monoamine oxidase - Basic and clinical frontiers.** Excerpta Medica, Amsterdam, Oxford, Princeton, 221 - 229.

LINNOILA, M., KAROUM, F. and POTTER, W.Z. (1982) Effect of low-dose clorgyline on 24-hour urine monoamine excretion in patients with rapidly cycling bipolar affective disorder. **Arch. Gen. Psychiatr. 39,** 513 - 516

LINNOILA, M., MILLER, T.L., BARTKO, J., and POTTER, W.Z. (1984) Five antidepressant treatments in depressed patients. Effects on urinary serotonin and 5-hydroxyindole-acetic acid output. **Arch. Gen. Psychiatry, 41,** 688 - 692

MAITRE, L., WALDMEIER, P., LAUBER, J. and BIECK, P. (1984) Urinary excretion of catecholamine metabolites, tyramine and phenylethylamine in human volunteers after prolonged treatment with CGP 11 305 A and tranylcypromine. In: **Monoamine Oxidase and Disease,** (TIPTON, K.F., DOSTER, P. and STROLIN-BENEDETTI, M), pp 117 - 126. Academic Press, London

SJOERDSMA, A., OATES, J.A., ZALTZMAN, P., UDENFRIEND, S. (1959) Identification and assay of urinary tryptamine: application as an index of monoamine oxidase inhibition man. **J. Pharmacol. Exp. Ther. 126, 217 -** 222.

WALDMEIER, P.C., ANTONIN, K.H., FELDTRAUER, J.-J., GRUNENWALD, Ch., PAUL, E. and BIECK, P. (1983) Urinary excretion of O-methylated catecholamines, Tyramine and Phenyl-Ethylamine by volunteers treated with tranylcypromine and CGP 11 305 A, **Eur. J. Clin. Pharmacol 25,** 361 - 368

WALDMEIER, P.C., BERNASCONI, R, FELNER, A.E., BAUMANN, P.A., TIPTON, K.F., and MAITRE, L. (1984) CGP 4718 A, a new antidepressant with two modes of action. 14. **C.I.N.P. congress,** P-592

WALDMEIER, P.C., TIPTON, K.F., BERNASCONI, R., FELNER, A.E., BAUMANN, P.A. and MAITRE, L. (1985) CGP 4718 A, a new potential antidepressant with a dual mode of action, **European Journal of Pharmacology, 170,** 79 - 89

STUDY OF THE SYNAPTOSOMAL CATECHOLAMINE UPTAKE IN POSTMORTEM BRAIN OF SCHIZOPHRENICS

L.Hetey, N.Haberland and R.Schwitzkowsky

Institute of Pharmacology, Humboldt University

1080 Berlin, C.-Zetkin-Str. 94, GDR

INTRODUCTION

Investigations in the neurochemical pathology of the dopamine (DA), noradrenaline (NA) and serotonin (5HT) systems in humans do not involve studies of functional parameters, as the high-affinity uptake and the stimulated release of the transmitters. This is probably based on the relative rapid decline of these parameters during postmortem storage in situ as well as their drastic decrease by tissue freezing in liquid nitrogen (LN, see below). As an alternative, binding studies were recently proposed as reliable probes for investigating the uptake complexes (see for example Javitch et al., 1984), but their specifity as well as their validity for the high-affinity uptake process is to be elucidated.

On the other hand, we investigated the uptake and release of above transmitters after freezing of brain tissue preincubated with the cryoprotective agent dimethylsulfoxid (DMSO; Haberland et al., 1984). Furthermore, the postmortem decline in situ was studied for the kinetic parameters of uptake (Km, Vmax) and for the K^+-stimulated release (Haberland and Hetey 1985). In this paper we give a survey in respect of these investigations and of uptake studies in postmortem brain regions of schizophrenic patients in comparison to matched controls.

MATERIALS AND METHODS

Experiments with rats were performed similarly as described earlier: synaptosomal DA uptake was estimated in the concentration range from 0.02 to 0.1 μM [^{3}H]DA (Hetey et al., 1985), 30 mM K^+-stimulated release of preloaded [^{3}H]DA on superfused synaptosomes (Hetey et al., 1983). The uptake of [^{3}H]NA and [^{3}H]5HT was determined in the same concentration range; NA uptake was shown recently to be independent from that of DA (Hetey and Zimmermann, submitted). The protein content of the synaptosomal fractions was estimated by Lowry's method.

Experiments with postmortem human brains were carried out under identical conditions.

The clinical evaluation of schizophrenia and the selection of controls was made by the cooperative psychiatrists and pathologists similar as described by Mackay et al. (1982). Schizophrenic patients (n = 6; 2 men and 4 Women, age at death: 62.2±6.5 years) have got haloperidol (1) or haloperidol and promethazine (2) (chlorpromazine equivalents: 410±108 mg/day); in 3 cases no neuroleptic medication had been given for at least the last year before death. In control patients, t.e. without neurologic or psychiatric disorder (n = 6; 3 men and 3 women, age at death: 69.8±5.8 years) cause of death was bronchopneumonia (2), acute heart failure (3) and myocardial infarction (1).

Both statistical analysis using the Wilcoxon-test and kinetic analysis according to Lineweaver and Burk was plotted on a Hewlett-Packard desk computer. All values represent means±S.E.M.

RESULTS

Freezing of rat brain slices or synaptosomes in LN resulted in a drastic decrease of the transmitter uptake and stimulated release. For example, pieces of the nc. accumbens yielded after freezing-thawing 22% and 10% of their native synaptosomal DA uptake and release capacity. However, after the preincubation of rat brain pieces with 5% DMSO before freezing in a two-step procedure (1:1.2 K/min to -25°C, 2: plunging into LN; Haberland et al., 1984) synaptosomal DA, NA and 5HT uptake and release is unchanged, e.g. DA uptake in nc. accumbens: Km = 97 ± 6% and Vmax = 96 ± 11% of controls (Haberland and Hetey, 1985), K^+-stimulated DA release: 101.2±6.6% of controls: 15.63±0.27 dpm-%, t.e.dpm of stimulated release in relation to total dpm in the synaptosomes (n = 4). The same is true for these parameters in human nc. accumbens: Km = 105 ± 9.5%, Vmax = 96.5 ± 11% of control DA uptake (Haberland and Hetey, 1985) and regarding the stimulated DA release: 96.2 ± 8.9 % of the release rate before freezing (controls: 7.7 ± 1.16 dpm-%, 10.5 ± 3 hours after death, n = 4). Keeping of rat heads in situ after decapitation at room or refrigerator temperature for prolonged intervals resulted in different alterations (see table 1). The Km values of DA uptake are relative constant, contrary, marked reduction of Vmax was observed resulting in an exponential decline at room temperature ($t_{1/2}$ = 30 hours), at 2 °C comparable changes were observed. On the other hand, K^+-stimulated DA release is very sensitive in respect to the temperature of preservation: at room temperature rapid exponential decrease was measured (initial $t_{1/2}$ near 9 hours), but surprisingly at 2 °C DA release remained unchanged (see table 1B).

Table 1: DA uptake (A) and K^+-stimulated DA release (B) in nc. accumbens synaptosomes after preservation of rat heads in situ at room temperature (near 22 °C) as well as at 2 °C after decapitation. * / ** $P < 0.05$ / 0.01; Controls: (A) Km = 0.104 μM; Vmax = 64.6 pmol/mg protein in 10 min; (B) 13.82 dpm-%.

(A) DA uptake	time/h/	Km (per cent of controls)	Vmax (per cent of controls)	n
control		100.0±11.5	100.0±15.6	5
22 °C	24	128.2±14.3	54.2±13.0**	5
	48	169.4±24.2*	30.5± 7.8**	4
2 °C	24	110.8±16.2	61.1±10.2**	4
	48	141.7±20.7*	39.2±11.6**	4

(B) DA release	time/h/	dpm-% (per cent of controls)	n
control		100.0 ± 6.6	12
22 °C	12	27.3 ± 2.4**	7
	24	18.2 ± 2.6**	5
2 °C	24	107.1 ± 11.2	6
	48	109.3 ± 7.4	4

Table 2: Kinetic parameters (Km, Vmax) of the synaptosomal DA and NA uptake in postmortem nc. accumbens and nc. caudatus caput of schizophrenic patients (S) and matched controls (C). Values are individually corrected regarding the postmortem delay (t.e. the time between death and freezing of tissue), the average delay was 21.6 ± 9.9 (S) and 22.4 ± 8.1 (C) hours; * / ** $P < 0.05/0.02$; Controls: nc. accumbens: DA - Km = 0.16 µM; Vmax = 47.42 pmol/mg protein in 10 min; NA - Km = 0.079; Vmax = 8.43; nc. caudatus: DA - Km = 0.138; Vmax = 57.17; NA - Km = 0.063; Vmax = 5.43;

Nc.accumbens		Km /per cent of controls/	Vmax
DA uptake:	C	100.0±16.1	100.0±21.3
	S	253.1±24.2**	270.9±25.7**
NA uptake:	C	100.0±16.6	100.0±19.6
	S	324.9±33.7**	345.1±67.5**
Nc.caudatus			
DA uptake:	C	100.0± 9.2	100.0±15.0
	S	201.4±17.2**	174.1±25.3*
NA uptake:	C	100.0±24.6	100.0±34.5
	S	261.9±33.0**	236.3±34.7*

The synaptosomal DA and NA uptake in postmortem nc. accumbens and nc. caudatus caput of schizophrenic patients in comparison to matched controls is summarized in table 2. In these nuclei significant increases of Km and Vmax of both DA and NA uptake could be demonstrated. On the other hand, the high-affinity uptake of 5HT was unchanged; in the frontal cortex (Area 11 by Brodmann) none of these transmitters has shown a significantly altered uptake (data not presented). The uptake values are individually corrected in respect to the postmortem delay based on model experiments with rat heads (see above). The comparable differences between the original and the corrected values were in all cases very similar, e.g. DA uptake in nc. accumbens was measured to 0.216 ± 0.036 μM and 31.9 ± 7.4 pmol/mg protein in 10 min (C) and 245.1±11.2 % and 281.7 ± 39.4% of them in schizophrenia ($P < 0.02$; compare table 2).

CONCLUSION

Postmortem brain tissue seems to be suitable for investigating functional neurochemical parameters, as the synaptosomal uptake and release of transmitters (at least of DA, NA and 5HT) after cryopreservation of tissue pieces preincubated with DMSO. For the pathobiochemical studies longer postmortem periods are acceptable, so for uptake studies in case of determination of kinetic parameter (Km), for investigating the stimulated release of transmitters after rapid postmortem cooling to near 0 ^{o}C. In all cases, comparison with carefully selected, reliable matched controls seems to be very important; our results underline that besides age, sex, manner and time of death a.o. (for review see Perry and Perry, 1983) similar postmortem delay and preservation temperature is very important at the selection of controls, than based on model experiments with rats (table 1) correction regarding this delay can only represent a rough approximation.

The alterations demonstrated in postmortem brains of schizophrenic patients give a first indication for changes of presynaptic functional parameters in schizophrenia. These alterations of the kinetic parameters of DA and NA uptake seem not to be primarily dependent on medication, as in the neuroleptic-free patients similar changes were shown. The increase of both DA and

NA uptake would underline the involvement of both transmission systems in the pathophysiology of schizophrenia (compare Iversen et al., 1983).

REFERENCES

Haberland N., Hetey L., Hackensellner H.A., and Matthes G. (1984) Characterization of the synaptosomal dopamine uptake from rat and human brain tissue after low temperature preservation. Probl. Haematol. Transfsn. Transplantn. 11, 277-281.

Haberland N., and Hetey L. (1985) Characterization of the synaptosomal dopamine uptake in studies with postmortem human brain, in: Dopamine Systems and their Regulation (Woodruff G.N., ed.), Macmillan Press Ltd., London, in press.

Hetey L., Schwitzkowsky R., and Oelssner W. (1983) Influence of psychotomimetics and lisuride on synaptosomal dopamine release in the nc. accumbens of rats, Eur. J. Pharmacol. 93, 213-220.

Hetey L., Haberland N., and Schulze R. (1985) Effects of psychotomimetics and lisuride on synaptosomal tyrosine hydroxylase activity and dopamine uptake in the nc. accumbens of rats. Biogenic Amines 2, in press.

Iversen L.L., Reynolds G.P., and Snyder S.H. (1983) Pathophysiology of schizophrenia - causal role for dopamine or noradrenaline? Nature 305, 577-578.

Javitch, J.A., Blaustein R.O., and Snyder S.H. (1984) [^{3}H]Mazindol binding associated with neuronal dopamine and norepinephrine uptake sites. Mol. Pharmacol. 26, 35-44.

Mackay A.V.P., Iversen L.L., Rossor M., Spokes E., Bird E., Arregui A., Creese I., and Snyder S.H. (1982) Increased brain dopamine and dopamine receptors in schizophrenia. Arch. Gen. Psychiatry 39, 991-997.

Perry E.K. and Perry R.H. (1983) Human brain neurochemistry - some postmortem problems. Life Sciences 33, 1733-1743.

A PHYSICOCHEMICAL APPROACH TO CHARACTERIZE [^{3}H]-TRYPTAMINE-BINDING-SITES IN HUMAN BRAIN

E.Kienzl[1], P. Riederer[1], K.Jellinger[1], N.Noller[2]

1) L.Boltzmann-Institute of Clin.Neurobiology, Lainz-Hospital,A-1130 Vienna,Austria, and

2) Institute of Physical Chemistry, Techn. University, Vienna, Austria

Small amounts of trace amines are able to modify neuronal responses to established aminergic and non-aminergic transmitters and may also have a function as neuromodulators. The presence of tryptamine (T) in nerve tissue and its structural similarity to indole- and catecholamine transmitters gave evidence for a possible function as neurotransmitter or neuromodulator. Depressant actions of T are generally more profound and prolonged than those avoked by equal applications of serotonin (S) (Aghajanian and Haigler, 1975) and they are unaffected by prior lesions of S-containing pathways. However, cells in cortex are depressed by T and excited by S (Jones,1982a). Electrophysiological studies confirmed a T-containing pathway arising in the median raphé nucleus and projecting to the cerebral cortex in rats (Jones,1982 b,c). T and S respond differently to antagonists and exhibit opposite effects on temperature (Cox et al.,1981). T also interacts with S in the "behavioral syndrome" when tryptophan plus monoamine oxidase (MAO) inhibitor is infused (Marsden and Curzon,1979). Iontophoretic studies revealed a reduction of substance P responses during ejection of T (Jones,1984). These studies as well as the finding of a high-affinity, saturable binding site, distinct from S1-receptors indicated possible specific T-recognition sites independent of the S-system (Kellar and Cascio,1982; Wood et al.,1984; Rommelspacher and Brüning, 1984). In membrane preparations of human post mortem frontal cortex we measured binding sites that are kinetically distinguishable and are referred to a high affinity (nM) or low affinity (μM) component on the basis of the relative magnitude of their apparent K_D-constants (Kienzl et al.,1984). The aim of the present study was to examine the properties of T-binding sites as well as a dynamic approach to discern the basic physiological dimensions of central processing and plasticity.

MATERIAL AND METHODS

Homogenates of the medial frontal cortex of human brain were studied after dissection 5 - 6 hours post mortem. The neuropathological examination using routine histological methods, did not show any abnormalities in controls. Other brain areas were dissected-out to evaluate the distribution of $[^3H]$-T binding sites within the brain. Hepatic encephalopathy (HE) showed characteristic histological changes, i.e. cerebral edema and increase of pathological astroglia of the Alzheimer type II;SDAT revealed a type III. Brain samples were stored frozen at -30°C. The tissue preparations and radio assay techniques have been recently described (Kienzl et al.1984) or were performed according the method of Kellar and Cascio (1982). To evaluate the time course of specific $[^3H]$-T binding at 37°C pargyline was omitted from buffer medium. For preparation of liver and brain mitochondrial binding fraction the tissue was homogenized with an Ultraturrax for 30 sec. in 0,25m sucrose-10 mM potassium - phosphate buffer, pH 7,5 (1:10 w/v). After centrifugation at 2100 rpm for 15 minutes to remove broken cells, nuclei and debris, the supernatant was centrifuged at 7600 rpm for 20 minutes to sediment a crude mitochondrial fraction and the histological examination was performed by electron microscopy. The pellet was resuspended in sucrose buffer and again centrifuged at 7600 rpm and transfered to the incubation medium (50 mM TRIS-HCl, pH 7,5 + 5,6 mM ascorbic acid + 20 μM pargyline).Correct definition of nonspecific binding was essential for the accurate analyses of the multiple components. To determine whether there was a single concentration of unlabeled T that could be used to define nonspecific T-binding over a wide range of concentrations, the inhibition of $[^3H]$-T binding by unlabeled T was evaluated at concentrations ranging from 10^{-5} to 10^{-3} M. A common plateau for the inhibition of binding occurred between the concentration of 100μM to 1mM where the competing drug, nonspecific [^{3}H]-T,,was not inhibited. 1 m T was chosen to define the nonspecific binding in the time course and binding characteristics at 37°C.

RESULTS

[^{3}H]-T binding sites in human brain are widely distributed with highest densities in caudate n., g. cinguli, amygdaloid N. and thalamus (Tab. 1).

Fig. 1 ^{3}H-tryptamine binding in human brain (homogenate)

BRAIN REGION	AGE (yrs)	POST MORTEM TIME (hours)	K_D (nM)	B_{max} (fmol/mg prot)	N
caudate n.	77±20	5,3±2,1	558± 76	12,0± 5,1	3
fr. cortex	60±8	6,1±0,6	249± 27	13,2± 2,0	4
g. cinguli	40/54	6/7	729/404	9,5/8,4	2
hippocampus	57/54	6/7	242/224	15,7/16,8	2
amygdaloid n.	56±16	6,0±0,0	564±109	7,8± 0,9	3
thalamus	52±10	6,3±0,4	335± 91	18,1± 5,5	3
cerebellum	70/80	5/9	77± 46	5,2/2,6	2

values are given as means ± s.e.m.; N = number of brains; all bindings assays were carried out in triplicate; [^{3}H]-T (39,4 Ci/mmol; NEN) varied between concentrations of 2,0 - 25 nM.
According to Kellar and Cascio,1982.

Binding studies performed according the method of Kellar ans Cascio (1982)at 0°C with 10 µM unlabeled T to define unspecific binding and 20 µM pargyline gave evidence for a saturable single affinity site with a finite number of binding sites (Tab. 2). The Hill-analysis (N ~ 1) reveals the theoretical consideration of interpretating the [^{3}H]-T binding system to be non-cooperative. However, specific binding determined in the presence of 1 mM nonradioactive T increases 2-fold in the B_{max} values as well as in the K_D- and Hill-coefficients. If pargyline was omitted from medium there occurred a shift to a low affinity capacity with a high B_{max} value. The low affinity component of [^{14}C]-T, measured at initial velocity at 37°C revealed only one single binding site.

Table 2 Physicochemical changes of tryptamine binding sites

LIGAND	INCUB. TIME (min)	CONC. T (unlabeled)	PARGYLINE MAO-I	T (°C)	B_{max} (fmoles/mg prot.)	K_D (nM)	HILL (nH)	ΔG^{o} $KJ.mol^{-1}$	N
$[^3H]$-T	50	1.10^{-5}	2.10^{-5}	0	$249 \pm 2,0$	$13 \pm 2,0$	$1,1 \pm 0,3$	-41,2	4
$[^3H]$-T	50	1.10^{-3}	2.10^{-5}	0	497 ± 60	$23 \pm 6,4$	$1,6 \pm 0,2$	-39,9	5
$[^3H]$-T	50	1.10^{-3}	0	0	2873	100	2,4	-36,6	1
$[^{14}C]$-T	7	5.10^{-5}	1.10^{-5}	37	71980 ± 8550	5030 ± 320	$1,0 \pm 0,1$	-27,7	4

In vitro effect of unlabeled T and pargyline on $[^3H]$-T and $[^{14}C]$-T binding in human cortex. The age of controls was $74 \pm 18,2$ years with a post mortem time of $6,5 \pm 0,9$ hours. Values are given as means ± s.e.m.
N = number of brains

[^{3}H]-T-binding at 0°C to liver mitochondria demonstrates a 2-fold increase in the maximal binding capacity, no change in dissociation constants and Hill-coefficient. At 37°C the time course of [^{3}H]-T binding (11,34 nM) in the mitochondrial fraction of brain frontal cortex raised up to values of 0,5 pmoles/mg protein in a time-interval of 6 hours, where there was a decrease in the liver fraction (Fig. 1)

Fig. 2 shows that T-binding is a temperature dependent process. At 0°C the complete equilibrium is reached after 90 min, whereas the time course at 37°C possesses greater initial velocity and binding capacity. Inhibition of binding of [^{3}H]-T (11,91 nM) at 37°C produced by 20 µM pargyline was compared by 20 µM clorgyline, 20 µM Lilly 51641 and 1mM Nialamide (Fig. 3). Important was the significant difference in the amount of inhibition of [^{3}H]-T binding produced by these concentrations of MAO-inhibitors. After 3 hours a three fold increase of specific [^{3}H]-T bound was found when pargyline was omitted from incubation medium. Ascorbic acid in the range of 5 mM causes additional catalytic effect on the association kinetics (Fig. 4). Different shapes of saturation curves could be demonstrated each in pathological material in post mortem frontal cortex of a patient with senile dementia of Alzheimer type and coma hepaticum (Fig.5).

Multiple states of binding properties in the presence of nucleotides could be evaluated. Guanine nucleotides (1 mM GTP) have the ability to decrease [^{3}H]-T binding (12,18 nM) as well as adenine nucleotides (1 mM ATP), but to a lesser extent ([^{3}H]-T : 11,973 nM).

A mixture of 100 µM cAMP, 1 mM ATP and 12,5 mM $MgCl_2$ in the binding assay (11,06 nM [^{3}H]-T) had a positive modulating effect on [^{3}H]-T binding (Fig. 6).

The [^{3}H]-T binding process with varying concentrations of T in the range from 1,57 up to 22,89 nM and from 10 min. to 6 hours reveals a non Michaelis-Menten kinetic and sigmoid characteristics (Fig. 6,7).

The Lineweaver-Burk plot gives evidence for positive cooperativity and non competitive inhibition during the binding process, probably causing the slow binding. Saturation was given at any period of time. 10 minutes incubation time revealed a B_{max} value of 1,35 pm/mg protein;

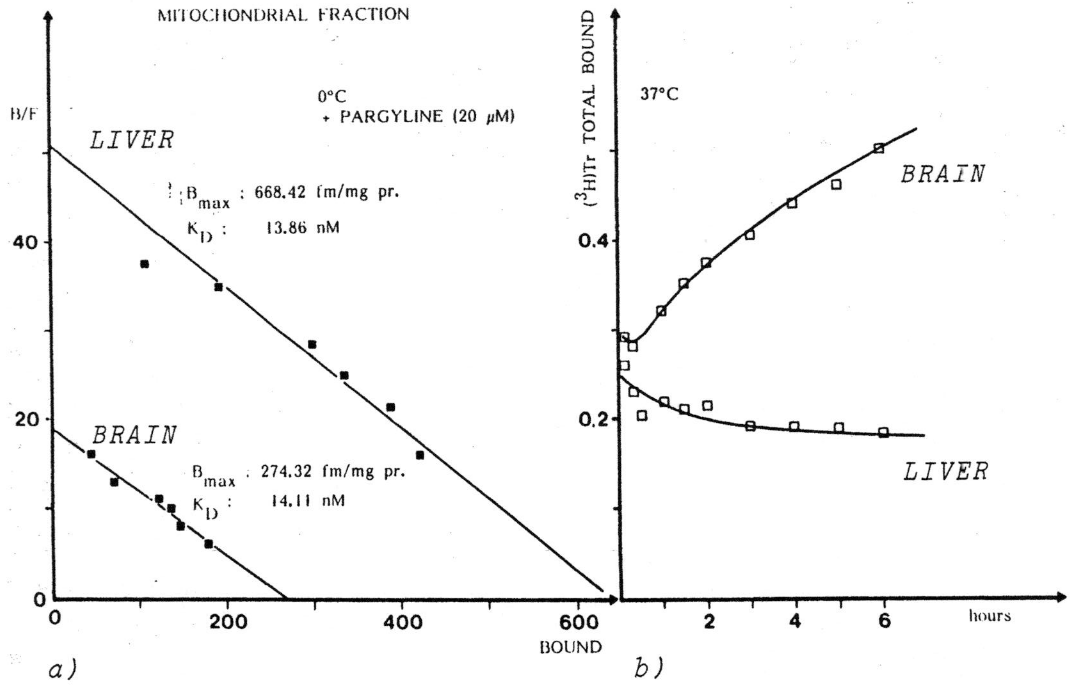

Figure 1:
a) [^{3}H] -T Scatchard analyses according the method of Kellar and Cascio (1982) of the mitochondrial fraction of liver and frontal cortex of control (90a, 6 hours post mortem) b) Time course of [^{3}H] -T total binding (11,34 nM).

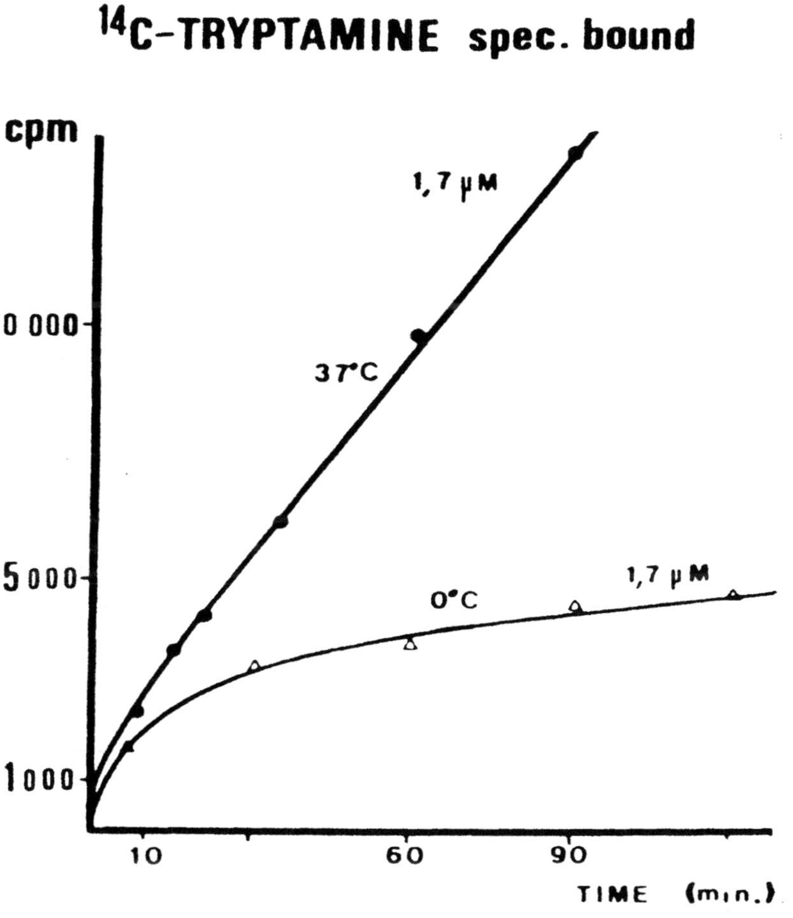

Figure 2:
Temperature influence at 0°C, respectively 37°C and saturation curve of [14C] -T binding (µM-range) to human frontal cortex of control tissue. 3 hours 3,34 pm/mg protein and 6 hours 6,67 pm/mg protein. A noncompetitive type of inhibition is demonstrated, the K_D value remained constant and was calculated to be 33,3 nM.

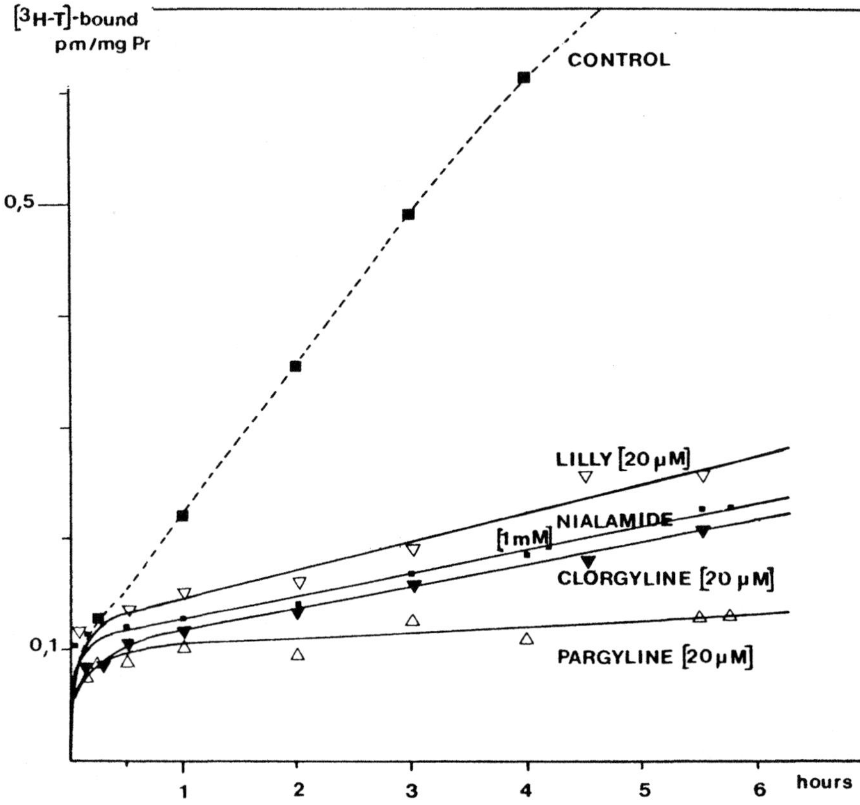

Figure 3:
Saturation curve and time course of [³H] -T binding (11,9146 nM [³H] -T at 37°C to human frontal cortex of control tissue in the presence of various MAO inhibitors. Lilly = Lilly 51641.

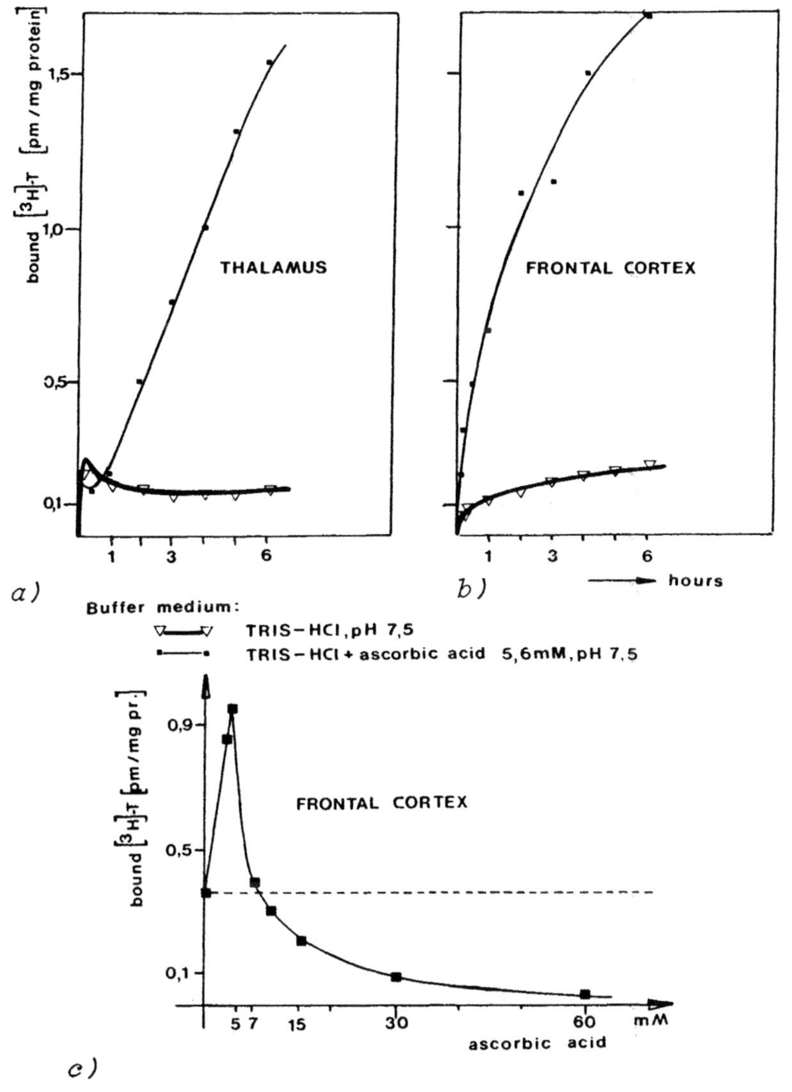

a,b) Saturation curve and time course of [^{3}H]-T binding (11,2nM) at 37°C to human brain tissue of control derived from thalamus and frontal cortex. c) Effects of various concentrations of ascorbic acid to the specific binding of [^{3}H] -T (12,01nM) in frontal cortex after an incubation period of 2,5 hours

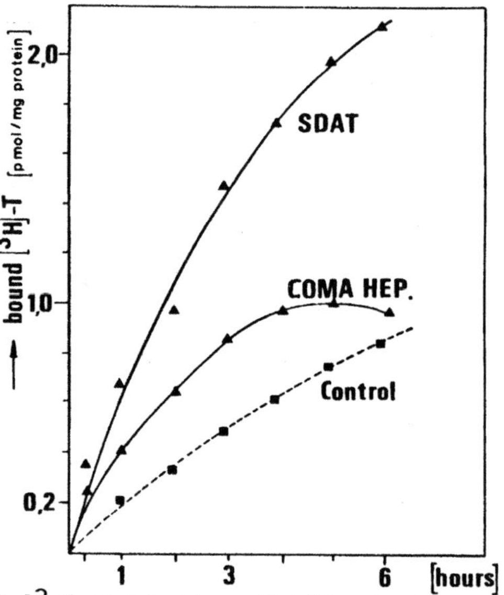

Figure 5:
Time course of [^{3}H]-T binding (11,03 nM) in post mortem frontal cortex of patients with senile dementia of Alzheimer type and coma hepaticum.

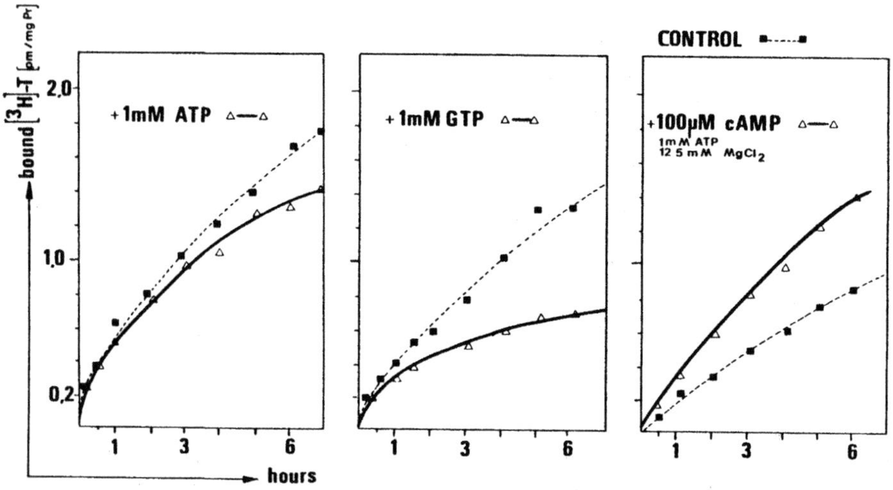

Figure 6:
Time course of [^{3}H]-T binding to human post mortem frontal cortex (11,97; 12,18 and 11,06 nM) in the presence of various nucleotides

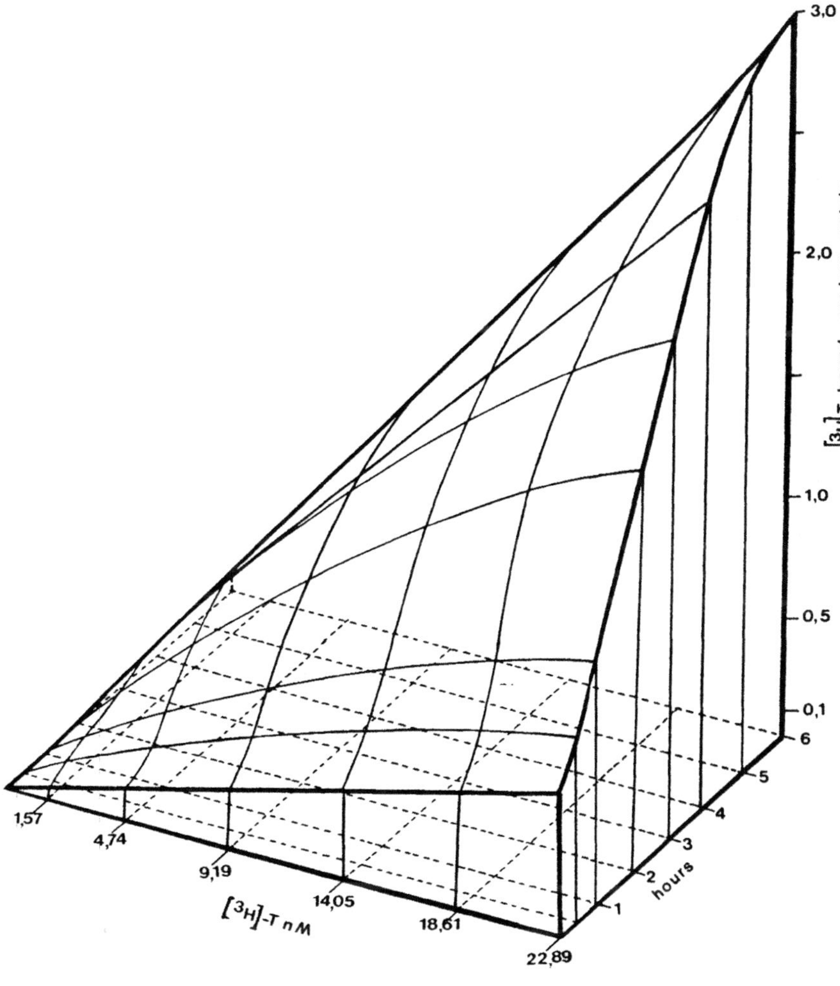

Figure 7a:
[^{3}H]-T binding to post mortem frontal cortex of control from 1,57 to 22,89 nM[^{3}H]-T; unlabeled T: 1 mM and in a time period from 10 min. up to 6 hours. Values were performed in duplicate.

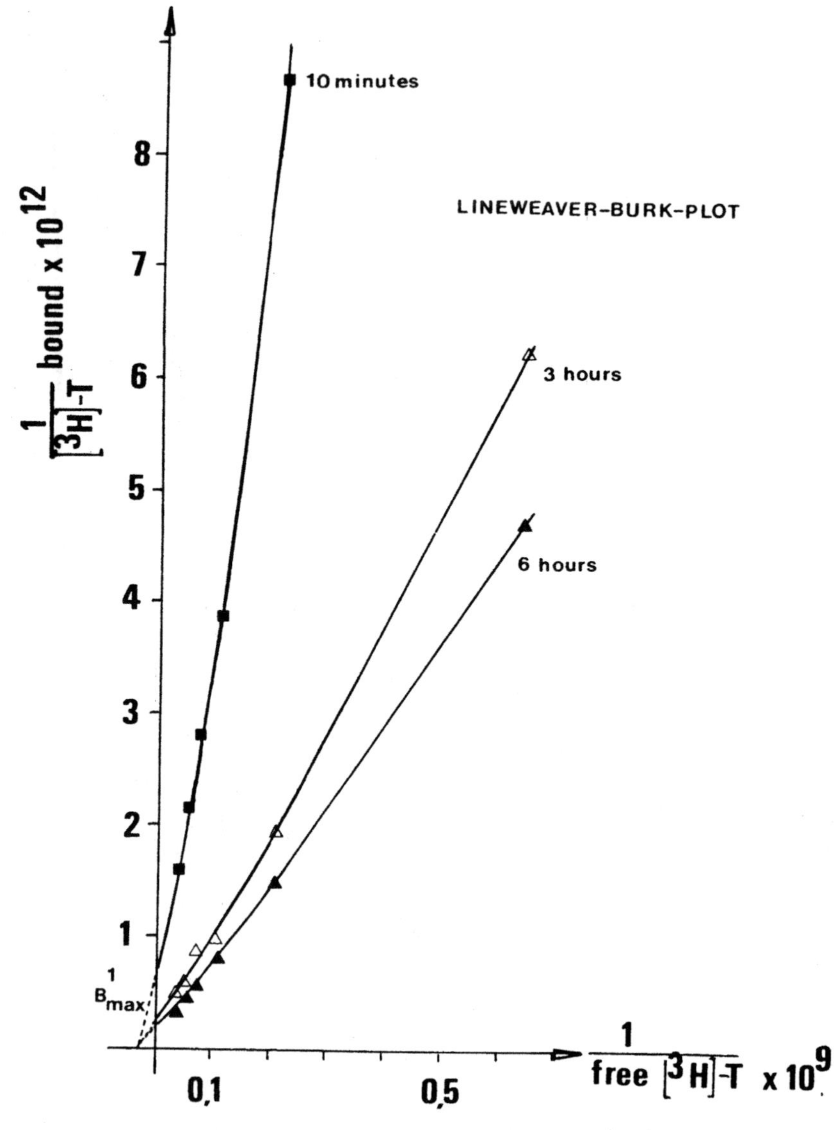

Figure 7 b
Lineweaver - Burk Plot

DISCUSSION

Questions concerning the types of T-binding sites in the central nervous system have developed into a complex issue. The significance of these binding sites involved in mediating modulating function has begun to be addressed.

T-protein-ligand complexes can be separated funtionally into two classes. In our findings, T-specific binding is characterized in relative terms, by a high affinity of the ligand and a low binding capacity at 0°C in the presence of MAO-inhibitors obeying simple Michaelis-Menten kinetics whereas a low affinity-state could be detected at 37°C in the central nervous system when the MAO-inhibitor was omitted from buffer medium. Probably the T-system therefore contains two specific interactions with a difference of order of magnitude between the two classes:
several types of binding experiments were conducted to determine whether multiple functions of the T-system were present. A subsequent increase in the maximal binding capacity is paralleled by an increase in the Hill-number if the concentration of unlabeled T to define unspecific binding is increased 100 fold from 1.10^{-5} to 1.10^{-3}M. If pargyline is omitted the calculation of the Hill-coefficient of 2,4 gave evidence for a complex T-binding system (Tab.2). The inhibition of the B_{max} value by 20 µM pargyline was about 91,36 %. In contrast to these findings in human brain, Cascio and Kellar (1983) found a decreased [^{3}H]-T binding capacity in rat brain by omission of pargyline and at 37° C, probably as result of the rapid oxidation of T by MAO. However, not only the neurotransmitter receptors, but also membrane enzymes have specific binding sites for ligands, substrates or inhibitors (Hamon and Pichat,1984; Parkinson and Callingham, 1980). To clarify the nature of these T-binding sites the assay according the method of Kellar and Cascio (1982) was repeated with liver as well as brain mitochondria fractions from human post mortem tissue (Fig. 1). Compared to the values in crude brain homogenate a 2,5 fold increase in the maximal binding capacity was measured in liver mitochondrial fractions; the K_D-value remained unchanged. Therefore no exclusive specificity for the T-binding site in the central nervous system was given.

Only three active classes of compounds were observed to displace [^{3}H]-T binding in rat frontal cortex: these were tryptamine analogues, ß-carbolines and phenethylamines, whereas antidepressants, serotonin-antagonists and agonists were inactive [IC_{50} >5000 nM] as well as the monoamines, apomorphine, dopamine, norepinephrine, amino acids like phenylalanine, kynuramine, isoquinolines, the hallucinogen mescaline and opiates (Wood et al.,1984). Harmaline (Ki-value 21 nM) was tested to be most effective in displacing [^{3}H]-T (Rommelspacher and Brüning,1984) but also a very useful tool studying the properties of the active site of MAO-A and for titrating the enzyme in tissue (Hamon and Pichat,1984). In assay conditions, similar to those of Kellar and Cascio (1982), 0°C and an incubation time of 2 hours, the specific binding to the enzyme was defined as the difference between total radioactivity bound minus that persisting in similar samples supplemented with 10 µM clorgyline. The K_D of the active enzyme site was in n-molar range, B_{max} about 7 pmoles/mg protein, similar to the values of [^{3}H]-T when pargyline was omitted. Furthermore, harmaline, harmine, tryptamine, serotonin, norepinephrine and quipazine acted as competitive inhibitors of [^{3}H]-harmaline binding and [^{3}H]-serotonin oxidative deamination whereas clorgyline, pargyline, Lilly 51641 and deprenyl acted irreversibly (Hamon and Pichat, 1984). Also in <u>in vivo</u> treatment with MAO-inhibitors such as pargyline or clorgyline specific binding sites for [^{3}H]-harmaline were destroyed but recovered some days later following treatment (Hamon and Pichat, 1984).

The distribution of T-binding sites in human brain reveals highest values in caudate n. and seems to correlate positively with MAO-activity.We might assume that in assay conditions described above, to a large part the T-binding-values probably reveal data of T-binding to MAO-enzyme remaining after inhibition of MAO (Tab. 1) (Kinemuchi et al., 1984). High MAO-activity therefore may prevent to measure the T-enzyme complex at 37°C. Indeed, no specific T-binding could be evaluated in both mitochondrial fractions (Fig. 1). Only the time course of total [^{3}H]-T binding (11,3 nM) at 37°C revealed a slow increase in the brain fraction, not detectable in peripheral tissue (Fig. 1).

The acceptor site of [^{3}H]-T at 37°C seems to be a functional unit rather than a classical pharmacological binding site. The increase in number of $[^{3}H]$-T molecules specifically bound is a temperature dependent process and nearly completely abolished in the presence of various MAO-inhibitors (Fig. 3). Maximal stimulation of the association kinetics at 5 mM ascorbic acid in the incubation medium reveals the necessity and dependence on a certain redox-potential probably connected with electron-transfer (Fig.4). In connection with MAO, known as flavoenzyme, such a formulation of a catalytic mechanism could be suggested (Fig.4 .). Preceeding a dissociation of an endogenous inhibitor bound to the active site in the same way as is substrate or conformational changes may cause that slow binding process which is saturable at any given moment. Obviously that association kinetics is modulated by cyclic nucleotides, cAMP is effector, GTP acts as negative modulator (Fig. 6). Positive cooperativity and allosteric properties might constitute the existence of a protein component of a mechanism by which regulating a long-term information may occur as well as direct effects on the genetic material in the cell nucleus that may lead to enzyme induction or long-term alterations according the suggested model of Jacob and Monod (Fig. 7). Alteration of these binding kinetics in pathological conditions like SDAT and coma hepaticum may confirm such a kind of regulating function (Fig. 5).

The effect of T on various receptors appears to be a direct one, only chemically addressed (Langer and Trendelenburg,1969). Recently, the influence of T in the rat caudal artery was found to be predominantly mediated by the same receptor as serotonin and that the differential inactivation of T by MAO-enzymes largely accounts for the different susceptibilities of serotonin and T to the antagonists examined (Bradley et al., 1985).

The T-binding kinetics can be seen of great potential complexity. The differential distribution of T-acceptor-sites well need new techniques to confirm T to possess a particular receptor system.

REFERENCES

Aghajanian, G.K. and Haigler, H.J. (1975) Hallucinogenic indoleamines: Preferential action upon presynaptic serotonin receptors. Psychopharm.Comm. 1, 619-629.

Bradley, P.B., Humphrey, P.P.A., and Williams, R.H. (1985) Tryptamine-induced vasoconstrictor responses in rat caudal arteries are mediated predominantly via 5-hydroxytryptamine receptors. Br.J.Pharmac. 84, 919-925.

Cascio, C.S. and Kellar, K.J. (1983) ^{3}H-Tryptamine: Characterization of binding sites in rat brain. Abs.Soc. Neurosci. 8, 783.

Cascio, C.S. and Kellar, K.J. (1983) Characterization of [^{3}H] tryptamine binding sites in brain. Europ.J. of Pharm. 95, 31-39.

Cox B., Davis, A., Juxon, V., Lee, T.F. and Martin, D.(1981) Core temperature changes following electrical stimulation of the midbrain raphe nuclei and intrahypothalamine injection of tryptamine in the rat. Br.J.Pharmacol. 74, 216 P.

Hamon, M. and Pichat, L. (1984) ^{3}H-harmaline: A useful tool for studying MAO A, in Monoamine oxidase and disease (K.F.Tipton, P.Dostert, M.Strolin-Benedetti eds). p 63-72, Academic Press.

Jones, R.S.G. (1982a) A comparison of the responses of cortical neurones to iontophoretically applied tryptamine and 5-hydroxytryptamine in the rat. Neuropharmac. 21, 209-214.

Jones, R.S.G. (1982b) Responses of cortical neurones to stimulation of the nucleus raphe medianus: A pharmacological analysis of the role of indolamines. Neuropharmac. 21, 511-520.

Jones, R.S.G. (1982c) Tryptamine modifies cortical neurone responses evoked by stimulation of nucleus raphe medianus. Brain Res. Bull. 8, 435-437.

Jones, R.S.G. (1984) Trace amine-peptide interactions on central neurones I. Tryptamine and substance P, in Neurobiology of the trace amines (A.A.Boulton, G.B.Baker, W.G.Dewhurst, and M.Sandler eds) Humana Press, Clifton, New Jersey, pp 321-326.

Kellar, K.J. and Cascio, C.S. (1982) [^{3}H]-Tryptamine: high affinity binding sites in rat brain. Eur.J.Pharmac. 78, 475-478.

Kienzl, E., Riederer, P., Jellinger, K., and H. Noller (1984) ^{14}C-tryptamine binding in Parkinson's disease and hepatic coma, in Neurobiology of the trace amines (A.A.Boulton, G.B.Baker, W.G.Dewhurst, M.Sandler eds) Humana Press, Clifton, New Jersey, pp 571-580.

Kinemuchi, H., Fowler, C.J., and Tipton, K.F. (1984) Substrate specificities of the two forms of monoamine oxidase, in Monoamine oxidase and disease (K.F.Tipton, P.Dostert, M.Strolin Benedetti eds), Academic Press,p 53-62

Langer S.Z., and Trendelenburg U. (1969) The effect of a saturable uptake mechanism on the slopes of dose-response curves for sympathomimetic amines and on the shifts of dose-response curves produced by a competitive agonist. J.Pharm.Exp.Therap. 167 (1), 117-142.

Marsden, C.A., and Curzon, A. (1979) The role of tryptamine in the behavioural effects of tranylcypromine + L-tryptophan. Neuropharmacology 18, 159-164.

Monod, J., Changeux, P., and Jacob, F. (1963) J.Mol.Biol. 6 p 306.

Parkinson, D. and Callingham, B.A. (1980) The binding of [^{3}H] pargyline to rat liver mitochondrial monoamine oxidase. J.Pharm.Pharmacol. 32, 49-54

Rommelspacher, H. and Brüning, G. (1984) Formation and function of tetrahydro-ß-carbolines with special reference to their action on [^{3}H]-tryptamine binding sites, in Progress in tryptophan and serotonin research (Schlossberger H.G., W.Kochen, B.Linzen, H.Steinhart eds), De Gruyter ,Berlin-New York; pp 209-216.

Warsh, J.J., Coscina, D.V., Godse, D.D. and Chan, P.W. (1979) Dependence of brain tryptamine formation on tryptophan availability. J.Neurochem. 32, 1191-1196.

Wood, P.L., Pilapil, C., La Faille, F., Nair, N.P.V. and Glennon, R.A. (1984) Unique [^{3}H] tryptamine binding sites in rat brain: distribution and pharmacology. Arch.int. pharmacodyn. 268, 194-201.

PHENYLETHYLAMINE METABOLISM BY POSTMORTEM HUMAN BRAIN AND LIVER PREPARATIONS

A.D. Mosnaim, M.E. Wolf, J. Chevesich, O.H. Callaghan, and P. Szanto

Chicago Med. School, N. Chicago Vet. Adm. Med. Center and Loyola Univ., N. Chicago, IL 60064

INTRODUCTION

In vivo and in vitro animal studies have elucidated in some detail the biochemical pathways involved in the central and peripheral biosynthesis and degradation of phenylethylamine (PEA) (Figs. 1 and 1a; ref. 1-4 and work therein cited). Human studies of the metabolism of this amine are few and mostly deal with the measurement of PEA and of its main metabolite, phenylacetic acid in urine, plasma and cerebrospinal fluid (4-6). Several investigators have suggested that the observed changes in the peripheral levels of these substances in a number of neuropsychiatric conditions may reflect variations in central PEA neuroactivity. On this rather weak evidence, it has been hypothesized that alterations in the brain PEA metabolism play a role in the pathophysiology of these disorders (1, 2,7). This communication presents the first evidence showing that in vitro human brain and liver preparations p-hydroxylate, N-methylate and N-acetylate PEA. It also reports on qualitative interspecie differences in the metabolism of PEA by these tissues. These preliminary results

This manuscript was written while Dr. A. Mosnaim and Dr. M.E. Wolf were Visiting Scientists at the Mailman Research Center, McLean Hospital and Harvard Medical School, Belmont, MA.

FIGURE 1

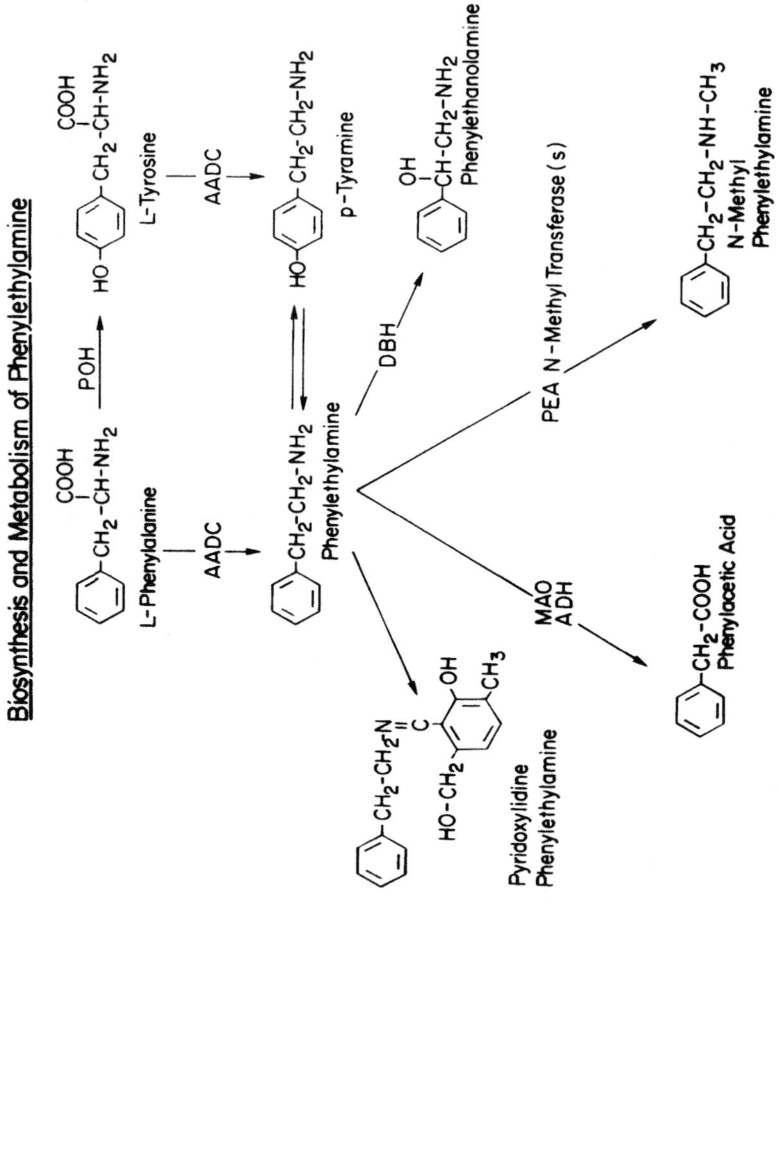
Biosynthesis and Metabolism of Phenylethylamine
L-Phenylalanine
POH
L-Tyrosine
AADC
AADC
Phenylethylamine
p-Tyramine
DBH
Phenylethanolamine
PEA N-Methyl Transferase (s)
N-Methyl
Phenylethylamine
MAO
ADH
Phenylacetic Acid
Pyridoxylidine
Phenylethylamine

FIGURE 1a

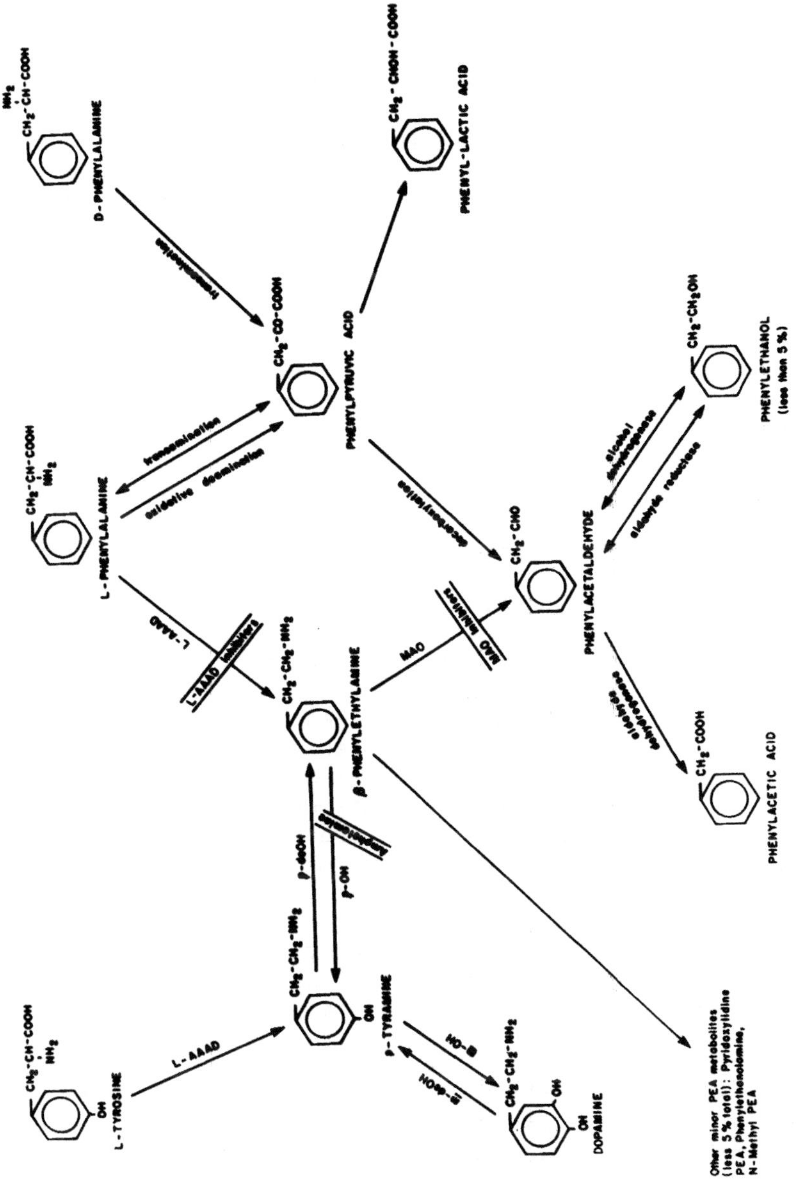
D-PHENYLALANINE
transamination
PHENYLPYRUVIC ACID
PHENYL-LACTIC ACID
L-PHENYLALANINE
transamination
oxidative deamination
decarboxylation
PHENYLACETALDEHYDE
alcohol dehydrogenase
aldehyde reductase
PHENYLETHANOL
(less than 5 %)
L-AAAD
L-AAAD inhibitors
β-PHENYLETHYLAMINE
MAO
MAO inhibitors
aldehyde dehydrogenase
PHENYLACETIC ACID
β-deOH
β-OH
L-TYROSINE
L-AAAD
p-TYRAMINE
DOPAMINE
Other minor PEA metabolites
(less 5 % total): Pyridoxylidine
PEA, Phenylethanolamine,
N-Methyl PEA

should encourage further research leading to the elucidation of the possible role of the "PEA metabolic pathway" in central mechanisms.

MATERIALS AND METHODS

The radiochemical purity of labeled substrates and standards, (1-^{14}C)PEA HCl, (1-^{14}C) p-hydroxy PEA HCl (p-tyramine, TRM) and custom made (NEN-Dupont) [^{3}H-(G)]-N-methyl PEA HCl, was checked by thin-layer chromatography (TLC) separation followed by radiochromatogram scanning (>95/pure) (for details see below and ref. 8). Other chemicals, including radioactive (^{3}H- N-Me) S-adenosyl-L-methionine (^{3}H-N-Me SAM) were of the highest purity commercially available and used without further purification. After obtaining informed consent from next of kin, the brain tissue of male patients (range 50 to 70 years of age) who expired at the North Chicago Veterans Administration Medical Center were rapidly removed, dissected out (4° cold room) into several brain areas and kept frozen (-80°C) until assayed for enzymatic activity. In some cases, a liver sample (∿100 g) was also stored for metabolic studies. Medical history and time elapsed between patient death and freezing of tissues were recorded for each subject. For animal studies, rabbits (white, male New Zealand, 2-3 Kg), rats (male Sprague-Dawley, 250-300 g) and mice (male Swiss albino, 20-25 g) were killed by decapitation, and brain and livers were removed separately (pooling of 2 or 3 animals per experiment) and assayed immediately (3). Human tissues (thalamus and putamen) were allowed to defrost in a cold room (4°C) before used. Samples were homogenized in ice-cold buffer (100 mM potassium phosphate pH 7.5, 10 mM $MgCl_2$, 0.01 mM $MnCl_2$; final mixture of 40% w/v), centrifuged (10,000 xg, 10 min, 40°C) and a fraction of the supernatant separated out to assay for methylating and acetylating activity. The rest was centrifuged again (100,000 xg, 60 min, 2°C) and the pellet, after resuspension in an equal volume of buffer, used as enzyme source for PEA hydroxylation (for details of incubation mixture see ref. 3 and Table 1). In the methylation studies, one serie of assays contained labeled PEA and "cold" SAM and a similar set of assays were added corresponding amounts of "cold" PEA and tritiated SAM. After incubation (37°C, 20 min), the reaction was stopped (equal volume ice-cold acetone) and the mixture centrifuged (2,500 x g, 2 min).

TABLE 1

PHENYLETHYLAMINE METABOLISM -[1] IN VITRO STUDIES

	p-Tyramine		N-Methyl PEA		N-Acetyl PEA	
	LIVER	BRAIN	LIVER	BRAIN	LIVER	BRAIN
RABBIT	+?	-	+++	+	++	-
MOUSE	-	-	+	-	+	-
RAT	++	-	-	-	-	-
HUMAN[2]	+	+	+	+	+	+

[1] Final incubation mixture (pH 7.5) contained: liver or brain 10,000 xg supernatant or 100,000 xg pellet (equivalent to 8% tissue homogenate) in 50 mM potassium phosphate buffer, 5 mM $MqCl_2$, 0.005 mM $MnCl_2$, 1mM NADPH, 0.01 mM pargyline and 0.01 mM 1-^{14}C PEA (~2×10^6 dpm). While the PEA hydroxylating enzyme(s) are mainly present in the microsomal fraction, the 100,000 xg supernatant contain the PEA methylating and acetylating enzymatic activities. NADPH is essential for tyramine formation and enhanced PEA N-methylation.

[2] Due to the problems in standardizing the time interval between autopsy and the in vitro biochemical studies it is difficult to make quantitative evaluations.

Products in the supernatnat were separated by TLC, visualized by autoradiography and identified by comparison to standards also visualized by autoradiography and by nynhydrin staining. In brief, in the studies of PEA hydroxylation, aliquots of the reaction's supernatant as well as appropriate blanks and standards, including PEA, TRM and N-Me PEA (5 μg each, spiked with the corresponding labeled amine) were spotted on cellulose-coated (0.1 mm thickness) glass plates and developed with a mixture of nitroethane: acetic acid:water (9.0:2.8:1.2) for 2h at 30°C (TLC system 1). The TRM containing fraction (R_f 0.51-0.61; PEA and N-MePEA R_f 0.75-0.80 and 0.79-0.83, respectively) was scrapped off the plate (Fig. 2) and spotted again (for standards, see above) on microcellulose-coated glass plates (0.25 mm thickness) and developed (nitroethane:acetic acid: water; 3:1:1; 2h at 37°C) (TLC system 2). TRM (R_f 0.23-0.28; PEA and N-Me PEA R_f 0.450-0.50 and 0.75-0.80) was scrapped off and counted (ref. 3 and 8). Aliquots of the reaction assaying the 10,000 x g supernatant were examined for presence of N-Me PEA and N-acetyl PEA. Samples and standards, see above, including N-acetyl PEA (visualized under UV light, 254 nm, ref. 8) were spotted on silica gel-coated glass plates (0.24 mm thickness, activated at 110°C for 40 min) and developed (acetone:1N ammonia; 10:3, 3-4h, room t°) (TLC system 3). N-Methyl PEA and N-acetyl PEA (R_f of 0.20-0.24 and 0.87-0.89; PEA and TRM R_f 0.62-0.68 and 0.54-0.61, respectively) were scrapped off the plate and respotted in TLC system 2 (see above). In this system, N-Me PEA, PEA and TRM show R_f of 0.75-0.80, 0.45-0.50 and 0.23-0.28, respectively. N-Methyl PEA was scrapped off the plate and counted. Fig. 3 illustrates the separation of these metabolites in TLC system 1 (ref. 8). In general, TLC plates were autoradiographed (Kodak Blue Brand X-ray films) for five to eight weeks or five to six months when using either (1-^{14}C) PEA or tritiated N-Me SAM, respectively before developed to visualize labeled PEA metabolites.

RESULTS

As can be seen from Table 1, under our experimental conditions there is considerable qualitative species differences in the metabolism of PEA by *in vitro* brain and liver tissue preparations. Whereas human 100,000 x g brain (thalamus and putamen) and liver pellet preparations *p*-hydroxylate PEA to form TRM, we failed to detect this reaction

FIGURE 2

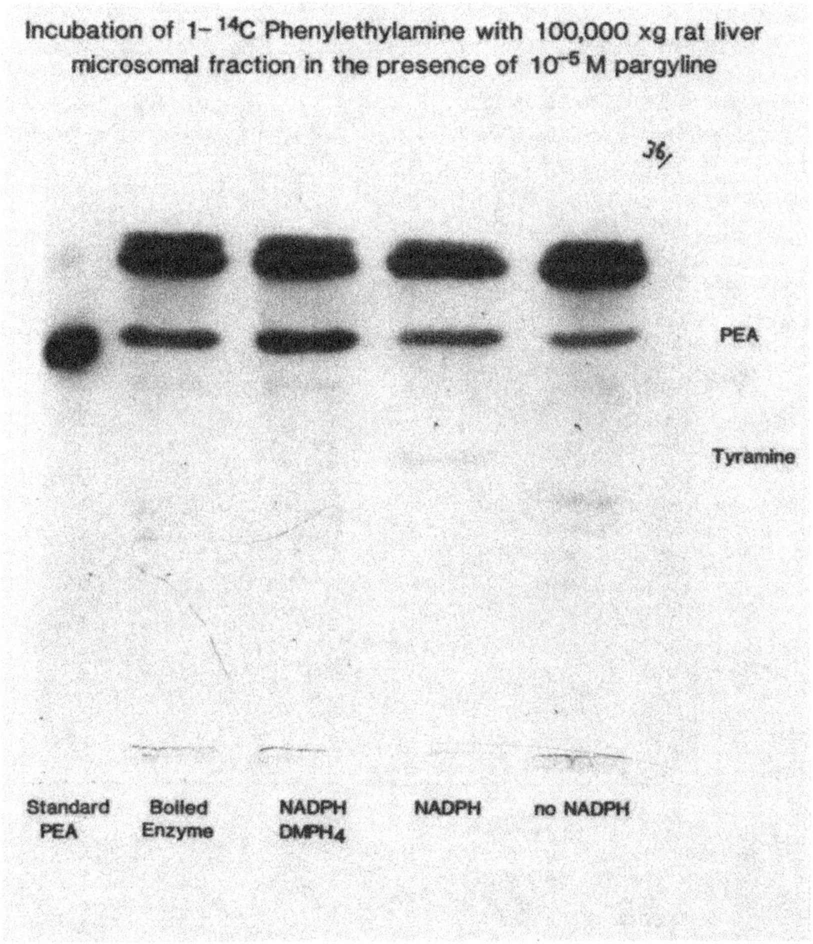

in either rabbit, mouse or rat brain tissues. Although rat liver actively converts PEA to TRM, the occurrence of this reaction is doubtful in the rabbit and does not appear to occur in mouse liver preparations. Human and rabbit brain and liver 10,000 x g supernatant N-methylate

FIGURE 3

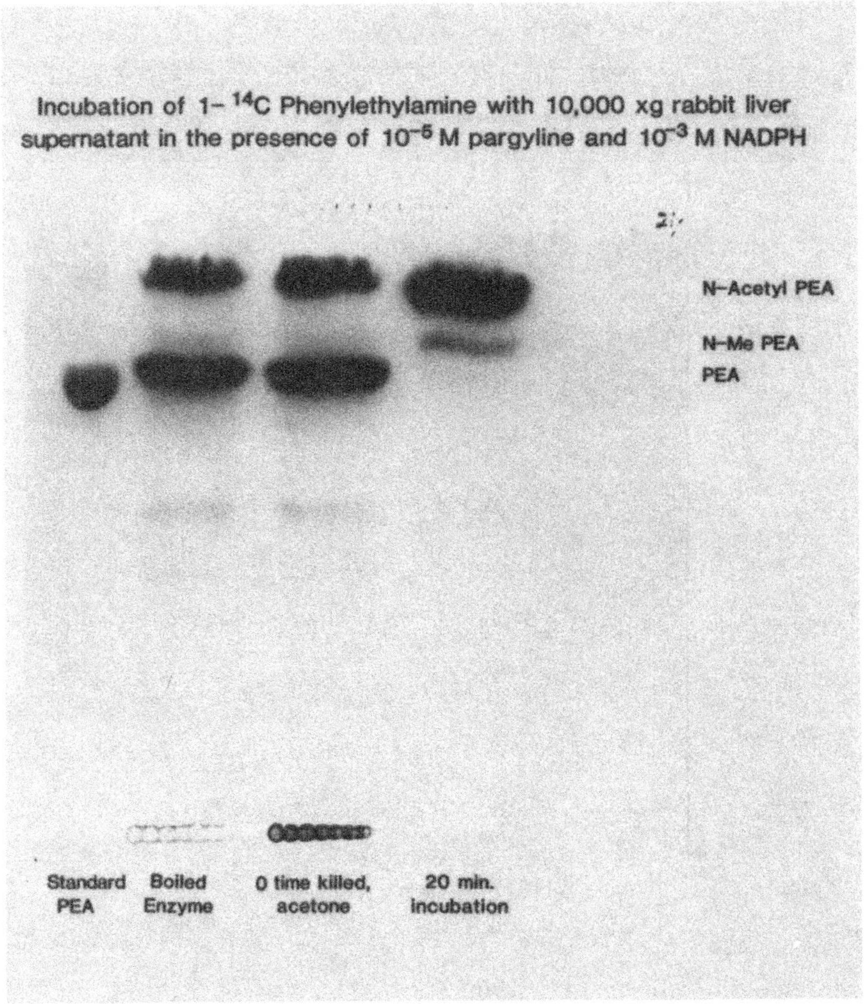

PEA, a reaction which also occurs in mouse liver but not in mouse or rat brain or rat liver preparations. Table 1 also shows the recovery of N-acetyl PEA after PEA incubation with human liver and brain 10,000 x g supernatant. This reaction, which also occurs in rabbit and mouse liver incubates, could not be detected in rat liver or rabbit, mouse or rat brain preparations. These findings, which are expressed only in a qualitative way, represent the results of at least four experiments (each done in duplicate) for each animal specie and tissue. Postmortem human studies were conducted in six brains and two liver tissues (each experiment was done in duplicate).

DISCUSSION

It has been proposed that alterations in the brain "PEA metabolic pathway" play a role in the pathophysiology of a number of neuropsychiatric conditions including schizophrenia, affective disorders, Parkinsonism, aggression and tardive dyskinesia (1,2,7,9). Most of the evidence supporting these suggestions are based on animal studies, as our knowledge of human brain PEA metabolism is very limited and mostly indirect, largely obtained from measuring this amine and some of its metabolites in urine, blood and cerebrospinal fluid (1,2,7,10). We now present evidence for the first time that postmortem liver and brain preparations (thalamus and putamen; chosen for their relatively high PEA content and evidence of physiological and pharmacological relevance) (1,2,9) of schizophrenic patients (n=4), medicated with neuroleptics for several years prior to death, and those of controls (neuroleptic-free, non-psychotic patients, n=2) p-hydroxylate, N-acetylate and N-methylate PEA. This latter reaction, which also takes place in in vivo rabbit brain (8,11) appears of particular interest as the administration (ip, mice and rats) of N-methyl PEA, a substance which easily crosses the blood-brain barrier (BUI 78, Mosnaim and Wolf, unpub. observ.), and is rapidly metabolized by monoaminooxidase, elicits behavioral (e.g., increased locomotor activity and stereotypies) and pharmacological effects (e.g., analgesia) similar to those of amphetamine and PEA itself (Mosnaim et. al. in ref. 9, see also ref. 12).

Results obtained from human samples are preliminary and expressed only qualitatively due to problems encountered in standardizing the procedure for tissue processing, e.g., the time interval between patient's death and actual obtention, dissection and freezing of the brain areas (and liver tissue) ranges from 2 to 10 h. The storage time, before actual biochemical determinations, also varied for the different samples (2 to 30 weeks). Comparison of these results with those obtained from animal studies emphasize the qualitative interspecie differences in the degradation of PEA; as these PEA metabolites could be responsible for some of these amine actions, one should stress the danger involved in drawing conclusions from animal experiments to explain the possible role of PEA in humans. The recovery of increased amounts of (^{14}C-) or (^{3}H-) N-Me PEA when using either 1-^{14}C amine and "cold" SAM or "cold" amine and ^{3}H-SAM, respectively, indicates that SAM plays a role in and enhances the methylation of PEA. However, this reaction is also enhanced (_in vitro_ rabbit brain studies) by a number of other folate derivatives known to act as methyl donors (13). Characterization of the human brain enzymatic system(s) responsible for the N-methylation of PEA and the study of its possible changes in mental disorders appears as a difficult but important future task. The possible physiological significance and pharmacological role of PEA hydroxylation to TRM deserves further consideration as this usually minor brain metabolic pathway has been postulated to play an important role in conditions involving a deficit of dopamine biosynthesis (see Fig. 1 and 1a, 14); the PEA---TRM---dopamine metabolic route could also act as a buffer-type regulator of dopamine availability at specific brain synapses (14). As an illustration of the complexities involved in trying to assess the mechanism(s) of action of PEA and metabolites, one should point out that in _in vitro_ rat liver preparations the endogenous N-ME PEA, as well as a number of other monomethylated PEAs, including α-Me PEA (amphetamine), inhibit PEA hydroxylation to TRM. Furthermore, the reaction is also inhibited to different degrees by a number of psychoactive drugs, including Iprindole and Alprazolam (Xanax) (Mosnaim et. al., unpub. observ.). Acetylation of PEA could be considered one of the excretion routes for this amine. It is important to stress that in all our studies the monoamine oxidase inhibitor pargyline was present in the incubation mixture; failure to add this substance results in rapid and extensive degradation of PEA to its main metabolite phenylacetic acid, making it technically very difficult to positively identify (autoradiography) the smaller amounts formed of other PEA metabolites. It should also be noted that results obtained from _in vitro_ animal experiments sometimes differ from _in vivo_ data, e.g., _in vivo_ but not _in vitro_ rat and rabbit p-hydroxylate PEA to TRM (this paper and refs. 1,2 and 11).

The present work establishes that selected postmortem human brain areas (and liver preparations) p-hydroxylate, N-methylate and N-acetylate PEA. It is clear that there exists a need for quantitative studies of PEA metabolism using several other relevant brain areas obtained under more rigorous experimental conditions. Our results show the existence of interspecie differences in the qualitative occurrence of these metabolic pathways (Table 1) and stress the need for caution in extrapolating observations from animal models to explain the possible role of alterations of PEA metabolism in human neuropathophysiology.

ACKNOWLEDGMENTS

We thank Ms. Elizabeth Murri and Barbara Smith for excellent help with this work. N-acetylphenylethylamine was a gift of Dr. Louise Hsu. This research was supported in part by the Veterans Administration, the National Migraine Foundation, and by the State of Illinois, Department of Mental Health and Developmental Disabilities.

REFERENCES

1. Mosnaim A. D. and Wolf M. E. (eds.) (1978) Non-catecholic Phenylethylamines, Part 1: Phenylethylamine: Biological Mechanisms and Clinical Aspects. Marcel Dekker, New York.
2. Mosnaim A. D. and Wolf M. E. (eds.) (1980) Non-catecholic Phenylethylamines, Part 2: Phenylethanolamine, Tyramines and Octopamine. Marcel Dekker, New York.
3. Callaghan O., Mosnaim A., Chevesich J., and Wolf M. (1984) The kinetics of hydroxylation of phenylethylamine, amphetamine and phenylalanine in rodent tissues. In: Neurobiology of Trace Amines, A. A. Boulton, G. B. Baker, W. G. Dewhurst and M. Sandler (eds.), Humana Press, Clifton, New Jersey.
4. Mosnaim, A. D., Wolf, M. E., and Madubuike, V. P. (1984) Presence and biosynthesis of phenylacetic acid in the rabbit brain. Biochem. Pharmacol. 33, 1993-1996.

5. Karoum F., Linniola M., Potter W. Z., Chuang L.-W., Goodwin F. K., and Wyatt R. J. (1982) Fluctuating high urinary phenylethylamine excretion rates in some bipolar affective disorder patients. Psychiat. Res. 6, 215-222.
6. Sandler M., Ruthven C. R. J., Goodwin B. L., Field H., and Matthews R. (1978) Phenylethylamine overproduction in aggressive psychopaths. Lancet 1269-1270.
7. Wolf M. E. and Mosnaim A. D. (1983) Phenylethylamine in neuropsychiatric disorders. Gen. Pharmacol. 14, 385-390.
8. Mosnaim A. D., Callaghan O. H., and Wolf M. E. (1981) Determination of noncatecholic phenylethylamines and monomethylated derivatives of phenylethylamine. J. Chromatog. 224, 481-487.
9. Wolf M. E. and Mosnaim A. D. (1984) Phenylethylamine and tardive dyskinesia. In: Neurobiology of the Trace Amines, A. A. Boulton, G. B. Baker, W. G. Dewhurst, and M. Sandler (eds.), Humana Press, Clifton, New Jersey.
10. Sabelli H. C., Fawcett J., Gusovsky F., Javaid J., Edwards J., and Jeffries H. (1983) Urinary phenylacetate: A diagnostic test for depression? Science 220, 1187.
11. Mosnaim A. D., R. Silkaitis, and M. E. Wolf (1980) Rabbit brain metabolism of phenylethylamine and tyramine: Drug effects. Life Sci. 27, 557-566.
12. Goudie A. J., D. Reid, and C. Demellweek (1984) Discriminative stimulus properties of beta-phenylethylamine in rats. Presented at E.N.A. Meeting, The Hague, September 1984 [Abstract published in Neuroscience Letters.]
13. Callaghan O. H. and Mosnaim A. D. (1979) In vitro methylation of phenylethylamine. Abstract, Seventh International Meeting of the Society for Neurochemistry, September 1979, Jerusalem, Israel.
14. Silkaitis R. P. and Mosnaim A. D. (1976) Pathways linking L-phenylalanine and 2-phenylethylamine with p-tyramine in rabbit brain. Brain Res. 114, 105-115.

INDEX

List of Abbreviations

ADR, Adrenaline
AMPH, Amphetamine
AUC, Area under the concentration/time curve
BP, Blood pressure
BZ, Benzylamine
DA, Dopamine
DMT, *N-N*-Dimethyltryptamine
DOPA, 3, 4-Dihydroxyphenylalanine
DOPAC, 3, 4-Dihydroxyphenylacetic acid
DSM III, Diagnostic and statistical manual of mental disorders
GC, Gas chromatography
GC–MS, Gas chromatography–mass spectrometry
HAMDS, Hamilton depression rating scale
HDS, *see* HAMDS
5-HIAA, 5- Hydroxyindoleacetic acid
m-HMA, *m*-Hydroxymandelic acid
m-HPA, *m*-Hydroxyphenylacetic acid
p-HPA, *p*-Hydroxyphenylacetic acid
HPLC, High-performance liquid chromatography
5-HT, 5-Hydroxytryptamine
HVA, Homovanillic acid
IAA, Indole acetic acid
iv, Intravenously
MA, Mandelic acid
MAO, Monoamine oxidase
MAO-A, Monoamine oxidase type A
MAO-B, Monoamine oxidase type B
MAOI, Monoamine oxidase inhibitors
5-MeODMT, 5-Methoxy-*N*, *N*-dimethyltryptamine
MHPG, 3-Methoxy-4-hydroxyphenylethylene glycol
MMPI, Minnesota multiple personality inventory
MOPEG, *see* MHPG
MS, Mass spectrometry
3-MT, 3-Methoxytyramine
NA, Noradrenaline
NE, Norepinephrine
NICIMS, Negative ion chemical ionization mass spectrometry
NMN, Normetanephrine
m-OA, *m*-Octopamine
p-OA, *p*-Octopamine
po, Per os (by mouth)
PAA, Phenylacetic acid
PE, *see* PEA
PEA, 2-Phenylethylamine
PHMA, *p*-Hydroxymandelic acid
PHPA, *p*-Hydroxyphenylacetic acid

PRL, Prolactin
PST, Phenolsulfotransferase
PTA, *p*-Tyramine
SBP, Systolic blood pressure
sc, Subcutaneous
SD, Standard deviation
SDAT, Senile dementia of Alzheimer type
SEM, Standard error mean
SSAO, Semicarbacide-sensitive amine oxidase
m-TA, *m*-Tyramine
p-TA, *p*-Tyramine
TCP, Tranylcypromine
THBC, Tetrahydro-beta-carbolines
TRY, Tryptamine
TY, *see* TYR
TYR, Tyramine
VMA, 3-Methoxy-4-hydroxymandelic acid